Introduction to
Optical Fiber
Communication Systems

The Oxford Series in Electrical and Computer Engineering

Adel S. Sedra, Series Editor, Electrical Engineering
Michael R. Lightner, Series Editor, Computer Engineering

Allen and Holberg, *CMOS Analog Circuit Design*

Bobrow, *Elementary Linear Circuit Analysis, 2nd Ed.*

Bobrow, *Fundamentals of Electrical Engineering, 2nd Ed.*

Campbell, *The Science and Engineering of Microelectronic Fabrication*

Chen, *Linear System Theory and Design, 3rd Ed.*

Chen, *System and Signal Analysis, 2nd Ed.*

Comer, *Digital Logic and State Machine Design, 3rd Ed.*

Cooper and McGillem, *Probabilistic Methods of Signal and System Analysis, 3rd Ed.*

Franco, *Electric Circuits Fundamentals*

Jones, *Introduction to Optical Fiber Communication Systems*

Krein, *Elements of Power Electronics*

Kuo, *Digital Control Systems, 3rd Ed.*

Lathi, *Modern Digital and Analog Communications Systems, 3rd Ed.*

McGillem and Cooper, *Continuous and Discrete Signal and System Analysis, 3rd Ed.*

Miner, *Lines and Electromagnetic Fields for Engineers*

Roberts, *SPICE, 2nd Ed.*

Santina, Stubberud, and Hostetter, *Digital Control System Design, 2nd Ed.*

Schwarz, *Electromagnetics for Engineers*

Schwarz and Oldham, *Electrical Engineering: An Introduction, 2nd Ed.*

Sedra and Smith, *Microelectronic Circuits, 4th Ed.*

Stefani, Savant, and Hostetter, *Design of Feedback Control Systems, 3rd Ed.*

Van Valkenburg, *Analog Filter Design*

Warner and Grung, *Semiconductor Device Electronics*

Wolovich, *Automatic Control Systems*

Yariv, *Optical Electronics in Modern Communications, 5th Ed.*

Introduction to
Optical Fiber
Communication
Systems

WILLIAM B. JONES, Jr.

Texas A & M University

New York   Oxford
OXFORD UNIVERSITY PRESS

Oxford University Press

Oxford   New York
Athens   Auckland   Bangkok   Bogota   Bombay   Buenos Aires
Calcutta   Cape Town   Dar es Salaam   Delhi   Florence   Hong Kong
Istanbul   Karachi   Kuala Lumpur   Madras   Madrid   Melbourne
Mexico City   Nairobi   Paris   Singapore   Taipei   Tokyo   Toronto   Warsaw

and associated companies in
Berlin   Ibadan

Published by Oxford University Press, Inc.,
198 Madison Avenue, New York, New York 10016
http://www.oup-usa.org

Oxford is a registered trademark of Oxford University Press

ISBN 0-19-510726-8

9 8 7 6 5 4 3 2

Printed in the United States of America
on acid-free paper

Preface

Optical fiber communication systems have moved very rapidly from the research labs into commercial application. When the attenuation inherent in the optical fiber was reduced to levels that made fiber economically attractive for long-haul communications, sources and detectors were ready and available for commercial applications. These first-generation systems were based on silicon devices operating at wavelengths in the 0.8–0.9 micron range. It was evident that fiber attenuation would be lower at longer wavelengths. Research on optical fibers with lower attenuation and larger bandwidths, as well as on lasers and photodetectors for wavelengths in the 1.3 micron range, was intensive. Second-generation systems, based on the 1.3 micron technology and improved optical fibers, are now state-of-the-art and third-generation systems, at 1.55 micron wavelengths, are on the horizon. Current research includes new fluoride fiber materials, with attenuation orders of magnitude smaller than is possible with silica fibers, and optoelectronic devices for use at the still longer wavelengths that will be attractive with the new fiber materials.

While the optical fibers, lasers, and photodetectors were making possible communication at very high data rates over increasingly long distances, research in communication technologies to take full advantage of the fiber and device technologies increased in scope and scale. As optical fiber communication system technologies have improved, an increasing variety of applications has become technically feasible and economically attractive.

This book is designed to prepare students, and perhaps others, to enter this fascinating, dynamic, and very important new field. It is suitable for use as a text for senior year college or first-year graduate courses. It should also be useful for self-study by practicing engineers and others with the necessary technical prerequisites and interests.

It is an *introductory* level textbook. Because it is introductory, many interesting and important aspects of the subject are left for later, more advanced studies. Because it is intended primarily as a textbook, the treatment of many topics is planned to flow smoothly from the prerequisite subjects upon which the subject matter of this book is built.

The prerequisites are the normal junior-level core courses in electromagnetic fields and semiconductor devices. Beyond these specific courses, it is assumed that students will have a level of maturity in electrical engineering subjects corresponding to first-term senior classification; subjects that are used include mathematics through differential equations, linear circuit analysis, Fourier and Laplace transforms, convolution, and electronic circuits. No communications prerequisites are expected; an introduction to modulation and other topics needed for later work is provided in Chapter 2.

The subject of this book is one of inherently broad scope. The selection of topics to be included, and those to be omitted, was based on the objectives of providing an overview of the principles and the potential of optical fiber communication systems and an understanding of the physical principles underlying the characteristics of component parts of the communication system. An engineering design point of view is maintained through most of the material.

Many instructors will choose to omit some topics and to supplement the text with notes on other topics. This should not present any unusual problems. In a field as dynamic as this, supplementing the textbook with notes on new results and placing greater emphasis on subjects of special interest to the instructor are highly desirable.

After completion of a one-semester course covering most of the book, the student should be able to read the technical literature with a moderate degree of understanding, to recognize the significance of new devices and new communication system techniques as they are developed and reported, to understand manufacturers' data sheets, and to design elementary optical fiber communication systems. It is the author's hope that many students will be inspired to pursue some of the topics through more specialized, advanced studies.

Acknowledgements

This book has evolved from the author's long standing interest in communications and his belief that optical communications is a subject that deserves a place in the electrical engineering curriculum. Although the words (and errors) are his, many people have helped to make this book possible. First and foremost are the students at the University of Florida and Texas A&M University whose interest in the subject and whose registration in Special Topics in Optical Communication Systems classes was the primary stimulation for the textbook project. Thanks are also due to faculty colleagues and to administrative officers at both of these Universities for their help in many ways. The treatment of several topics was influenced by discussions with Dr. Bob Biard, Chief Scientist, Honeywell Optoelectronics Division, and with Professors Henry Taylor, Bob Nevels, John Eknoyan, Don Parker and Mark Weichold, all members of the Electrical Engineering faculty at Texas A&M University. Technical reviewers included Antonio Arroyo, University of Florida; Ted Batchman, University of Virginia–Charlottesville; Larry Bowman, Brigham Young University; John Buck, Georgia Institute of Technology; David Giri, University of California–Berkeley; Steven Hill, Clarkson College of Technology; H.C. Hsieh, Iowa State University; H. Hsu, Ohio State University; Mysore Lakshminarayana, California State Polytechnic University; Robert Paknys, Clarkson College of Technology; Joseph Plunkett, California State University–Fresno; Mehdi Shadaran, University of Texas–El Paso; and M.E. Van Valkenburg, University of Illinois–Urbana. Their insightful comments and constructive suggestions have been useful and are sincerely appreciated. Special thanks are due to Mrs. Sybil Liebhafsky for her careful review of the entire manuscript. I would like to thank my editor, Deborah L. Moore, at Holt, Rinehart and Winston for her guidance and encouragement, and Chuck Wahrhaftig, Senior Project Manager, for producing this book. Finally, words are not adequate to express my gratitude to my wife for her patience, support, and active assistance in many, many ways.

William B. Jones, Jr.

Contents

Introduction to
**Optical Fiber
Communication Systems**

CHAPTER

Introduction

The development of the laser and the optical fiber has brought about a revolution in communication system design. In 20 years, optical fiber communications has evolved from a remote possibility for the distant future to a powerful, practical, and widely used communication technology.

In the 1960s, the laser evolved from a laboratory curiosity to become a versatile and widely applied family of devices and systems. In the 1970s, the optical fiber was developed from an idea with some promise to a proven communication channel capable of carrying high data rates with low attenuation over distances far exceeding those used in coaxial and microwave systems. In the 1980s, optical fiber communication system technologies are achieving a broad and still expanding range of successful commercial applications and are expanding the horizons of optical fiber systems capabilities.

The future is difficult to envision. Progress in all aspects of optical fiber communication technologies has exceeded forecasts. Current research in optical fibers, lasers, photodetectors, and communication systems all hold promise of the continuing evolution of new capabilities and new applications.

This book is intended as an introduction to this dynamic and exciting new field. It will cover most of the principal components of the optical fiber communication system. Its basic strategy is to provide an overview of the field rather than a study in depth of some component or aspect of the system. Its specific objectives include preparing the student to read the current technical literature, to design elementary optical communication systems, and to study in greater depth some of the topics introduced here.

1.1 Communication Systems

First and foremost, the optical fiber communication system is a communication system. Its purpose is to transfer information from some source to a distant user; the key words are *information* and *distance*. The optical fiber system has many of the same considerations, and same limiting factors, as any other communication system. Therefore, we will begin by defining some of the terms and some of the broad design questions that must be considered.

1.1.1 *Information*

An electrical, acoustic, or optical signal conveys information only to the extent that the received signal has some measurable characteristic that is unknown to the receiver until the signal is received and processed. In one important class of communications, the amplitude of a sinusoidal wave is caused to change in an unpredictable manner. By measuring the amplitude, the receiver can learn what the transmitter is sending. When the "information" is a speech or other audio waveform, it is possible to reconstruct a similar audio signal at the receiver. The communication system has performed its function successfully if the reconstructed signal is an accurate and useful replica of the original audio signal that was the input signal to the transmitter.

For some purposes, a more precise definition of information is useful. A quantitative definition of information is based on the concept of making random selections from a finite group of possibilities. Consider the selection of a sequence of letters from a four-letter alphabet: *A*, *B*, *C*, and *D*. When we select one letter, it is one from a set of four possibilities. If we select two, three, or four letters, we can convey more information; an intuitive measure indicates that with four letters we can convey four times as much information as with one letter. But with two letters there are sixteen (4^2) different possible messages, *AA*, *AB*, . . . , *DC*, *DD*; with three letters there are 64 possible combinations, and with four letters, 256 combinations. This line of reasoning leads to a logarithmic measure of information, $I = \log n$, where n is the total number of possible messages.

The most common unit for information is based on the logarithm to the base 2; the unit is the "bit," a term familiar to computer users. The choice of one letter from the four-letter alphabet represents $\log_2 4 = 2$ bits of information. Two letters give $\log_2 16 = 4$ bits; three letters, 6 bits; four letters, 8 bits; thus satisfying our feeling that the amount of information should be proportional to the length of the message.

Complications begin to arise with this definition when all possible choices are not equally likely. The definition can be extended to accommodate unequal probabilities of the various possible choices by casting it in terms of probability theory. We will not pursue the matter further here.

1.1.2 *Telecommunication*

The second key word in defining the objectives of a communication system was *distance*. The word telecommunication, using the Greek *tele-*, which means "at a distance," is popular with some communication system engineers because it distinguishes their business from many other uses of the word communications.

Telecommunication requires some means for sending the information to be conveyed to the distant point. Electrical signals propagating on wire transmission lines and radio waves propagating in space have long been suitable ways for covering the necessary distances. An important class of telecommunication system is characterized by sending computer tapes by air freight; if the quantity of data is large and a delay of one day in transmission is acceptable to the user, then air freight may be the best, that is, most cost-effective, system. Radio, coaxial cable, and optical-fiber communication systems are "real-time" systems; the delay between transmission and reception of information-bearing signals is very small, at least in comparison with air freight.

Every communication system has a limited information capacity. The capacity can be expressed in terms of a maximum information rate, in bits per second, or in terms of bandwidth. The system designer can usually increase the information rate by sacrificing distance and, perhaps, cost. One figure of merit for these systems is the product of the information capacity and the maximum distance over which it can operate. For example, an optical fiber is specified by the manufacturer as being capable of 200 (Mb/s)-km operation. It should be possible to use this fiber in systems carrying 100 megabits per second over distances up to 2 kilometers, 4 Mb/s over 50 km, etc. To transmit 100 Mb/s over 50 km with this fiber would require $\frac{100 \cdot 50}{200} = 25$ separate transmission paths. The 25 paths might be placed in tandem, each carrying 100 Mb/s a distance of 2 km, with a repeater (receiver and transmitter) placed at intervals of not more than 2 km; or 25 fiber paths could be placed in parallel, each carrying 4 Mb/s for the full 50 km distance. We will see in later chapters the reasons for the (Mb/s)-km limitation on communication capacity and some ways the system designer can make trade-offs between information rate and transmission distance.

1.1.3 *Component Parts of a Communication System*

For the purposes of design, the communication system can be broken down into five component subsystems. These are shown in Figure 1.1.1. Two of them, the source and destination (of the information to be transmitted) are associated primarily with the user of the system. They would normally be described to the system designer through specifications that define the information rate, the form

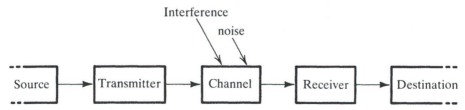

Figure 1.1.1 *Block diagram of a communication system*

of the input from the source to the communication system, and the form in which the output must be provided at the destination. The remaining three parts, the transmitter, the channel, and the receiver, constitute the communication system itself. The design problem is to select and/or design system components that will provide the communication capacity required by the user and to assure that the interconnections between each subsystem and the adjacent subsystems are compatible.

The transmission channel can be a pair of electrical conductors, an optical waveguide, free space through which microwave transmission is feasible, or an air cargo plane. The communication channel between two fixed points on the surface of the earth can use coaxial cable or optical fiber as waveguides. The channel between two airplanes is restricted to free-wave propagation. The attenuation, per km, and the bandwidth or bit rate of the channel are important constraints. The inevitable noise and interference that will be added to the signal during its transmission through the channel is another most important characteristic of the channel. Considerations such as these will help to identify some possible choices for the transmission medium as unsuitable for a particular system design and others as potentially satisfactory.

The transmitter must accept the information-bearing signals from the source and adapt them as necessary for transmission through the channel. Adapting to the channel consists of generating an electrical, optical, or other signal that the channel requires and modulating that signal so that the receiver can recover the information being sent. The interface between the transmitter and the channel is an important aspect of the system design. For example, in an optical system, the physical size of the transmitting devices must be similar to the diameter of the fiber in order that the lightwave can be coupled into the fiber for transmission.

The receiver must be capable of receiving a weak signal from the channel, separating the desired signal from the unavoidable noise and interference, recovering information from the signal, and converting it into a form acceptable to the user.

Each of the three major communication system component parts must have the capability of operating at the communication rates required for the overall system. Each has some requirements relating to the power level or signal amplitudes needed. There are also such considerations as reliability, initial and continuing costs, and the ease with which technicians can repair and maintain the system after it is placed into operation.

1.2 Optical Communications

1.2.1 *The Evolution of Optical Fiber Communications* [1.1]

The first attempt to develop an optical communication system was probably made by Alexander Graham Bell, who was issued a patent in 1880 for a "photophone," a system for transmitting voice over distances of several hundred meters.

The photophone was based on modulating reflected sunlight by causing the mirror to vibrate. The receiver was a photocell. Although Bell's apparatus apparently worked, the photophone was not a commercial success.

The invention of the laser in 1960 offered a new opportunity for optical communications. The search for a suitable transmission medium was promptly launched. Among the candidates were line of sight between two telescopes facing each other and several kinds of waveguides that used periodic or continuous focusing elements. Dielectric waveguides were studied theoretically but were not useful because of their high losses—of the order of 1000 dB/km.

Research on materials for optical transmission made dramatic progress. Losses were reduced from 1000 dB/km to 20 dB/km in 1970, and 4 dB/km in 1973. By using longer wavelengths, transmission losses were further reduced to 2 dB/km in 1974, 0.5 dB/km in 1976, and 0.2 dB/km in 1979.

Research and development of lasers and photodetectors also received intensive attention. The first semiconductor lasers were announced in 1962. Subsequent research on materials, device structures, and manufacturing technologies led to lasers with reliability, stability, and power levels that were useful for the first-generation commercial systems in the late 1970s. The prospects for lower fiber losses at longer wavelengths led to intensive research for lasers and photodetectors at 1.3 μm, and later 1.55 μm, wavelengths. These efforts, too, have been successful. Commercial systems that use the 1.3-μm technologies appeared in the early 1980s; 1.55-μm systems have been demonstrated in research laboratories and are expected to be commercially useful in the later 1980s.

The capacity for data communications and the distances over which the data can be transmitted have continued to increase. Commercial long-distance systems being installed in the mid- and late 1980s use the "second-generation" 1.3-μm technology and transmit data at 256 Mb/s on each of several fibers over distances of 50 km between repeaters.

The first commercial long-distance optical-fiber telephone system was put into service by the American Telephone and Telegraph Company (AT&T) in 1983 [1.2]. Today most major cities in the United States are linked by optical fiber systems. In addition to AT&T, most of the long-distance telecommunications companies either have or are installing optical fiber networks. A transatlantic optical-fiber system will be put into commercial service in 1988. A transpacific system, linking California, Hawaii, and Japan, is planned for installation in 1989.

Beyond the technologies on the immediate horizon and the systems proposed or envisioned, several other opportunities can be dimly seen and still more expected. New fiber materials and structures can greatly increase the available bandwidth and decrease the attenuation. Coherent communication techniques can improve receiver sensitivity and make available more sophisticated and powerful modulation methods. Somewhere, however, there are limits. Some researchers predict that optical fiber technology may outstrip the capacity of the electronic devices to supply and receive data at rates that the optical fiber can handle; there may be an "electronic barrier" to higher data rates. Some pragmatists question our need for or ability to utilize some of the new and emerging technologies. Although there are legitimate questions about what the future holds in store, there is no doubt that optical fiber communication is already a major technical and commercial success and that continuing growth of the optical communications industry is assured.

1.2.2 *Optical Fiber*

It is the optical fiber that stimulated the revolution in optical communications. When the attenuation was reduced to 10 dB/km, optical fiber communications became technically and commercially feasible. The optoelectronic[1] and other electronic devices necessary for the first-generation communication systems were already available. Prototype communication systems were being field tested in a remarkably short time. Research on still better fiber materials and more suitable electronic devices was promptly launched.

The optical fiber is a dielectric waveguide. Current fibers are made of silica materials, basically glass, but very high-quality glass. Other materials are under consideration for future use. A light wave lauched into the fiber at one end can propagate with very low attenuation, currently as low as 0.2 dB/km, making possible transmission over fiber paths of more than 100 kilometers without intermediate amplification.

Two factors limit the usefulness of the optical fiber for communications. The first is attenuation, mentioned above. There is always a maximum transmitter power level and a minimum useful received power level. The difference between these is the total transmission loss, which is due primarily to fiber attenuation. The fiber attenuation, expressed in dB/km, limits the maximum distance between the transmitter and the receiver. For example, if the transmitter can deliver 1 mW into the fiber and the minimum useful received power is 1 μW, then the total attenuation cannot exceed 1000, or 30 dB. If the average attenuation of the optical fiber is 1.5 dB/km, then the maximum length of fiber is 20 km.

The other limiting factor is bandwidth. The optical fiber will have a maximum bandwidth for signals to be transmitted through the fiber without distortion. The bandwidth limits the rate at which the signal can change its intensity or other signal parameter, and thus the rate at which information can be transmitted. For high-data-rate digital communication systems, the bandwidth rather than attenuation tends to be the factor that limits the transmission distance.

1.2.3 *Optical Fiber Communication Systems*

The principal parts of the optical fiber communication system are the transmitter, the fiber, and the receiver. Each of these parts has limited capabilities

---

[1] Optoelectronic devices are electronic devices whose electrical characteristics are influenced by optical waves. Similarly, electrooptic devices are optical devices whose optical characteristics are influenced by electric or magnetic fields.

with respect to both the intensity and the bandwidth of the signals it can handle without distortion.

The optical transmitter usually consists of a forward-biased semiconductor junction diode. It can be either a light-emitting diode or a semiconductor laser. In either case, the output light intensity can be a linear function of the diode current, making intensity modulation feasible. Because the diameter of the semiconductor light source is comparable to that of the optical fiber, coupling the light wave from the semiconductor into the optical fiber with reasonable efficiency is possible.

The optical receiver is a photodetector, in most cases a reverse-biased semiconductor junction diode. The light wave incident on the photodetector can deliver energy to electrons in the semiconductor crystal, exciting some of them into the conduction band. The high electric field of the reverse-biased junction will sweep these conduction electrons out of the junction region and cause a current to flow in the external circuit. This output current is proportional to the incident optical field intensity and therefore to the current that was used to modulate the intensity at the transmitter.

The study of optical fiber communications must start with the study of the principles of communication systems so that we can understand the functions that the optical communication system must provide and the physical limitations that nature and the real world impose. We can then study the component parts of the optical system to learn the capabilities of each and the limitations that each part imposes on the total system performance. Finally, we can examine the total system; the selection of suitable transmitter, optical fiber, and photodetector component parts; and ways for exploiting the capabilities and overcoming the limitations of these parts.

1.3 Optics

Classical optics, as such, will not be treated in this book. Many topics from classical optics are an essential part of optical communications technology and will be included in the appropriate sections.

Introductory courses and textbooks on optics will normally introduce the electromagnetic wave theory of light waves and derive some of the characteristics of optical phenomena on this basis. Reflection and refraction, interference phenomena, and optical properties of thin films can be studied as electromagnetic wave phenomena. The electromagnetic wave theory of light wave propagation is necessary for the study of optical fibers.

Another important aspect of optics, the generation and absorption of light, is necessary for the study of transmitting and receiving devices for optical communication systems. This subject is included in many optics texts, as well as in texts and courses on semiconductors and atomic physics. Our treatment of this aspect of optics will be similar to that in semiconductor electronics courses. It is based on concepts from atomic physics but does not make any direct use of quantum mechanics.

Optical devices, including lenses, filters, prisms, interferometers, gratings, and devices based on electrooptic effects, are important in optical systems, including some communication systems. However, they are not discussed in this book except by reference to places where such devices might be used.

A variety of good books on optics is available. Several are listed as references [1.3] to [1.6]. The first three are suitable for an introduction to the subject. The last is a well-recognized reference, more thorough and rigorous than most introductory level books.

1.4 Units and Symbols

Much of the material in this book is in the form of mathematical developments and equations. The symbols in the equations represent physical quantities, physical constants, and other quantities that influence the behavior of the phenomenon being studied. Care in the interpretation of the symbols and in handling the units associated with the symbols is important for the understanding and correct use of the results.

1.4.1 *Units*

The standards for units and symbols will be IEEE Standards [1.12], incorporating SI units, in cases where these standards are applicable. In many cases, IEEE Standards identify alternative units for a quantity that can be used in different contexts; in such cases, the unit chosen is in common use in the current literature on optical communications. In many cases, two or more different units may be used; for example, energy may be expressed in joules, J, or in electronvolts, eV, depending on context.

It is expected that the reader is familiar with the units being used and that conversion between alternative units for the same quantity will present no difficulty.

1.4.2 *Symbols*

IEEE Standards define letter and graphical symbols for many of the quantities to be used. We will follow IEEE Standards as matter of first choice, but must, in some cases, recognize that standard practice in optical communications and closely related disciplines is different. In such cases current usage, especially when widely practiced, will take precedence over the IEEE definition.

In a few cases, the proper choice of a symbol is not clear. The fact that we are dealing with a subject that has deep roots in both electrical engineering and classical optics, where different symbols are used for frequency, for example, presents a particularly difficult dilemma. We will use f to represent frequency, although expressing the energy of a photon as hf rather than $h\nu$ will be controversial with some readers.

One caution to students seems in order. Symbols are used to represent specific quantities. It is unavoidable that the same symbol will be used in different places to represent different quantities. If an equation is seen in terms of the symbols themselves rather than in terms of the quantities they represent, unnecessary confusion can result. The symbol "*e*" can represent a time-varying electromagnetic field, the charge on an electron, or the base for natural logarithms. The symbol "*α*" has different meanings. The symbol "*k*" might represent Boltzmann's constant, a wave propagation constant, or a summation index. It is possible in almost all cases to determine from context the proper interpretation for such symbols.

CHAPTER

Communication Systems

This chapter will introduce some general principles of communication systems. This background provides the context in which the components of optical communication systems will be considered. The student who has a background in communications will find it easy reading, but should review this chapter for the perspective it provides and for some definitions of terminology and symbols to be used in subsequent chapters.

A long-range communication system relies upon the use of a "carrier," a signal that can be transmitted over a long distance with minimum attenuation and distortion. In a light-wave communication system, the carrier is an electromagnetic wave with wavelength in the optical region of the electromagnetic spectrum, and the channel can be an optical fiber. The information to be transmitted over the system is contained in an input signal, $x(t)$, which modulates the carrier for transmission through the system.

Among the considerations in the design or analysis of the communication system are the manner in which the information to be transmitted is associated with the carrier, the bandwidth of the transmitted signal, and the noise and distortion that are inevitably associated with the received signal. In this chapter, we will introduce these considerations as background for the study of the major components of optical communication systems in Chapters 3 through 8. We return to the systems point of view in the final chapter.

2.1   **Continuous-Wave Modulation**

Modulation is the operation by which an information-bearing signal, $x(t)$, is attached to the sinusoidal carrier wave. For the light wave, we will consider that the electric field vector E is the carrier to be modulated; the magnetic field vector and the power can be found from the E field. The carrier is assumed to have the form $e(t) = E \cos(\omega_c t + \theta)$. This carrier has three parameters that can be controlled: amplitude, frequency, and phase. If one of these parameters is made to vary in proportion to $x(t)$, the carrier is said to be modulated by $x(t)$. A demodulator at the receiver can recover $x(t)$ from the received signal.

We will treat two classes of modulation: continuous-wave modulation and pulse modulation. For our purposes, continuous-wave (CW) modulation is modulation imposed on a sinusoidal carrier as discussed above. Pulse modulation uses discrete pulses rather than a continuous sine wave as the carrier; pulse code modulation (PCM) is perhaps the best known of the pulse modulation methods.

2.1.1   *Amplitude and Intensity Modulation*

With amplitude modulation (AM), the amplitude E takes the form $E[1 + mx(t)]$, where m is a modulation index. We restrict $mx(t)$ to the range $\{-1 \leq mx(t) \leq 1\}$ so that the resultant amplitude can never become negative.

The schematic diagram of Figure 2.1.1 illustrates the modulation of the amplitude of a sinusoidal carrier by using a variable-gain amplifier, with $x(t)$ controlling the gain. At the output of this amplifier, the amplitude carries a replica of the modulating signal. A demodulator at the receiver can recover the $x(t)$. In a light-wave system, the intensity of the light generated by a light source can be controlled by a modulating signal, thus producing amplitude, or intensity, modulation of the output lightwave.

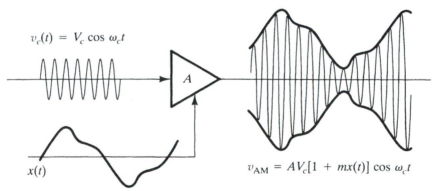

Figure 2.1.1 *Amplitude modulator. This figure illustrates one method for producing amplitude modulation. The gain of the amplifier is controlled by the modulating signal $x(t)$; the carrier amplification is $A[1 + mx(t)]$.*

When the modulating signal is a sinusoid, $\cos(\omega_m t)$, the AM wave has the form

$$e_{AM}(t) = E[1 + m\cos(\omega_m t)]\cos(\omega_c t)$$

$$= E\cos(\omega_c t) + mE\cos(\omega_m t)\cos(\omega_c t)$$

$$= E\cos(\omega_c t) + \frac{mE}{2}[\cos(\omega_c + \omega_m)t + \cos(\omega_c - \omega_m)t] \qquad (2.1.1)$$

The modulated wave has components at three different frequencies: the carrier frequency, ω_c, and modulation terms at $(\omega_c - \omega_m)$ and $(\omega_c + \omega_m)$. The frequencies due to ω_m are called modulation sidebands. If $x(t)$ consists of the sum of several sinusoids and ω_m is the highest frequency of these components of $x(t)$, we can conclude that ω_m represents the bandwidth of $x(t)$. The bandwidth of the modulated carrier wave is $2\omega_m$, twice that of the modulating signal.

In a more general representation, the form of $x(t)$ is not specified except that it is assumed to have a Fourier transform, $X(\omega)$, which is band-limited to a bandwidth B, that is,

$$X(\omega) = 0 \quad \text{for } |\omega| > \omega_B, \quad \omega_B = 2\pi B$$

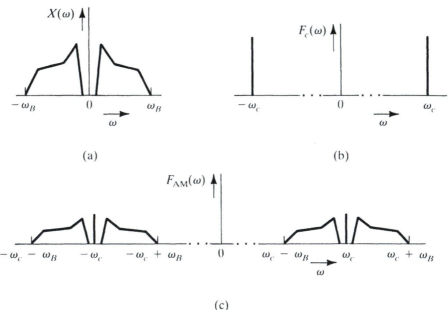

(a)

(b)

(c)

Figure 2.1.2 *Spectrum of an AM wave. (a) Spectrum of the modulating signal. It has bandwidth $\omega_B = 2\pi B$ and is centered at zero frequency. (b) Spectrum of the unmodulated carrier. (c) Spectrum of the amplitude-modulated carrier. The spectrum of the modulating signal is translated from base-band to the carrier frequency ω_c. See Equation (2.1.2).*

The Fourier transform of $E[1 + mx(t)] \cos(\omega_c t)$ is then [1.7, p. 124][1.8, p. 116]

$$F_{AM}(\omega) = \frac{E}{2}\{\delta(\omega - \omega_c) + mX(\omega - \omega_c) + \delta(\omega + \omega_c)$$

$$+ mX(\omega + \omega_c)\} \tag{2.1.2}$$

Since $X(\omega)$ is zero for $|\omega| > \omega_B$, the spectrum of the AM signal extends from $(\omega_c - \omega_B)$ to $(\omega_c + \omega_B)$; the bandwidth, in hertz, is $2B$. The spectrum of the AM wave is illustrated in Figure 2.1.2.

To transmit the AM signal without distortion will require a transmission channel that will pass all frequencies within this spectrum with the same attenuation and time delay. If either the relative amplitudes or the relative phase angles among the carrier and sidebands are disturbed, the modulation carried by the envelope will be distorted.

EXAMPLE 2.1

In an amplitude-modulation wave, define the modulating signal to be $x(t) = 3 \sin \omega_1 t + \sin 3\omega_1 t$ and the modulation index $m = 0.2$. The carrier frequency is 2 MHz and the fundamental frequency of the modulating signal is 15 kHz. The amplitude of the carrier is 3 kV/m and the power in the unmodulated carrier is 10 mW. Calculate the amplitude and frequency of each term in the spectrum of the modulated wave, the bandwidth of the modulated wave, and the percentage of the total power in the carrier and each pair of sidebands.

The equation for the modulated wave is

$$e(t) = 3[1 + mx(t)] \cos \omega_c t \text{ (kV/m)}$$
$$= 3 \cos \omega_c t + 0.9[\sin(\omega_c + \omega_1)t - \sin(\omega_c - \omega_1)t]$$
$$+ 0.3[\sin(\omega_c + 3\omega_1)t - \sin(\omega_c - 3\omega_1)t]$$

There are five terms, with frequencies 1.955, 1.985, 2.000, 2.015, and 2.045 MHz. The bandwidth is $2.045 - 1.955 = 0.090$ MHz or 90 kHz.

The amplitudes of the terms in this equation for $e(t)$ are 3, 0.9, and 0.3 kV/m. The carrier amplitude, 3 kV/m, is unaffected by the presence of modulation. There are two pairs of sidebands, with amplitudes of 0.9 and 0.3 kV/m.

The power in each term is proportional to the amplitude squared:

Frequency	Amplitude	Amplitude squared	Power (mW)	Percent of total power
ω_c	3.0	9.0	10.0	83.3
$\omega_c \pm \omega_1$	0.9	0.81	1.8	15.0
$\omega_c \pm 3\omega_1$	0.3	0.09	0.2	1.7
			12.0	100.0

The power at the carrier frequency, 10 mW, is unaffected by the presence of modulation. The total power at the sideband frequencies is 2.0 mW. The total power is increased by 20 percent due to the presence of the modulation sidebands.

Intensity modulation is similar to AM in principle but differs from it in one important detail. The average power of the transmitted wave, rather than its amplitude, varies in proportion to $x(t)$.

$$P_o(t) = P_c[1 + mx(t)]\ (W) \tag{2.1.3}$$

Light-wave systems will normally use intensity, rather than amplitude, modulation.

The spectrum of the intensity-modulated wave can be found by taking the Fourier transform of the electric field phasor. Since the power P_c is proportional to the square of the amplitude E_c, the amplitude has the form

$$E(t) = E_c\left[1 + mx(t)\right]^{\frac{1}{2}} \tag{2.1.4}$$

Thus, $[1 + mx(t)]^{\frac{1}{2}}$ is the waveform that must modulate the carrier amplitude if the intensity is to vary in proportion to $x(t)$. If $x(t)$ is again assumed to be a sinusoid, it is clear that the waveform of $E(t)$ is not sinusoidal.

To find the Fourier transform of this $E(t)$ we can expand the $[1 + mx(t)]^{\frac{1}{2}}$ in an infinite power series:

$$[1 + mx(t)]^{\frac{1}{2}} = 1 + \frac{1}{2}[mx(t)] - \frac{1 \cdot 1}{2 \cdot 4}[mx(t)]^2 + \frac{1 \cdot 1 \cdot 3}{2 \cdot 4 \cdot 6}[mx(t)]^3 + \dots \tag{2.1.5}$$

When $x(t) = M\cos(\omega_m t)$, the higher powers of $mx(t)$ will give corresponding harmonics of ω_m. The spectrum thus has infinite bandwidth. However, when the magnitude of mM is small, only a few terms are required for a good description of the nonsinusoidal waveform. The spectrum is then approximated by one of finite bandwidth; when mM is sufficiently small, the bandwidth is the same as for an AM wave.

EXAMPLE 2.2

The intensity of a light wave carrier is modulated by a signal $mx(t) = 0.8\cos\omega_1 t$. Calculate the amplitude spectrum of this intensity-modulated wave.

The equivalent amplitude-modulation expression is

$$e(t) = E\left[1 + 0.8\cos\omega_1 t\right]^{\frac{1}{2}}\cos\omega_c t$$

The power series expansion of the amplitude would give an infinite series. We will use only the first three terms.

$$[1 + 0.8 \cos \omega_1 t]^{\frac{1}{2}} = 1 + 0.4 \cos \omega_1 t - 0.08 \cos^2 \omega_1 t + \ldots$$
$$\approx 1 + 0.4 \cos \omega_1 t - 0.04 - 0.04 \cos 2\omega_1 t$$
$$= 0.96 + 0.4 \cos \omega_1 t - 0.04 \cos 2\omega_1 t$$

The equation for the modulated carrier is

$$e(t) = E [0.96 + 0.4 \cos \omega_1 t - 0.04 \cos 2\omega_1 t] \cos \omega_c t$$
$$= 0.96E \cos \omega_c t + 0.2E[\cos (\omega_c - \omega_1)t$$
$$+ \cos (\omega_c + \omega_1)t] - 0.02E[\cos (\omega_c - 2\omega_1)t + \cos (\omega_c + 2\omega_1)t]$$

To determine whether this $e(t)$ is or is not an adequate approximation, we can square the amplitude series and compare the result with $1 + 0.8 \cos \omega_1 t$, the desired intensity modulation.

$$(0.96 + 0.4 \cos \omega_1 t - 0.04 \cos 2\omega_1 t)^2 = 1.002 + 0.752 \cos \omega_1 t$$
$$+ 0.003 \cos 2\omega_1 t - 0.016 \cos 3\omega_1 t$$
$$- 0.001 \cos 4\omega_1 t$$

The amplitude of the fundamental (desired) frequency term, 0.752, is approximately 6 percent smaller than its value. For some purposes a more accurate approximation (more terms in the power series expansion) may be needed. Omission of these additional terms, by band-limiting, for example, represents distortion.

2.1.2 Frequency Modulation

The sinusoidal carrier can be expressed as $E_c \cos(\alpha)$, where $\alpha = \omega t + \theta$. This phase angle includes one term, ωt, that depends on the frequency and another, θ, that is a phase displacement from ωt. Two different ways for modulating α are to modulate the frequency, ω, or the phase, θ. Frequency modulation (FM) and phase modulation (PM) are two aspects of angle modulation. We will restrict our attention to FM.

The instantaneous frequency is defined as the time derivative of the phase angle.

$$\omega(t) = \frac{d\alpha}{dt} \tag{2.1.6a}$$

$$\alpha(t) = \int \omega(t) \, dt$$
$$= \omega_{co} t + \int \Delta\omega \, x(t) \, dt + \theta \tag{2.1.6b}$$

where $\Delta\omega$ is the radian-frequency deviation from the carrier frequency per unit of $x(t)$. If $x(t) = \cos(\omega_m t)$, then

$$\alpha(t) = \omega_{co}t + \frac{\Delta\omega}{\omega_m}\sin\omega_m t + \theta \qquad\qquad (2.1.7a)$$

$$\omega(t) = \omega_{co} + \Delta\omega\cos\omega_m t \qquad\qquad (2.1.7b)$$

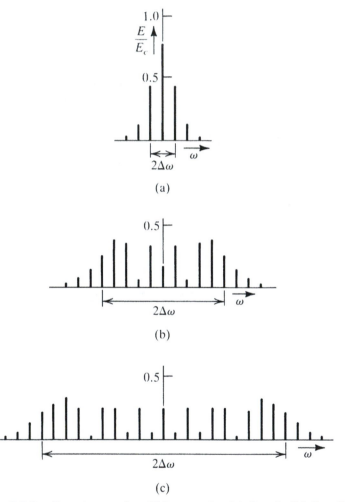

(a)

(b)

(c)

Figure 2.1.3 *Spectrum of an FM wave for (a) $\beta = 1$, (b) $\beta = 5$, and (c) $\beta = 10$. The center frequency is ω_c. Each line in this line spectrum represents one sideband of the FM spectrum; the spacing between sidebands is ω_m. The range over which the actual frequency varies is $2\Delta\omega$. The modulation index, $\beta = \Delta\omega/\omega_m$, is equal to the number of sidebands contained within $\Delta\omega$. The amplitude of the sidebands of order greater than β decreases with increasing order. See Equation (2.1.10).*

The frequency-modulated carrier wave is now

$$e(t) = E_c \cos \alpha(t)$$
$$= E_c \cos [\omega_{co}t + \beta \sin (\omega_m t) + \theta] \tag{2.1.8}$$

where $\beta = (\Delta\omega/\omega_m)$ is the modulation index. The angle θ, an arbitrary phase shift, will be set equal to zero.

This $e(t)$ is not sinusoidal since the angle is not increasing uniformly with time. We can expand it into a Fourier series to find the bandwidth of $e(t)$. We do this by first using the identity for the cosine of the sum of two angles

$$e(t) = E_c [\cos (\omega_{co}t) \cos (\beta \sin (\omega_m t)) - \sin (\omega_{co}t) \sin (\beta \sin (\omega_m t))] \tag{2.1.9}$$

The $\cos (\beta \sin (\omega_m t))$ and $\sin (\beta \sin (\omega_m t))$ can each be expanded into an infinite series [1.9, formula 818] to give, finally

$$\begin{aligned}
e(t) = E_c \{ &J_o(\beta) \cos (\omega_c t) \\
&- J_1(\beta) [\cos (\omega_c - \omega_m)t - \cos (\omega_c + \omega_m)t] \\
&+ J_2(\beta) [\cos (\omega_c - 2\omega_m)t + \cos (\omega_c + 2\omega_m)t] \\
&- J_3(\beta) [\cos (\omega_c - 3\omega_m)t - \cos (\omega_c + 3\omega_m)t] + \ldots \}
\end{aligned} \tag{2.1.10}$$

The coefficients $J_n(\beta)$ are Bessel functions. They can be evaluated using tables or power series [1.10].

The spectrum of this FM wave is shown in Figure 2.1.3. It consists of an infinite number of sidebands, spaced at intervals ω_m, on each side of the carrier frequency. It is evident in Figure 2.1.3, and it is true in general, that most of the significant sidebands lie within a range only slightly greater than the range over which the carrier frequency actually varies. Only two or three pairs of sidebands beyond the actual frequency-deviation range are needed to include most of the energy in the complete, infinite-series representation of $e(t)$.

2.2 Pulse-Code Modulation

2.2.1 Principles of Pulse Modulation

There are a variety of pulse-modulation techniques, each of which modulates some parameter of a sequence of pulses. Among the pulse parameters that can be modulated are the amplitude, duration, spacing, or position within a fixed time interval.

There is a major difference between pulse-modulation systems and the sinusoidal carrier systems described previously. With the sinusoidal carriers, the parameter being modulated is a continuous variable. In principle, we can examine the carrier wave continuously and determine the value of the modulating signal, $x(t)$, at any time. In a pulse system, the pulses are transmitted at discrete times. We can determine the values for $x(t)$ at these discrete times, but we cannot determine directly from the modulated pulses values for $x(t)$ at times between pulses. Pulse systems are therefore called discrete communication systems; sinusoidal-carrier systems are called continuous communication systems.

In a discrete communication system we take samples of $x(t)$ at regular intervals and transmit only these sample values. It can be shown that if we take samples at a rate that is greater than twice the highest frequency component of $x(t)$, then we can reconstruct the original continuous $x(t)$ completely and accurately [1.11]; the $x(t)$ must be band-limited, and the sampling rate must be greater than twice the bandwidth. For example, a speech signal, band-limited so that it contains no frequencies higher then 3.5 kHz, can be transmitted through a discrete communication channel using 8000 samples per second.

2.2.2 *Pulse-Code Modulation*

The pulse-carrier system we will examine is pulse code modulation (PCM). In PCM, each sample of $x(t)$ is quantized into a finite number of discrete amplitude levels. These sample values are then coded, usually into a binary number, and the code is transmitted using pulse amplitude modulation. In binary PCM, each pulse carries one bit of information. The number of pulses per sample depends on the accuracy with which we choose to code the value of each sample. For speech signals, coding each sample into 128 different possible amplitudes can give high-quality speech transmission; this would require seven bits per sample. We could thus transmit speech through a discrete communication system with 56 kbits per second.

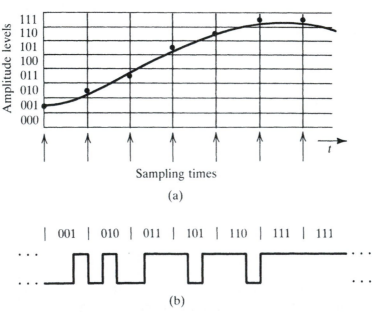

Figure 2.2.1 *Sampling a continuous waveform for PCM coding.* (a) *Quantized samples representing a continuous $x(t)$.* (b) *A binary coded pulse sequence representing the samples of* (a).

Figure 2.2.1 illustrates a continuous-time signal $x(t)$ that is sampled at uniform intervals and quantized into eight levels. It is evident that the quantized samples, represented by the dots on or near the $x(t)$ curve, do not provide an exact representation of $x(t)$. However, if the sampling interval is sufficiently short and the sampling levels sufficiently close, the error in this representation can be made arbitrarily small.

2.2.3 *Bandwidth*

Every communication system has a finite bandwidth. A signal with spectral width larger than the system bandwidth cannot be transmitted through the system without distortion. It is important in system design to know the bandwidth required for satisfactory transmission of the signals to be used, the transmission characteristics of the system at all frequencies in the signal spectrum, and the effects of the transmission characteristics on the signals. We will examine here the spectrum required for PCM signals and the effects of inadequate system bandwidth on the PCM signals. In later chapters, we will see how various parts of the optical communication system influence the system's frequency response and its overall effective bandwith.

The bandwidth required for the baseband PCM signal is determined by the pulse shape and the pulse rate. The spectra of several pulse shapes are shown in Figure 2.2.2. These spectra extend to infinity, but they can be limited to a bandwidth that includes little beyond the first zero of the spectrum. One would expect, from examining these spectra, that a bandwidth of $2/T$ Hertz might be adequate for transmitting such pulses. It will be shown that this bandwidth is more than is needed for a PCM system.

It is not necessary that the shape of the transmitted pulse be preserved. In PCM systems, the shape of the pulse in the receiver is important only to the extent that shape influences our ability to decide whether a binary *1* or *0* was transmitted. The system designer must compute or estimate the bandwidth of the system, determine the effect of this bandwidth on the pulse shape, and confirm that the resulting received pulse shape is satisfactory.

We can calculate the effect of a low-pass filter on a rectangular pulse of duration T by finding the response of the filter to a unit step and subtracting from this a similar response delayed by T seconds. The step response of an ideal low-pass filter is shown in Figure 2.2.3a.[1] The step response of a realizable low-pass filter, a three-pole maximally flat (Butterworth) filter, is shown in Figure 2.2.3b.[2] The characteristic of the step response of greatest interest is the "rise

---

[1] It is convenient to normalize results such as these in terms of a time-bandwidth product. Here it is the bandwidth, B, multiplied by the time, t. Thus $Bt = 0.5$ may represent a $B = 0.5$ MHz and $t = 1$ μs, or $B = 5$ kHz and $t = 0.1$ ms. It is convenient to use simple integers, such as $B = 1$, to study the effect of bandwidth on pulse shape. When we have selected a suitable bandwidth, we can then unnormalize by adjusting B and t to realistic values while maintaining the Bt product constant.
[2] The transfer function of the three-pole Butterworth filter having bandwidth ω_B is $1/[(s + \omega_B)(s^2 + \omega_B s + \omega_B{}^2)]$.

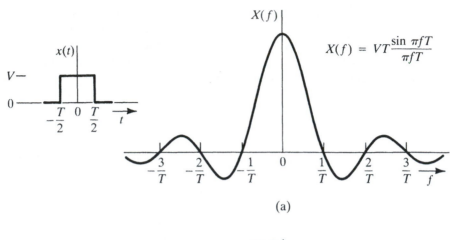

(a)

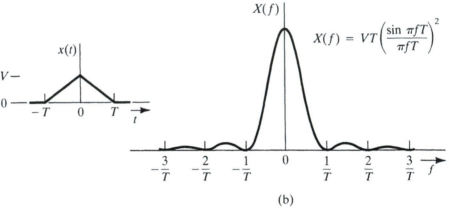

(b)

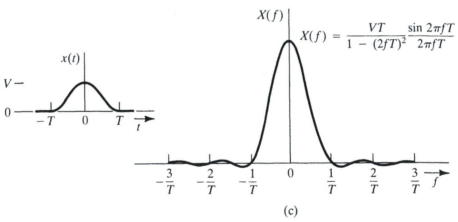

(c)

Figure 2.2.2 *Spectra of* (a) *rectangular pulse,* (b) *triangular pulse, and* (c) *raised cosine pulse. For each pulse, T is the duration between half-amplitude points.*

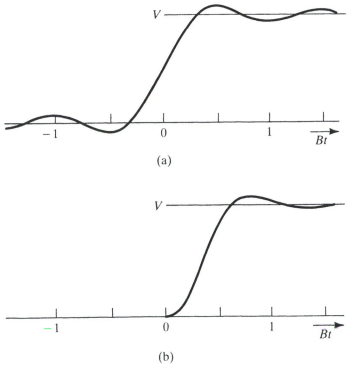

Figure 2.2.3 *Step response of (a) an ideal low-pass filter with cutoff frequency B, and (b) a three-pole maximally flat (Butterworth) filter with a 3 dB bandwidth B. The ideal filter, with an anticipatory transient of infinite duration, is unrealizable. The Butterworth filter is realizable. By increasing the number of poles in the Butterworth filter, its cutoff characteristic can be made to approach the ideal sharp cutoff.*

time," the time required for the voltage or current variable to rise from its initial value to its final value. Since most such responses approach their final values in some exponential manner, the rise time, strictly speaking, will be infinite. As we attempt to find a more realistic definition of rise time, it is apparent that no definition is clearly superior to all others. However, it is clear from Figure 2.2.3, that something in the range $1/2B < t_r < 1/B$ is appropriate.

Figure 2.2.4 shows the effect of filters of several bandwidths on a rectangular pulse of duration T; the filter is the three-pole Butterworth filter of Figure 2.2.3b. When $BT < 0.5$, the peak amplitude of the pulse is less than the value it could reach with larger bandwidths. When $BT > 1$, the flat top of the rectangular pulse begins to appear; since the flat top is not necessary for PCM demodulation, these larger bandwidths are not needed.

If the bandwidth is too small, the duration as well as the shape of the pulse will be affected. Smaller bandwidths will result in slower rise times and

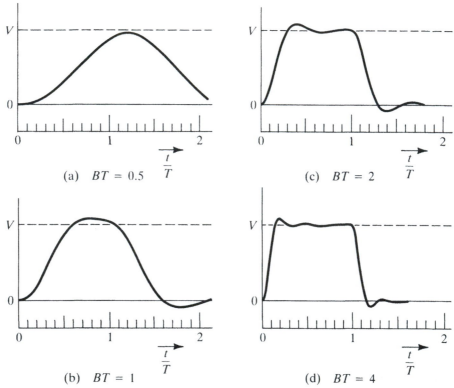

Figure 2.2.4 *Effects of bandwidth on the shape of a rectangular pulse. The input to the filter is in each case a rectangular pulse of duration T. The filter is the three-pole Butterworth filter of Figure 2.2.3(b). For $BT \geq 0.5$, the output pulses reach their full amplitudes before beginning to decay. For $BT > 1$, the output pulses have approximately the same shape and duration as the input pulse.*

longer-duration pulses. This is shown in Figure 2.2.5 for the three-pole Butterworth filter.

For input pulse shapes other than rectangular, we can find the output pulse shape by convolution of the input pulse with the impulse response of the filter. The duration of the impulse response depends on the bandwidth, as is indicated in Figure 2.2.6; the fact that the duration of the impulse response increases as the bandwidth decreases is true of essentially any filter. The amount of pulse stretching will be approximately equal to the duration of the impulse response. Input pulses that are long compared to the impulse response will not be stretched significantly; a short input pulse will produce a response that is approximately equal to the impulse response

The rate at which pulses can be transmitted is limited by the pulse duration and by the necessity to control the overlap between successive pulses. A specific time interval equal to or greater than the pulse duration can be assigned to

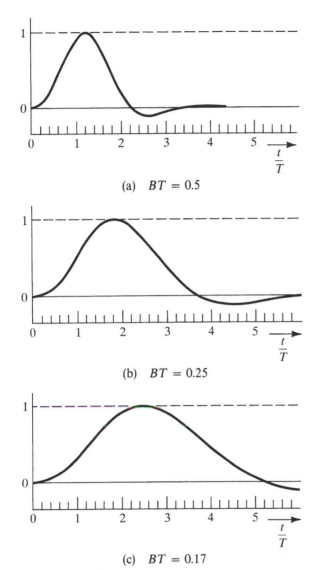

(a) $BT = 0.5$

(b) $BT = 0.25$

(c) $BT = 0.17$

Figure 2.2.5   *Effects of bandwidth on pulse duration. The input pulse and the filter are the same as for Figure 2.2.4. The shape of the output pulse resembles that of the impulse response; the duration of the output pulse is longer than that of the input and is approximately proportional to 1/B. These output pulses have been normalized to have the same peak amplitudes.*

each pulse. If, in the receiver, a pulse extends beyond its assigned time interval into the following interval, interference between adjacent pulses (intersymbol interference) may result in the receiver not being able to distinguish between them. Errors in binary *1/0* decisions will result.

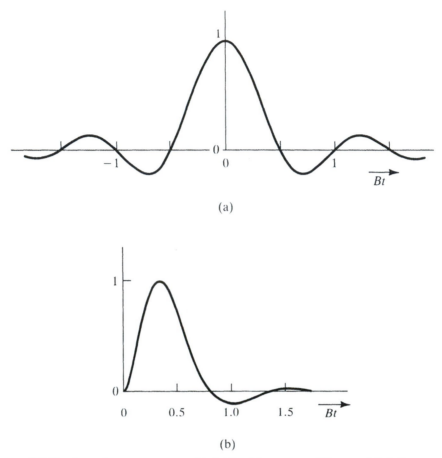

(a)

(b)

Figure 2.2.6 *Impulse response of (a) ideal low-pass filter and (b) three-pole Butterworth filter. The time scales are normalized to $1/B$.*

The effects of intersymbol interference can be seen in the "eye diagram," commonly used in system design for this purpose. It consists of the superposition of all possible symbol-to-symbol transitions (*1-0, 1-1, 0-1,* and *0-0*). An eye diagram for the case $BT = 0.5$ is shown in Figure 2.2.7; this diagram corresponds to the pulse shape shown in Figure 2.2.4a and for a pulse repetition period equal to the pulse duration. This diagram shows clearly the time at which the most reliable decision can be made, the margin for error, and the levels at which the decision threshold could be placed. We will refer again to the eye diagram in considering the effects of noise on the reliability of binary decisions in the receiver.

We can conclude by examining Figure 2.2.7 that $BT = 0.5$ should be adequate for a PCM system. If the pulse duration T is determined by the signaling rate, then $B = 1/2T$ is the bandwidth required. For example, we can transmit 10^6 bits per second using pulses having $T = 1$ μs and $B = 500$ kHz. As we reduce the bandwidth below $1/2T$, the "eye" would begin to close and binary decisions would be less reliable.

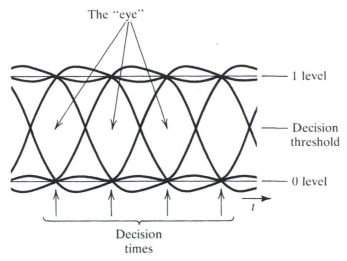

Figure 2.2.7 *Eye diagram. This diagram is derived from the step-response waveform of Figure 2.2.3(b) with the pulse duration equal to the pulse repetition period T and the bandwidth equal to 1/2T. The pulse shape may resemble any of those of Figure 2.2.4, depending on how many successive 1's are transmitted without an intervening 0. The output due to a single pulse would resemble that of Figure 2.2.4a*

For transmission through a light-wave PCM system, the continuous $x(t)$ is first sampled and quantized to produce a sequence of pulses that represents $x(t)$. These pulses then become the modulating signal for the sinusoidal light-wave carrier. The bandwidths discussed in this section are for the baseband PCM signal before it is used to modulate the light-wave carrier. This determines the highest modulating frequency, which is then used to find the transmission system bandwidth required for AM or FM. Inadequate transmission system bandwidth will have the same effects on pulse shape and intersymbol interference as would inadequate bandwidth at baseband.

EXAMPLE 2.3

A signal $x(t)$ is band-limited to frequencies in the range 0 to 3.5 MHz. It is to be sampled at $8 \cdot 10^6$ samples per second and quantized into 64 amplitude levels.

(a) Calculate the data rate and the bandwidth required for binary PCM transmission of this signal.

$\log_2 64 = 6$(bits/sample)

Data rate $= 6$(bits/sample) $\cdot\ 8 \cdot 10^6$(samples/second)
$= 48 \cdot 10^6$ bits/second

The signaling period is $T = \frac{1}{48} \,\mu s$

For $BT = 0.5, B = \dfrac{1}{2T} = 24 \cdot 10^6 \,\text{Hz} = 24 \,\text{MHz}$

(b) This binary PCM signal is to be transmitted through an intensity-modulated light wave system. What is the minimum acceptable bandwidth of this light wave transmission system?

The light wave carrier must pass a spectrum twice as wide as the spectrum of the baseband signals. Thus, bandwidth > 48 MHz. An intensity-modulated light wave system can transmit 48 Mbits per second if the system bandwidth is at least 48 MHz.

2.3 Noise

Noise in a communication system is represented as an undesired, randomly varying voltage or current that is present along with the desired signal in the receiver. A noise current can be characterized by its statistical parameters such as the average noise current and the average squared noise current. We can calculate its probable values or the range of most-probable values, but we cannot calculate the specific value the noise current will have at some future time as we could with a deterministic signal.

Figure 2.3.1 shows waveforms such as we might find by recording (a) a noise current and (b) a signal-plus-noise current. The effect of adding noise to the signal is to cause a degree of uncertainty about the exact value of the signal current. It is clear that if the signal is much larger than the noise, the uncertainty caused by the noise may be small and of no real consequence. If the noise is much larger than the signal, it may not be possible even to know that a signal is present.

We will examine the principal types of noise sources and the ways in which they can be represented in communication receiver circuits. In later chapters, specific noise sources and the effects of noise in communication systems will be treated in more detail.

2.3.1 *Thermal Noise*

The electrons in a resistor are always in motion; the velocities and directions of motion are random. The intensity of this motion is proportional to the temperature. The effect of this motion of electrons (and perhaps other charges) is to produce a randomly varying voltage or current that is measurable and that can cause measurable voltages and currents in other parts of the circuit in which the resistor is connected.

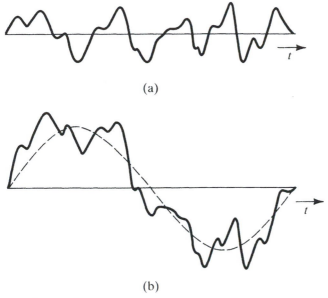

(a)

(b)

Figure 2.3.1 *Effect of adding a noise voltage to a signal voltage. (a) Noise voltage waveform. (b) Signal plus noise; S/N = 8 dB.*

This phenomenon can be represented in the circuit diagram by placing a voltage source, v_{th}, in series with the resistor or a current source, i_{th}, in parallel with it. The mean-squared values of the sources are:[3]

$$\langle v_{th}{}^2 \rangle = 4kTRB \quad (V^2) \tag{2.3.1a}$$

or

$$\langle i_{th}{}^2 \rangle = 4kTGB \quad (A^2) \tag{2.3.1b}$$

where

$k = 1.38066 \cdot 10^{-23}$ J/°K (Boltzmann's constant)

T = temperature, in degrees Kelvin

R = resistance in ohms

G = conductance in siemens $\left(G = \dfrac{1}{R} \right)$

B = bandwidth, in hertz

---

[3] Angle brackets, $\langle \ \rangle$, indicate the average value of the variable enclosed.

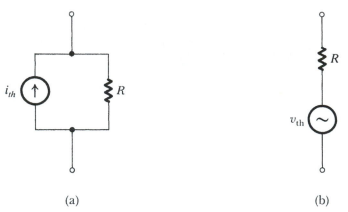

(a)   (b)

Figure 2.3.2 *Thermal noise sources.* (a) *Current source.* (b) *Voltage source.*

The equivalent circuits for the resistance and its associated noise sources are shown in Figure 2.3.2.

2.3.2 *Shot Noise*

The random emission of electrons from the cathode of a thermionic diode produces a current with average value I but with fluctuations about this average value. The fluctuations can be explained on the basis of random emission times and the discrete quantity of charge on the electron. The fluctuations are represented by a noise current, called shot noise, that is added to the average current. The total current can be expressed

$$i(t) = I + i_{sh}(t) \ (A) \tag{2.3.2}$$

The mean-squared value of the shot noise current is given by

$$\langle i_{sh}^2 \rangle = 2qIB \ (A^2) \tag{2.3.3}$$

where

$q = 1.60219 \cdot 10^{-19} C$ (the electronic charge).

I = average current, in amps

B = bandwidth, in hertz

Most electronic devices produce shot noise that can be calculated from Equation (2.3.3). In a photodiode, the average current, I, is proportional to the intensity of the incident light. Fluctuations about this average value are represented by Equation (2.3.3). The equivalent circuit for the photodiode would have two current sources, one for the signal current, I, and one for the shot

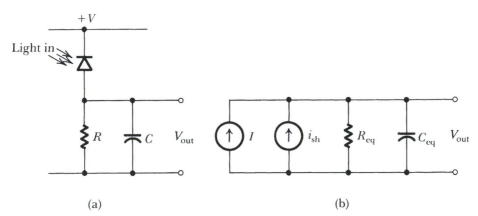

Figure 2.3.3 *Photodetector circuit including shot noise source.* (a) *Schematic circuit diagram.* (b) *Equivalent circuit showing signal and shot noise current sources.* R_{eq} *includes* R_L *and the equivalent resistance of the photodiode in parallel.* C_{eq} *includes* C, *the capacitance of the photodiode, and other parasitic capacitances.*

noise current, i_{sh}. Figure 2.3.3 shows a schematic photodiode circuit and its equivalent circuit.

2.3.3 *Total Noise*

The magnitude of a noise source is represented by the mean-squared value of a noise current or voltage.

To combine the effects of two or more independent noise sources in one circuit we must add mean-squared voltages or currents from the various sources. For a circuit in which a thermal noise source and a shot noise source are in parallel, the total noise current, i_n, is given by

$$\langle i_n^2 \rangle = \langle i_{th}^2 \rangle + \langle i_{sh}^2 \rangle \, (A^2) \tag{2.3.4}$$
$$= (4kTG + 2qI)B \tag{2.3.5}$$

Whenever two or more independent noise currents, or voltages, are added, the sum must be on a mean-squared basis.

2.3.4 *Signal-to-Noise Ratio*

A most important parameter in the analysis of communication systems is the signal-to-noise ratio in the receiver. This ratio is defined:

$$\frac{S}{N} = \frac{\text{Signal power}}{\text{Total noise power}} \tag{2.3.6}$$

It is usually expressed in dB. Thus

$$\frac{S}{N} = 10 \log \frac{P_s}{P_n} \ (\text{dB}) \tag{2.3.7a}$$

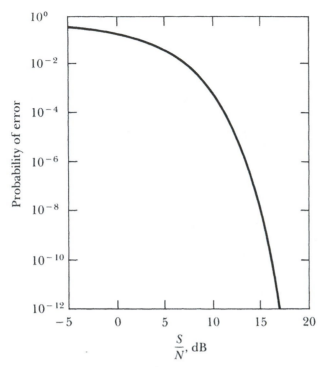

Figure 2.3.4   *Probability of error in binary decisions as a function of signal-to-noise ratio. Figure 9.1.4 shows a similar graph, plotted on a larger scale.*

or

$$\frac{S}{N} = 20\log\frac{V_s}{V_n} \text{ (dB)} \qquad\qquad \textbf{(2.3.7b)}$$

The quality of service provided by the communication system is directly related to this S/N. Figure 2.3.4 shows a graph that relates the probability of error in a digital communication system to the signal-to-noise ratio. Current practice in such systems often calls for a probability of error to be 10^{-9} or less. Figure 2.3.4 indicates that this will require S/N of 16 dB or higher.

EXAMPLE 2.4

In a photodiode circuit such as is shown in Figure 2.3.3, the load resistor is 100 ohms and the bandwidth is 100 MHz. The photodiode produces 0.5 amperes of signal current per watt of incident optical power: $I_s = 0.5\ P_i$. Find the signal-to-noise ratio in the output of the photodetector.

The *S/N* can be found from Equation (2.3.7b). The signal voltage developed across the load resistor is $V_s = I_s R$, and the noise voltage is $V_n = R[\langle i_{th}^2 \rangle + \langle i_{sh}^2 \rangle]^{\frac{1}{2}}$. Because in taking the ratio of voltages, the resistance R will cancel, we can use the ratio of signal and noise currents.

$$
\begin{aligned}
\langle i_{sh}^2 \rangle &= 2qI_s B \\
&= 2 \cdot 1.60219 \cdot 10^{-19} \cdot 0.5 P_i \cdot 10^8 \\
&= 1.60 \cdot 10^{-11} P_i \ (\mathrm{A}^2)
\end{aligned}
$$

$$
\begin{aligned}
\langle i_{th}^2 \rangle &= 4kTGB \\
&= 4 \cdot 1.38066 \cdot 10^{-23} \cdot 300 \cdot 0.01 \cdot 10^8 \\
&= 1.656 \cdot 10^{-14} \ (\mathrm{A}^2)
\end{aligned}
$$

The total mean-squared noise current in the load resistor, R, is

$$
\begin{aligned}
\langle i_n^2 \rangle &= \langle i_{sh}^2 \rangle + \langle i_{th}^2 \rangle \\
&= 1.60 \cdot 10^{-11} P_i + 1.656 \cdot 10^{-14} \ (\mathrm{A}^2)
\end{aligned}
$$

The signal current is $I_s = 0.5 \ P_i$. The signal-to-noise ratio is

$$
\frac{S}{N} = 10 \log \frac{I_s^2}{\langle i_n^2 \rangle}
$$

$$
= 10 \log \frac{(0.5 P_i)^2}{1.6 \cdot 10^{-11} P_i + 1.656 \cdot 10^{-14}} \ (\mathrm{dB})
$$

To find P_i for a given *S/N*, we could solve a quadratic equation, or we could plot *S/N* versus P_i. If we need a signal-to-noise ratio of 16 dB or better, we must require that the input optical power be at least 1.62 μW or higher. Note that in this example and with an input power of a few microwatts, the total noise is essentially equal to the thermal noise; the contribution of shot noise to the total is negligible.

2.4 System Design

System engineering, or system design, involves many considerations. The starting point is usually some specifications describing the objectives of the system, its performance parameters, or the needs the system should serve; from these are determined data rates, bandwidth requirements, and minimum received S/N. Specifications may also include constraints such as cost, operating environment, reliability, and lifetime. The design team will review the alternative technologies that appear to be applicable and evaluate the more attractive options. A preliminary design of the major system components and subsystems will define further, more detailed, specifications and may identify questions and problems overlooked in the earlier design studies. The detailed design is then begun.

Although the subject of this book is optical communication systems, most of the chapters will treat the devices and subsystems that are required for such systems. Lasers, photodiodes, and optical fibers all have characteristics that can limit the rate at which information can be transmitted through the communication system and the distance over which the system can function effectively. A thorough understanding of components that make up the system is necessary before a successful design can be executed. We will return to the total system point of view in the last chapter.

PROBLEMS

2.1 In an amplitude-modulated sinusoidal-carrier system, the carrier has amplitude of 1 kV and frequency of 100 MHz. The modulation index is $m = 0.15$ and the modulating signal is $x(t) = 3 \cos \omega_1 t + 2 \cos 2\omega_1 t + \cos 3\omega_1 t$, where $\omega_1 = 2\pi \cdot 2$ MHz. The average power in the unmodulated carrier is 10 kW.
   (a) Calculate all of the frequencies present in the spectrum of this amplitude-modulated wave, the amplitude of each, the power in the carrier and each pair of sidebands, and the total bandwidth of the AM signal.
   (b) Determine the peak amplitude and the peak instantaneous power of this AM signal.

2.2 An unmodulated optical carrier has an amplitude of 10 kV/m. If the modulating signal is $x(t) = 5 \cos \omega_1 t + 3 \cos 2\omega_1 t + \cos 3\omega_1 t$, what is the maximum value that the modulation index m can have when the maximum field intensity is limited to (a) 17 kV/m? (b) 25 kV/m?

2.3 Calculate the effect on the fundamental-frequency amplitude of including one additional term (the cubic term) in the power series expansion in Example 2.2.

2.4 An intensity-modulated wave has a carrier power level of 1 W and modulation $mx(t) = 0.5 \cos \omega_1 t + 0.2 \cos 3\omega_1 t$. Use the first three terms in the power series to calculate the spectrum of the intensity-modulated wave.

2.5 Estimate the bandwidth required for a frequency-modulated wave when the modulating signal is a single sinusoid of frequency 500 kHz and $\beta =$ (a) 10, (b) 5, (c) 1, (d) 0.1.

2.6 Estimate the bandwidth required for a frequency-modulated wave when the modulating signal has a spectrum extending from 100 Hz to 20 kHz and $\beta = 10$.

2.7 An instrumentation system samples 15 different variable parameters, each at 1 ms intervals. Each sample is quantized into 100 amplitude levels. What is the total data rate, in bits per second, required to transmit all of these data?

2.8 A video surveillance system has a baseband spectral width of 3.5 MHz. It is quantized for transmission through a binary PCM system using 32 amplitude levels. What will be the minimum data rate, in bits per second, required for transmission of this video signal?

2.9 (a) Use Laplace transform methods to find the step response of a three-pole Butterworth low-pass filter. The transfer function, $H(s)$, for this filter is given in a footnote in Section 2.2.3. Sketch the output voltage versus time and compare with Figure 2.2.3b.

(b) Sketch the step response of a three-pole Butterworth low-pass filter having a 3-dB bandwidth of 3.8 MHz.

(c) Sketch the magnitude of the frequency response, $H(j\omega)$, of the filter of part (b) above. Normalize this response to unity (0 dB) at $\omega = 0$.

2.10 The input to a three-pole Butterworth low-pass filter consists of two rectangular pulses, each 1 μs duration, with a 1 μs interval between the end of the first pulse and the start of the second. Plot the output of the filter when the bandwidth is (a) 1.0 MHz, (b) 0.5 MHz, (c) 0.25 Mhz.

2.11 A PCM communication system uses intensity modulation to transmit binary data at the rate 1.0 Mbit/second. The modulating waveform is a rectangular pulse that has amplitude 1 or 0 to represent a binary symbol. Each pulse has duration $T = 1$ μs.

(a) Sketch the pulse waveform representing the binary data sequence *010011101*.

(b) If the received signal is to correspond to that of the eye diagram of Figure 2.2.7, what is the minimum bandwidth for transmitting these data using an intensity-modulated light wave?

2.12 (a) What is the rms noise voltage produced in a 1000-ohm resistance when its temperature is 50°C (note °C, not °K) and the effective bandwidth is 75 MHz?

(b) A 5000-ohm resistance is connected in parallel with the 1000-ohm resistance of (a). What is the resulting rms noise voltage?

2.13 The current flowing in a semiconductor diode is 30 mA (dc). If no further information is given, what can be said about the shot noise produced in this diode?

2.14 The signal current in a photodetector such as that of Figure 2.3.3 is 0.1 mA (dc). The load resistance is 300 ohms; its temperature is 350°K. The bandwidth of the RC load is 1 GHz. What is the signal-to-noise ratio at the output of the photodetector ?

2.15 In the photodetector defined in Problem 2.14, the signal current of 0.1 mA is produced by an input optical power of 0.25 mW. The signal current is proportional to the input optical power; $I_s = 0.4$ P_i. Since the S/N is higher than required for good performance, the system should function adequately with lower input power. What power would be required for an output S/N of 20 dB?

CHAPTER

Light and Electromagnetic Waves

Light wave propagation is an electromagnetic phenomenon. It is governed by the same equations as are used for microwaves and other electromagnetic waves. In this chapter, we will review the wave propagation equations, derived from Maxwell's equations. In the following chapter, these results will be applied to study optical fibers and their characteristics.

3.1 The Nature of Light

In attempting to describe the nature of light we face an apparent contradiction. On the one hand, light waves are known to behave like any other electromagnetic wave; the physical laws describing propagation, reflection and refraction, and attenuation, developed and refined in the hundred years since Maxwell's equations were introduced, accurately describe the behavior of light waves. On the other hand, some optical phenomena are defined and studied in terms of photons rather than waves. The discovery of the photoelectric effect showed beyond any doubt that the electromagnetic wave model was not adequate for describing the absorption of light by a photosensitive material.

In some cases we rationalize that when there are very many photons they are no longer distinguishable as individual particles but behave collectively as a continuum—an electromagnetic wave. This point of view is acceptable for studying propagation phenomena but cannot explain phenomena such as the interaction of light waves, or photon beams, with the semiconductor material in a photodiode.

The wave versus particle dilemma can be addressed in a more formal, and more productive, way using techniques beyond the scope of this introductory book. Quantum mechanics explains some of the properties of photons in terms of wave packets. Much progress has been made in reconciling the dilemma of the wave versus the particle. Quantum mechanics also calls on the uncertainty principle to define some questions as unanswerable.

In this book, we will use both the electromagnetic wave and the photon concepts, each in the places where it best matches the phenomena we are studying. It is comforting to know that although these two models for light wave phenomena are different, we can be assured that they are not really in conflict [2.1].

3.2   Maxwell's Equations

Maxwell's equations provide the basis for the development of equations for electromagnetic wave propagation in dielectric materials and other optical phenomena that will be needed in subsequent chapters. It is assumed that the student has some prior familiarity with electromagnetic fields. The material in this section is included for reference and to provide a familiar starting point for the developments to follow.

Maxwell's equations in differential form are:

$$\mathbf{\nabla} \times \mathbf{E} = -\frac{\partial \mathbf{B}}{\partial t} \tag{3.2.1a}$$

$$\mathbf{\nabla} \times \mathbf{H} = \mathbf{J} + \frac{\partial \mathbf{D}}{\partial t} \tag{3.2.1b}$$

$$\mathbf{\nabla} \cdot \mathbf{B} = 0 \tag{3.2.1c}$$

$$\mathbf{\nabla} \cdot \mathbf{D} = \rho \tag{3.2.1d}$$

There are, in addition, the three constitutive relationships:

$$\mathbf{D} = \epsilon \mathbf{E} \qquad \mathbf{B} = \mu \mathbf{H} \qquad \mathbf{J} = \sigma \mathbf{E} \tag{3.2.2}$$

where ϵ, μ, and σ are properties of the medium.

We usually assume that ϵ, μ, and σ are not functions of time and that the medium is linear, homogeneous, and isotropic; ϵ, μ, and σ are therefore constant and uniform throughout the medium. However, there are some circumstances under which these assumptions will not be valid. In such cases we must reexamine the consequences of these assumptions; one such case is the graded-index fiber in which the dielectric constant is a function of the radial distance from the axis of the fiber.

When **E** and **H** are sinusoidal functions of time, the time dependence can be represented with the exponential exp($j\omega t$). Equations (3.2.1a) and (3.3.1b) then become

$$\nabla \times \mathbf{E} = -\mu\frac{\partial \mathbf{H}}{\partial t} = -j\omega\mu\mathbf{H} \tag{3.2.3a}$$

$$\nabla \times \mathbf{H} = (\sigma + j\omega\epsilon)\mathbf{E} \tag{3.2.3b}$$

These are vector equations. For the present, we will use rectangular coordinates with unit vectors, $\mathbf{a}_x$, $\mathbf{a}_y$, and $\mathbf{a}_z$, designating the directions along the three coordinate axes. The electric field vector, for example, is $\mathbf{E} = \mathbf{a}_x E_x + \mathbf{a}_y E_y + \mathbf{a}_z E_z$. By equating corresponding components on the right and left sides, Equations (3.2.3) can each be written as three scalar equations. For example, the x components of Equation (3.2.3b) give

$$\frac{\partial H_z}{\partial y} - \frac{\partial H_y}{\partial z} = (\sigma + j\omega\epsilon)E_x \tag{3.2.4}$$

When electromagnetic fields appear at the interface between two different materials, Maxwell's equations require that certain relationships must exist between the fields on the two sides of the interface. These are the boundary conditions. They are:

1. The tangential components of **E** are equal on the two sides of the interface.
2. The difference between the normal components of **D** is equal to the charge density.
3. The difference between the tangential components of **H** is equal to the current density.
4. The normal component of **B** is continuous across the interface.

The boundary conditions at the interface between two dielectric materials, where there is no free charge and no current, are:

1. Tangential components of **E** and **H** are continuous across the interface.
2. Normal components of **D** and **B** are continuous across the interface.

These boundary conditions are used to write equations relating the fields on each side of the interface. They will be used later in this chapter to study the reflection of electromagnetic waves from the interface between two dielectric materials.

Poynting's theorem describes power flow in a region containing electric and magnetic fields. The form of Poynting's theorem we will use is

$$\mathbf{S} = \mathbf{E} \times \mathbf{H} \qquad\qquad (3.2.5a)$$

S is a vector representing the instantaneous power density with magnitude and direction defined by the cross product of **E** and **H**; the direction will be normal to the plane in which the **E** and **H** vectors lie. Its units are W/m^2. The spatial relationships between **E**, **H**, and **S** are illustrated in Figure 3.2.1.

For fields with sinusoidal variation with time, the average power density is

$$\mathbf{S}_{\mathbf{avg}} = \tfrac{1}{2}\text{Re}\{\mathbf{E} \times \mathbf{H}^*\} \qquad\qquad (3.2.5b)$$

where the magnitudes of **E** and **H** are the peak values of the sinusoids and the phase angle between the sinusoids is assumed to be zero; the asterisk indicates the time-domain complex conjugate. If the phase angle between **E** and **H** is ϕ, the average power density given by Equation (3.2.5b) will include an exp $(j\phi)$ term. If this phase angle is $\pi/2$, the power is imaginary. Imaginary power means that the instantaneous power, given by Equation (3.2.5a), alternates in direction with net power flow equal to zero.

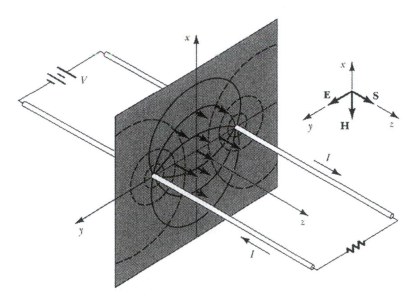

Figure 3.2.1 *The **E**, **H**, and **S** fields in a dc transmission line. If the polarity of V, and therefore the direction of I, were reversed, the **E** and **H** vectors would be reversed in direction but the direction of **S** would remain the same.*

EXERCISE 3.1

A coaxial transmission line has a constant V between the inner and outer conductors and a direct current I flowing in them. The magnitudes of the **E** and **H** fields are

$$E = \frac{V}{r \ln\left(\frac{b}{a}\right)} \quad \text{and} \quad H = \frac{I}{2\pi r} \quad a < r < b$$

where a is the radius of the inner conductor and b is the inner radius of the outer conductor. Both E and H fields are zero for $r < a$ and $r > b$. Show, by integrating the Poynting vector over the area where the fields are not zero, that the total power is VI watts.

3.3 The Wave Equation

3.3.1 *Derivation of The Wave Equation*

The wave equation, developed from the time-dependent Maxwell's equations, describes the propagation of an electromagnetic wave through a uniform medium. By taking the curl of Equation (3.2.3a) and then substituting for $\mathbf{V} \times \mathbf{H}$ from Equation (3.2.3b), we can write

$$\begin{aligned}
\mathbf{V} \times \mathbf{V} \times \mathbf{E} &= -j\omega\mu\mathbf{V} \times \mathbf{H} \\
&= -j\omega\mu(\sigma + j\omega\epsilon)\mathbf{E} \\
&= (\omega^2\mu\epsilon - j\omega\mu\sigma)\mathbf{E}
\end{aligned} \qquad (3.3.1)$$

This is further reduced, using the vector identity

$$\mathbf{V} \times \mathbf{V} \times \mathbf{E} = [\mathbf{V}(\mathbf{V} \cdot \mathbf{E}) - \mathbf{V}^2\mathbf{E}] \qquad (3.3.2)$$

and the assumption of a linear, homogeneous, isotropic, charge-free region. In such a region, $\mathbf{V} \cdot \mathbf{E} = 0$ and

$$\mathbf{V} \times \mathbf{V} \times \mathbf{E} = -\mathbf{V}^2\mathbf{E} \qquad (3.3.3)$$

By substituting this into (3.3.1) we can show that

$$\mathbf{V}^2\mathbf{E} - \gamma^2\mathbf{E} = 0 \qquad (3.3.4a)$$

where $\gamma^2 = -\omega^2\mu\epsilon + j\omega\mu\sigma$.

By following the same procedure, a similar equation can be developed for the **H** field.

$$\mathbf{\nabla}^2\mathbf{H} - \gamma^2\mathbf{H} = 0 \qquad\qquad (3.3.4b)$$

Equations (3.3.4) are the wave equations. They can be used in any orthogonal coordinate system. Each can be expressed as three scalar equations. For example,

$$\frac{\partial^2 E_y}{\partial x^2} + \frac{\partial^2 E_y}{\partial y^2} + \frac{\partial^2 E_y}{\partial z^2} - \gamma^2 E_y = 0 \qquad\qquad (3.3.5)$$

3.3.2   Solution of The Wave Equation

We assume a uniform, unbounded plane wave. Without loss of generality we can orient a rectangular coordinate system so that $E_x = E_z = 0$. Thus $\mathbf{E} = \mathbf{a}_y E_y$, and the vector Equation (3.3.4a) reduces to one scalar Equation (3.3.5). We further assume that propagation is in the z direction. The plane of the wave is then the x,y plane. Since it is uniform and unbounded in the x,y plane, its partial derivatives with respect to x and y are both zero. Equation (3.3.5) then reduces to

$$\frac{\partial^2 E_y}{\partial z^2} - \gamma^2 E_y = 0 \qquad\qquad (3.3.6)$$

The solution to this equation has the form

$$E_y = A \exp(pz) \qquad\qquad (3.3.7)$$

which satisfies (3.3.6) when $p = \pm\gamma$. Then the solution becomes

$$E_y = A_1 \exp(\gamma z) + A_2 \exp(-\gamma z) \qquad\qquad (3.3.8)$$

The two solutions represent waves traveling in the $-z$ and $+z$ directions, respectively. In some cases, both of these waves are present. For our purposes, however, we will consider only the positive traveling wave. Thus

$$E_y = A \exp(-\gamma z) \qquad\qquad (3.3.9)$$

Let $\gamma = (-\omega^2\mu\epsilon + j\omega\mu\sigma)^{\frac{1}{2}} = \alpha + j\beta$. Then

$$E_y = A \exp\left[-(\alpha + j\beta)\right] = A \exp(-\alpha z)\exp(-j\beta z) \qquad\qquad (3.3.10)$$

By restoring the time-dependence term, $\exp(j\omega t)$, we get the complete expression for E as a function of time and distance.

$$E_y(t,z) = A \exp(-\alpha z) \exp\left[j(\omega t - \beta z)\right] \qquad\qquad (3.3.11)$$

The amplitude A is found from the power or from other constraints imposed for specific problems.

3.3.3 *Propagation Parameters*

The magnitudes of the **E** and **H** field vectors are related through the properties of the medium. For the plane wave considered above, Equation (3.2.3a) has only the x components:

$$-\frac{\partial E_y}{\partial z} = \gamma E_y = -j\omega\mu H_x \tag{3.3.12}$$

Let the conductivity $\sigma = 0$; then $\gamma = j\omega(\mu\epsilon)^{\frac{1}{2}}$. The ratio of electric to magnetic field vectors is then

$$\frac{E_y}{H_x} = -\left(\frac{\mu}{\epsilon}\right)^{\frac{1}{2}} \tag{3.3.13}$$

The magnitude of this ratio is the intrinsic impedance η. The negative sign indicates that the **H** vector is directed in the negative x direction.

The intrinsic impedance of space is $(\mu_0/\epsilon_0)^{\frac{1}{2}} \approx 377$ ohms. In a dielectric material, $\eta = 377/\epsilon_r^{\frac{1}{2}} = 377/n$. n is the index of refraction, to be defined in Equation (3.3.18).

The power density in the plane wave can be found using Poynting's theorem, Equation (3.2.5b). In the plane wave, with $\mathbf{E} = \mathbf{a}_y E_y$ and $\mathbf{H} = \mathbf{a}_x H_x$, the power flow is in the z direction and its average magnitude is $E_y H_x/2$. Since $E_y = \eta H_x$, $|S| = E_y^2/2\eta$.

$E_y(t,z)$ of Equation (3.3.11) has magnitude $A \exp(-\alpha z)$ and phase $(\omega t - \beta z)$. The propagation constant γ is a complex number with units m^{-1}. The real part α is the attenuation constant and the imaginary part β is the phase constant. Together they describe the behavior of the electric field vector as it moves through space. In the case represented here, propagation is in the z direction.

Propagation through space can be examined by tracing a point of constant phase as t and z change.

$$(\omega t - \beta z) = \text{constant} \tag{3.3.14}$$

The time derivative of this constant phase angle is

$$\frac{d}{dt}(\omega t - \beta z) = \omega - \beta\frac{dz}{dt} = 0 \tag{3.3.15}$$

We interpret dz/dt as the velocity, v, of the wavefront,

$$v = \frac{dz}{dt} = \frac{\omega}{\beta} \tag{3.3.16}$$

and

$$v = \frac{\omega}{\beta} = \frac{\omega}{\omega(\mu\epsilon)^{\frac{1}{2}}} = \frac{1}{(\mu\epsilon)^{\frac{1}{2}}} \qquad (3.3.17)$$

In free space $\mu = \mu_0$, $\epsilon = \epsilon_0$.

$$v = c = \frac{1}{(\mu_0\epsilon_0)^{\frac{1}{2}}} \approx 3 \cdot 10^8 \ (\text{m/s})$$

In other media $\mu = \mu_0\mu_r$, $\epsilon = \epsilon_0\epsilon_r$, and

$$v = \frac{c}{(\mu_r\epsilon_r)^{\frac{1}{2}}} = \frac{c}{n} \qquad (3.3.18)$$

where $n = (\mu_r\epsilon_r)^{\frac{1}{2}}$ is the index of refraction. In dielectric materials, $\mu_r = 1$ and $n = \epsilon_r^{\frac{1}{2}}$.

As a phase front propagates with velocity v, the time variable appears in the form $\omega(t - z/v)$. The field vectors at z are called "retarded" with respect to those at $z = 0$ because variations at z are delayed in time, that is, retarded, by the propagation time z/v. Similarly, $(t - z/v)$ can be called retarded time.

The factor $(\omega t - \beta z)$ represents the variation of the phase angle of the field vectors with time and distance. If z is fixed at some arbitrary value $z = z_1$, then βz_1 is a constant phase difference between the field at $z = z_1$ and the field at $z = 0$. At this point, $(\omega t - \beta z_1)$ gives the phase as a function of time. The period T is the time required for the phase angle at z_1 to go through one complete cycle, or 2π radians; $\omega T = 2\pi$ and $T = 2\pi/\omega$. Similarly, $(\omega t_1 - \beta z)$ is the phase angle as a function of distance at a fixed time t_1. The wavelength λ is the distance along the direction of propagation required for the phase angle at time t_1 to go through 2π radians; $\beta\lambda = 2\pi$ and $\lambda = 2\pi/\beta$.

Since

$$\beta = \frac{2\pi}{\lambda} = \frac{\omega}{v} = \frac{2\pi f}{v}$$

then

$$v = f\lambda = \frac{c}{n}$$

and

$$\lambda = \frac{v}{f} = \frac{c}{nf} = \frac{\lambda_0}{n} \qquad (3.3.19)$$

The velocity of propagation in free space is c, and the velocity in a material having index of refraction n is c/n. If the wavelength in free space is $\lambda_0 = c/f$, the wavelength in a material of index n is λ_0/n. The wavelength in free space will usually be written λ, without subscript. A subscript will be used to designate the wavelength in a specific material, region, and so forth.

In Equation (3.3.11), the x,y,z coordinate axes were oriented so as to be aligned with the **H**, **E**, and **S** vectors. In a more general frame of reference, the direction of propagation is represented by a unit vector **n**, normal to the wavefront, and the propagation constant is a vector $\mathbf{k} = \beta\mathbf{n} = \omega(\mu\epsilon)^{\frac{1}{2}}\mathbf{n}$; both **n** and **k** will have x, y, and z components. In a plane wave propagating in the **n** direction, **E** and **H** will be normal to each other and to **n**; each of these field vectors will have x, y, and z components.

Consider the wave of Figure 3.3.1 with components of the propagation vector **k** in both the x and the z directions. The **E** vector has only a y component. The vector propagation constant $\mathbf{k} = \beta\mathbf{n} = \mathbf{a}_x k_x + \mathbf{a}_z k_z$.

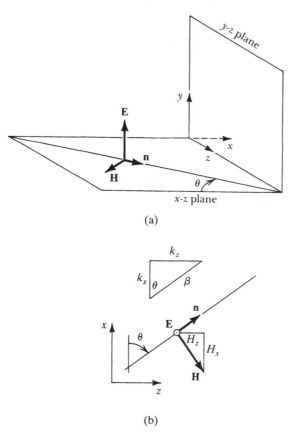

(a)

(b)

Figure 3.3.1 *Field vectors representing a plane wave with the wavefront normal to the x-z plane. The direction of propagation has x and z components.*

$$k_x = \beta \cos \theta = \omega(\mu\epsilon)^{\frac{1}{2}} \cos \theta \tag{3.3.20a}$$

$$k_x = \beta \sin \theta = \omega(\mu\epsilon)^{\frac{1}{2}} \sin \theta \tag{3.3.20b}$$

Equation (3.3.11) can now be expressed

$$E_y(t,x,z) = A \exp(j\omega t) \exp(-jk_x x) \exp(-jk_z z) \tag{3.3.21}$$

The components of the propagation vector are related by

$$k_x{}^2 + k_z{}^2 = \beta^2 = \omega^2 \mu\epsilon \tag{3.3.22}$$

For the wave of Figure 3.3.1, in which **E** has only a y component, the **H** field vector can be calculated using Equation (3.2.3a).

$$-\mathbf{a}_x \frac{\partial E_y}{\partial z} + \mathbf{a}_z \frac{\partial E_y}{\partial x} = -j\omega\mu[\mathbf{a}_x H_x + \mathbf{a}_y H_y + \mathbf{a}_z H_z] \tag{3.3.23}$$

The partial derivatives can be found by using Equation (3.3.21) for $E_y(t,x,z)$. The vector equation is thus reduced to three scalar equations that give the relationship between E_y and the components of **H**.

$$-jk_x E_y = -j\omega(\mu\epsilon)^{\frac{1}{2}} \cos \theta \, E_y = -j\omega\mu H_z$$

$$jk_z E_y = -j\omega(\mu\epsilon)^{\frac{1}{2}} \sin \theta \, E_y = -j\omega\mu H_x \tag{3.3.24}$$

$$0 = -j\omega\mu H_y$$

from which we find

$$H_x = -\frac{E_y}{\eta} \sin \theta$$

$$H_y = 0 \tag{3.3.25}$$

$$H_z = \frac{E_y}{\eta} \cos \theta$$

The directions for the components of **H** in these equations are consistent with those shown in Figure 3.3.1.

The velocities of the components of the wave can be found from the propagation constants. Recall that $v = \omega/\beta = \omega/|k|$. If we apply this relationship to the components of **k**, we find

$$k_x = \beta \cos \theta = \frac{\omega}{v} \cos \theta$$

and

$$v_x = \frac{\omega}{k_x} = \frac{v}{\cos \theta} \qquad\qquad \textbf{(3.3.26a)}$$

Similarly,

$$v_z = \frac{v}{\sin \theta} \qquad\qquad \textbf{(3.3.26b)}$$

These velocities are faster than that of the wavefront moving in the direction of **n** and can be faster than *c*. Each can be visualized as the velocity of a point of constant phase on the wavefront. They are examples of *phase* velocities. Refer to Figure 3.3.1: the phase velocity v_z is the velocity of the intersection of a plane of constant phase with the interface between the two regions. It can be seen there that as θ approaches zero, the phase velocity increases without limit. The velocity of a plane wave in an infinite medium, $v = c/n$, is its phase velocity.

EXERCISE 3.2

A light wave with $\lambda = 1.3 \ \mu m$ is propagating in a material having $\epsilon_r = 2.1$ and $\mu_r = 1.0$. (a) Find the index of refraction, the wavelength, the velocity of propagation, and the intrinsic impedance. (b) For $E_{max} = 1 \ kV/m$, find H_{max} and the average power density.

Answers
(a) 1.45, 0.897 μm, $2.07 \cdot 10^8$ m/s, 260 ohms.
(b) 3.85 A/m, 1.93 kW/m^2.

3.3.4 *Group Velocity*

The velocity of propagation found above was derived on the basis of the motion of a plane of constant phase. It is the phase velocity and is, in some cases, the only velocity of significance. However, note that it can be a function of frequency. In that case, when each component of a multifrequency signal travels with its own unique velocity, the concept of a group velocity becomes important.

In waveguides, where the wave is not an infinite plane wave but is confined by the sides of the waveguide, the propagation constant β can be a function of the frequency. It is also possible that the μ or ϵ of the material in which the wave propagates may vary with frequency. In either case, the phase and group velocities will not be equal. A medium in which the velocity of propagation is a function of frequency is called a dispersive medium.

When an electromagnetic wave is modulated, the sidebands constitute additional frequencies that must propagate with the wave in order for the modulation to be propagated with the wave. When the sidebands propagate with phase velocities different from that of the carrier, the phase relationships between the carrier and the sidebands change as the wave moves forward. The effects of this phenomenon can be seen by examining an AM signal

$$e_{AM} = E\,(1 + m \cos \omega_1 t) \cos \omega_c t \quad (\text{at } z = 0)$$

$$= E \cos \omega_c t + E\,\frac{m}{2} \{\cos (\omega_c - \omega_1)t + \cos (\omega_c + \omega_1)t\}$$

$$= E \operatorname{Re} \left[\exp (j\omega_c t) + \frac{m}{2} \{\exp \left[j(\omega_c - \omega_1)t \right] \right.$$

$$\left. + \exp \left[j(\omega_c + \omega_1)t \right] \} \right] \tag{3.3.27}$$

The three components of this wave propagate in the z direction with different values for the phase propagation constant β. If we assume that the variation of β with ω in the neighborhood of ω_c is linear, then $\Delta\beta$ is proportional to $(\omega - \omega_c) = \Delta\omega$. Here $\Delta\omega = \omega_1$ and $\Delta\beta = \omega_1 d\beta/d\omega$. The $(\omega t - \beta z)$ terms for the three components are

$$(\omega_c t - \beta_c z) \quad \text{for the carrier,}$$

and

$$(\omega_c \pm \Delta\omega)t - (\beta_c \pm \Delta\beta)z \quad \text{for the sidebands.}$$

After propagating a distance z through this dispersive medium, the equation for e_{AM} will be

$$e_{AM}(t,z) = E \operatorname{Re} \left[\exp \{j(\omega_c t - \beta_c z)\} + \frac{m}{2} \exp \{j[(\omega_c - \Delta\omega)t \right.$$

$$\left. - (\beta_c - \Delta\beta)z]\} + \frac{m}{2} \exp \{j[(\omega_c + \Delta\omega)t - (\beta_c + \Delta\beta)z]\} \right] \tag{3.3.28}$$

The order of the terms in the exponents can be rearranged as follows:

$$[(\omega_c - \Delta\omega)t - (\beta_c - \Delta\beta)z] = [(\omega_c t - \beta_c z) - (\Delta\omega t - \Delta\beta z)]$$

$$[(\omega_c + \Delta\omega)t - (\beta_c + \Delta\beta)z] = [(\omega_c t - \beta_c z) + (\Delta\omega t - \Delta\beta z)]$$

The equation for the AM signal can now be reduced to

$$e_{AM}(t,z) = E[1 + m \cos (\Delta\omega t - \Delta\beta z)] \cos(\omega_c t - \beta_c z) \tag{3.3.29}$$

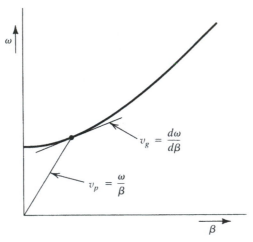

Figure 3.3.2 *An ω-β diagram illustrating phase and group velocities.*

The modulation term travels with the group velocity

$$v_g = \frac{dz}{dt} = \frac{\Delta\omega}{\Delta\beta} = \frac{d\omega}{d\beta} \tag{3.3.30}$$

The carrier velocity remains $v_p = \omega/\beta$, the phase velocity found earlier. Figure 3.3.2 illustrates graphically the relationships between v_p, v_g, and the ω-β curve.

An alternative and more general treatment of the group velocity is based on a wave packet of finite duration propagating in a dispersive medium. Let the wave packet be represented by an envelope, $x(t)$, of duration T and a Fourier transform, $X(\omega)$. The signal to be propagated through the dispersive medium is designated $y(t,z)$ with Fourier transform $Y(\omega,z)$. At $z = 0$, this is $y(t,0) = x(t) \cos \omega_c t$, and its Fourier transform is

$$Y(\omega,0) = \tfrac{1}{2} X(\omega - \omega_c) + \text{complex conjugate.} \tag{3.3.31}$$

The propagation constant is $\gamma = \alpha + j\beta$ where β is a function of frequency; it can be represented by the first two terms of a Taylor series,

$$\beta = \beta_c + \frac{d\beta}{d\omega} \Delta\omega + \text{higher order terms} \tag{3.3.32}$$

where $\Delta\omega = 2\pi(f - f_c)$. Propagation at frequency $\omega = \omega_c + \Delta\omega$ is given by

$$\exp\left[-(\alpha + j\beta)z\right] = \exp\left[-(\alpha + j\beta_c)z\right] \exp\left[-j\frac{d\beta}{d\omega} \Delta\omega z\right] \tag{3.3.33}$$

The spectrum at z is

$$Y(\omega,z) = Y(\omega,0) \exp\left[-(\alpha + j\beta_c)z\right] \exp\left[-j\frac{d\beta}{d\omega}\Delta\omega z\right] \tag{3.3.34}$$

The inverse Fourier transform is

$$y(t,z) = F^{-1}[Y(\omega,z)] = \int_0^\infty Y(\omega,z) \exp(j\omega t) \frac{d\omega}{2\pi}$$

$$= \int_0^\infty X(\omega - \omega_c) \exp\left[-(\alpha + j\beta_c)z\right] \exp\left[-j\frac{d\beta}{d\omega}\Delta\omega z\right]$$

$$\cdot \exp[j\omega t] \frac{d\omega}{2\pi} + \text{c.c.} \tag{3.3.35}$$

This can be rearranged to give

$$y(t,z) = \exp[-\alpha z] \exp[j(\omega_c t$$

$$- \beta_c z)] \int_0^\infty X(\Delta\omega) \exp\left[j\Delta\omega\left(t - \frac{d\beta}{d\omega}z\right)\right] \frac{d(\Delta\omega)}{2\pi} + \text{c.c.} \tag{3.3.36}$$

The integral is recognized as an inverse Fourier transform. Then

$$y(t,z) = \exp[-\alpha z] \exp[j(\omega_c t - \beta_c z)] x\left(t - \frac{d\beta}{d\omega}z\right) + \text{c.c.}$$

$$= \exp[-\alpha z] x\left(t - \frac{d\beta}{d\omega}z\right) \cos(\omega_c t - \beta_c z) \tag{3.3.37}$$

Since the time variable in the envelope has the form of retarded time, $(t - z/v)$, the $x(t,z)$ envelope can be said to have velocity $d\omega/d\beta$. This is the group velocity. The carrier-frequency wave moves with phase velocity $v_p = \omega_c/\beta_c$ and the envelope (the packet) with the group velocity v_g. If this waveform could be viewed on an oscilloscope, the sinusoidal carrier wave would appear to move forward through the envelope.

The group velocity can be thought of as the velocity with which an information spectrum or an energy packet is propagated. A photon can be considered to be a wave packet containing a discrete quantity of energy; photons travel at the group velocity. In a uniform nondispersive medium, the group and phase velocities are equal.

Equation (3.3.26b) gives the phase velocity in the direction parallel to the interface to be $v_p = v_z = v/\sin\theta$. Examination of Figure 3.3.1 can show that the component of velocity of the wavefront parallel to the interface is $v\sin\theta$. The energy of the wave moves in the z direction with this velocity; this corresponds to the group velocity. The product of group and phase velocities, in this case, is v^2. This relationship between v_p and v_g is true in many cases but is not true in all cases.

EXERCISE 3.3

A plane wave with wavelength 1.3 μm travels in a material having $\beta = 0.3 \, (\omega - \omega_0)^{\frac{1}{2}}$, where $\omega_0 = 8.66 \cdot 10^{14}$ s^{-1}. (a) Calculate the phase and group velocities. (b) Calculate the index of refraction n.

Answer

(a) $2 \cdot 10^8$ m/s, $1.61 \cdot 10^8$ m/s.

(b) 1.5.

3.3.5 *Dispersion*

The expressions developed above are based on the use of only the first two terms in the Taylor series expansion of $\beta(\omega)$. The second term in this series corresponds to the group velocity and the first to the phase velocity; higher order terms were neglected. Since the ω-β curve normally shows significant curvature, that is, the first derivative is not constant, we should examine the third term in the Taylor series for $\beta(\omega)$.

To interpret the effect of this nonlinear term, consider the propagation delay, τ, of a wave propagating in a dispersive medium.[1] The information-bearing spectrum of a wave traveling with group velocity v_g over a distance of one meter will be delayed by $\tau/L = 1/v_g$. If v_g is frequency-dependent,

$$\frac{d\tau}{d\omega} = \frac{d}{d\omega} \frac{1}{v_g} = \frac{d}{d\omega} \frac{d\beta}{d\omega} = \frac{d^2\beta}{d\omega^2} \; (\text{s}^2/\text{m}) \tag{3.3.38}$$

If the signal has a spectrum extending from $(\omega_c - \omega)$ to $(\omega_c + \omega)$, then the difference in propagation time of parts of this signal at the opposite extremes of the spectrum will be

$$\frac{\Delta\tau}{L} = \frac{d\tau}{d\omega} \Delta\omega = \frac{d^2\beta}{d\omega^2} 2\omega \; (\text{s}/\text{m}) \tag{3.3.39}$$

The dispersion in the time of arrival of the signal is a form of distortion. It limits the rate at which data can be transmitted through the medium.

---

[1] A more thorough treatment of dispersion and the development of the results summarized here can be found in reference [2.2, Section 6.5].

This dispersion can be related to the frequency dependence of the index of refraction. Since $\beta = \omega n/c$, then

$$\frac{d\beta}{d\omega} = \frac{d}{d\omega}\frac{\omega n}{c} = \frac{1}{c}\left[n + \omega\frac{dn}{d\omega}\right] \tag{3.3.40}$$

The expression in brackets is defined as the *group index*, N, giving the relationship

$$v_g = \frac{c}{N} \tag{3.3.41}$$

We can then express the dispersion as

$$\frac{d^2\beta}{d\omega^2} = \frac{d}{d\omega}\frac{1}{v_g} = \frac{d}{d\omega}\frac{N}{c} = \frac{1}{c}\cdot\frac{dN}{d\omega} \tag{3.3.42}$$

Dispersion is thus shown to be proportional to the first derivative of the group index.

It is often preferable to express dispersion in terms of wavelength rather than frequency. To do so, we can differentiate β with respect to λ.[2]

$$\frac{d\beta}{d\omega} = -\frac{\lambda}{\omega}\frac{d\beta}{d\lambda} = \frac{1}{c}\left[n - \lambda\frac{dn}{d\lambda}\right] = \tau \ (\text{s/m}) \tag{3.3.43}$$

and

$$D = \frac{d\tau}{d\lambda} = \frac{1}{c}\left[\frac{dn}{d\lambda} - \lambda\frac{d^2n}{d\lambda^2} - \frac{dn}{d\lambda}\right] = -\frac{\lambda}{c}\frac{d^2n}{d\lambda^2} \ (\text{s/m}^2) \tag{3.3.44}$$

The units normally used for material dispersion are ns/(km · nm) rather than s/m^2. To find the dispersion of propagation time, in ns, multiply this dispersion parameter D by the distance in km and by the spectrum width in nm.

It will sometimes be useful to express D in terms of $d^2\beta/d\omega^2$ rather than n and λ. To do so, we note that

$$\frac{d\tau}{d\lambda} = \frac{d\tau}{d\omega}\cdot\frac{d\omega}{d\lambda} = -\frac{\omega}{\lambda}\cdot\frac{d^2\beta}{d\omega^2} = -\frac{2\pi c}{\lambda^2}\cdot\frac{d^2\beta}{d\omega^2} \tag{3.3.45}$$

The dispersion coefficient D is thus defined in either of two forms

$$D = -\frac{\lambda}{c}\cdot\frac{d^2n}{d\lambda^2} = -\frac{2\pi c}{\lambda^2}\cdot\frac{d^2\beta}{d\omega^2} \tag{3.3.46}$$

---

[2] Note that this equation shows an alternative definition of the group index:

$$N = \left[n - \lambda\frac{dn}{d\lambda}\right]$$

Dispersion can arise from two major sources, material dispersion and waveguide dispersion. Material dispersion results when the dielectric constant, and therefore the index of refraction, is a function of frequency; material dispersion is considered further in Section 3.4. Waveguide dispersion results when the propagation constants for the waveguide are functions of frequency. In optical fibers, both forms of dispersion must be considered; waveguide dispersion and total dispersion in optical fibers are considered in Chapter 4.

EXAMPLE 3.1

An optical fiber transmission system uses an amplitude-modulated carrier having a wavelength of 1.3 μm and dispersion of 20 ps/(km $\cdot$ nm). This system is to be used to transmit a binary PCM signal with a data rate of 500 Mbits/s. The length of the fiber is 150 km. What is the dispersion in the propagation time?

Solution:

Assume that the bandwidth at baseband is 500/2 MHz. The total bandwidth of the modulated light wave is therefore twice the baseband spectrum, or 500 MHz. The frequency of the carrier is $c/\lambda = 3 \cdot 10^8/(1.3 \cdot 10^{-6}) = 2.3 \cdot 10^{14}$ Hz. The width of the spectrum of the modulated carrier is 500 MHz; the fractional spectral width, $\Delta f/f$, is $500 \cdot 10^6/(2.3 \cdot 10^{14}) = 2.17 \cdot 10^{-6}$. The spectral width expressed in terms of wavelength is $\Delta\lambda/\lambda = \Delta f/f$.

$$\Delta\lambda = 2.17 \cdot 10^{-6} \cdot 1.3 \cdot 10^{-6} \text{ (m)} = 2.82 \cdot 10^{-3} \text{ (nm)}$$

$$\Delta\tau = 20 \cdot 150 \cdot 0.00282 = 8.46 \text{ ps}$$

3.4 Material Dispersion

A dielectric material is one in which the dielectric properties are the most significant ones and other properties, such as conductance and magnetic permeability, have relatively small effects. The magnetic properties are usually taken to be the same as in vacuum. The conductance or other loss mechanisms are such that a relatively small fraction of the energy flowing through the material is absorbed.

In optical materials, the dielectric constant ϵ can be complex and is often a function of frequency. Since the index of refraction n, the propagation constant β, the phase and group velocities, and other parameters of interest all depend on ϵ, they too will be functions of frequency.

The frequency dependence of the index of refraction and the effects of this dependence on optical signal propagation constitute material dispersion as defined in the previous section. In this section, we will examine the nature and magnitude of material dispersion in some of the materials used in optical fibers.

3.4.1 *Complex Dielectric Constant*

When $\sigma = 0$, the propagation constant γ, defined through Equation (3.3.4a), is given by $\gamma^2 = -\omega^2 \mu \epsilon$. When σ has some small value, its effect can be represented as follows:

$$\gamma^2 = -\omega^2 \mu \epsilon + j\omega\mu\sigma = -\omega^2 \mu \epsilon \left(1 - j\frac{\sigma}{\omega\epsilon} \right) \tag{3.4.1}$$

The effect of σ is thus represented as a small imaginary part in a complex ϵ, where $\epsilon = \epsilon - j\sigma/\omega$.

For small σ, using the approximation $(1 - \sigma)^{\frac{1}{2}} = (1 - \sigma/2)$, the propagation constant can be expressed

$$\gamma = \alpha + j\beta = j\omega(\mu\epsilon)^{\frac{1}{2}} + \frac{\sigma}{2}\left[\frac{\mu}{\epsilon}\right]^{\frac{1}{2}} \tag{3.4.2}$$

Thus, a small conductance σ causes α to be greater than zero, representing attenuation of the electromagnetic wave as it propagates in this material. The equations above indicate that a negative imaginary part for the complex dielectric constant corresponds to a positive attenuation constant.

There are other ways for ϵ to become complex. In general, if $\epsilon = \epsilon - j\epsilon'$, then $n = n - jn'$ where, for $\epsilon' \ll \epsilon$, $n = (\epsilon/\epsilon_0)^{\frac{1}{2}}$ and $n' = n\epsilon'/(2\epsilon)$. Since $\beta = \omega n/c$, then for complex n we can show that $\alpha = \omega n'/c$.

3.4.2 *Dispersion In Dielectric Materials*

The physical basis for the dielectric constant and index of refraction can be understood by studying the manner in which an electromagnetic field interacts with the electric charges on an atomic scale and some of the consequences of this interaction. We will examine a simple model of this interaction and from this model explain the origins and the nature of dispersion in dielectric materials. The results that follow are summaries, not complete derivations, of the relevant results. For more complete treatment of the subject, references [2.3, Vol. 1, Chapter 31] and [2.4, Section 5.4] are suggested.

An electric charge q in an electric field E will have a force qE exerted on it. The electric charges in an atom will experience such forces and will, if free to move, be displaced from their equilibrium positions. If the field has the form $E \exp(j\omega t)$, the charges in the atom will oscillate about their equilibrium positions. In writing the equations of motion, we assume when the charge is moved from its equilibrium position there will be a restoring force that is proportional to the displacement and that there will be a damping force proportional to its velocity. With these assumptions, the equation of motion is

$$\ddot{x} + \gamma\dot{x} + \omega_0{}^2 x = qE \exp\left[j\omega t\right] \tag{3.4.3}$$

The motion of the charge is

$$x(t) = -\frac{\frac{q}{m} E \exp\left[j\omega t\right]}{(\omega_0{}^2 - \omega^2) + j\gamma\omega} \tag{3.4.4}$$

This is a complex number. The phase angle of $x(t)$ indicates that the sinusoidal motion of $x(t)$ is not in phase with the E field that causes it.

The displacement of charges under the influence of an electric field is the physical basis of the dielectric constant. Since the displacement is complex, the dielectric constant and the index of refraction should be expected to be complex, at least when the frequency is within the general region where atomic resonances exist.

An equation for the complex dielectric constant, derived from the atomic model defined above, is

$$\epsilon_r = 1 + \frac{\frac{Nq^2}{2\epsilon_0 m}}{(\omega_0{}^2 - \omega^2) + j\gamma\omega} \tag{3.4.5}$$

where N is the number of electrons per unit volume having this specific characteristic resonance.

When the frequency of the electromagnetic wave is near a resonance, the energy gained by the oscillating charges accumulates and can become large enough that the structure of the atom is changed. For example, when the oscillating charge is an electron, it may acquire enough energy to escape from the forces that bind it to the nucleus. This is in effect what happens when an electron in a semiconductor is excited into the conduction band. However, the resonance phenomena are not the aspects of the motion that we are concerned with here. Since our present interest is in the propagation of light waves in optical fibers, it is the behavior of the material far from resonance that concerns us. For a thorough discussion of the behavior near a resonance, see reference [2.4].

In most materials there are several atomic resonances, each with a characteristic resonant frequency and damping constant. In most materials there will be more than one type of atom. In the calculation of dielectric constant, each resonance will be represented by a term of the form of (3.4.5). The total effect is the sum of the contributions of all of the resonances near the frequency of interest.

$$\epsilon_r = n^2 = 1 + \frac{q^2}{\epsilon_0 m} \sum_k \frac{N_k}{(\omega_k{}^2 - \omega^2) + j\gamma\omega} \tag{3.4.6}$$

When the frequency is near a resonant frequency, the $(\omega_k{}^2 - \omega^2)$ term is small and the $j\gamma\omega$ term becomes significant. When the frequency is not near a resonance, the $j\gamma\omega$ term is relatively very small and can sometimes be neglected. Our immediate objective is to show the manner in which the index of refraction varies with frequency. For that purpose, a small phase angle that results from the $j\gamma\omega$ term is not significant and can be dropped. The equation for the index of refraction thus becomes

$$n^2 = 1 + \frac{q^2}{\epsilon_0 m} \sum_k \frac{N_k}{(\omega_k{}^2 - \omega^2)} \tag{3.4.7a}$$

where N_k is the number of atoms and ω_k is the resonant frequency of the kth resonance. Since in optical fiber design we more often use wavelength rather than frequency, it will be convenient to write n in the form

$$n^2 = 1 + \sum_k \frac{G_k \lambda^2}{(\lambda^2 - \lambda_k{}^2)} \tag{3.4.7b}$$

where G_k includes N_k and physical constants.

This last equation is known as the Sellmeier dispersion formula. It has been established that for frequencies (wavelengths) not near a resonance, only a few terms in the summation are required for a reasonable characterization of n. For optical fiber materials, three terms, with values for G_k and λ_k determined empirically, are sufficient [2.5].

A graph showing the nature of equation (3.4.7a) is shown in Figure 3.4.1. If we look at the region between the two resonances, it can be seen, from either the graph or the equation, that the slope, $dn/d\omega$, is positive. Since β is proportional to n, $d\beta/d\omega$ is also positive. It is also apparent that the slope is not constant. Dispersion, as defined and discussed in Section 3.3.5, is present.

The index of refraction for three optical fiber materials is plotted as a function of λ in Figure 3.4.2. It is clear that the index is a function of λ and that $dn/d\lambda$ is negative; the negative $dn/d\lambda$ is consistent with the positive $dn/d\omega$ deduced from Figure 3.4.1 or Equation (3.4.7).

The curves of Figure 3.4.2 can be represented by the Sellmeier formula, Equation (3.4.7b). The parameters, G_k and λ_k, determined from measured data, are given in Table 3.4.1. The Sellmeier formula, with three terms, can represent

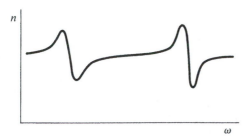

Figure 3.4.1 *Index of refraction as a function of frequency. This is a schematic diagram, not to scale, illustrating the nature of the variation of n in the vicinity of and between two resonant frequencies.*

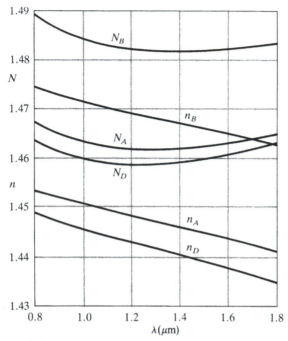

Figure 3.4.2 *The index of refraction, n, and group index, N, as functions of wavelength for three optical fiber materials. The material dispersion, D_m, is proportional to the slope of the N curve. These curves are calculated using the Sellmeier formula and coefficients from Table 3.4.1.*

the measured index of refraction with accuracy of 0.00002; this is comparable to the accuracy of carefully measured data and should therefore be adequate.

Material dispersion can be calculated using the Sellmeier formula. Material dispersion versus wavelength, for the three materials illustrated in Figure 3.4.2, is shown in Figure 3.4.3. For each of these materials, there is one wavelength for which the material dispersion is zero. Note that the dispersion is proportional to the derivative of the group index, $dN/d\lambda$. The dispersion is negative in the region where the slope of the N curve is negative, and the dispersion zero occurs at the wavelength where the slope is zero.

Table 3.4.1. Sellmeier coefficients for several optical fiber materials.

		G_1	λ_1	G_2	λ_2	G_3	λ_3
A	Quenched SiO_2	0.696750	0.069066	0.408218	0.115662	0.890815	9.900559
B	$13.5GeO_2:86.5\ SiO_2$	0.711040	0.064270	0.451885	0.129408	0.704048	9.425478
C	$9.1P_2O_5:90.0SiO_2$	0.695790	0.061568	0.452497	0.119921	0.712513	8.656641
D	$13.3B_2O_3:86.7SiO_2$	0.690618	0.061900	0.401996	0.123662	0.898817	9.098960
E	$1.0F:99.0SiO_2$	0.691116	0.068227	0.399166	0.116460	0.890423	9.993707
F	$16.9Na_2O:32.5B_2O_3:50.6SiO_2$	0.796468	0.094359	0.497614	0.093386	0.358924	5.999652

From Fleming, J. W., "Material Dispersion in Light-guide Glasses," Electronics Letters, vol. 14, pp. 326–328, 1978 [2.5].

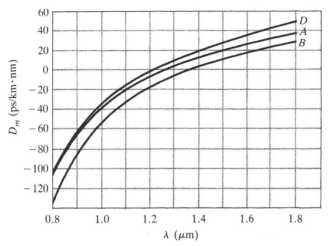

Figure 3.4.3 *Dispersion in optical fiber materials. These curves show dispersion for the three materials represented in Figure 3.4.2.*

3.5 Reflection and Refraction at a Dielectric Interface

When an electromagnetic wave is incident upon any discontinuity in the medium, some energy will be reflected from the discontinuity. In some cases, the discontinuity is due to random irregularities or imperfections in an otherwise uniform medium. In such cases the reflected energy is lost from the propagating wave and serves no useful purpose. Energy reflected from random irregularities is called "scattered" energy and represents one type of attenuation. In other cases, the discontinuity is carefully designed and fabricated to produce a desired result. The case to be studied in this section is reflection from the interface between two dielectric materials having different dielectric constants. The results will be the basis for the study of dielectric waveguides, including optical fibers.

Figure 3.5.1 shows the geometry of the reflection problem to be analyzed. The interface is a plane of infinite extent, and the incident wave is a plane wave. The index of refraction of medium 1, through which the incident wave approaches, and that of medium 2 are unequal; the interface is therefore a discontinuity from which a reflection is expected. Some of the energy of the incident wave will continue into medium 2; this constitutes the refracted, or transmitted, wave. The incident, reflected, and transmitted waves are indicated by the subscripts i, r, and t, respectively. We will refer to the wave in medium 2 as the transmitted wave, although refracted wave would be an acceptable and commonly used alternative.

3.5.1 *Snell's Law and the Fresnel Equations*

The rectangular coordinates are oriented with the $y - z$ plane parallel to the interface and the x axis normal to it. The direction of the incident wave is

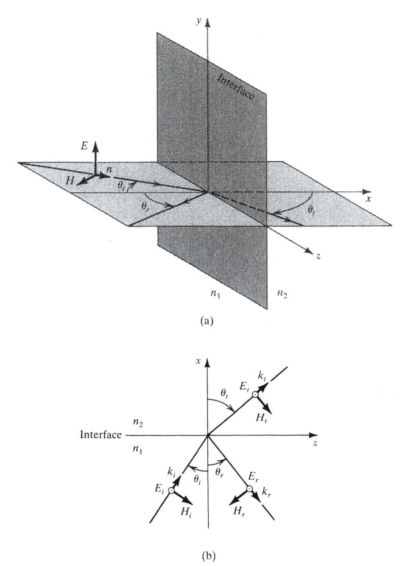

(a)

(b)

Figure 3.5.1 *Reflection and refraction of a plane wave at a plane dielectric interface. The geometry and the incident wave are the same as those in Figure 3.3.1. The $x = 0$ plane is the interface between two regions having different indices of refraction.*

at an angle θ_i from the normal to the interface. The **E** vector is in the y direction, parallel to the interface.[3] The **H** vector will then have both x and z components; H_z is parallel to the interface and H_x is normal to it. The propagation vector

---

[3] This is a transverse-electric, or TE, wave. A wave with its magnetic field vector parallel to the reference plane is a TM wave. A plane wave of infinite extent, with no boundary conditions and no reflected or refracted waves, is a TEM wave.

$\mathbf{k}_i$ lies in the x-z plane. For the $\mathbf{E}$ and $\mathbf{H}$ vectors we can write:

$$\mathbf{E} = \mathbf{a}_y E_y$$

$$\mathbf{H} = \mathbf{a}_x H_x + \mathbf{a}_z H_z \qquad (3.5.1)$$

$$E_x = 0 \qquad E_z = 0 \qquad H_y = 0$$

The problem to be solved is to find the $\mathbf{E}$ and $\mathbf{H}$ fields of the reflected and transmitted waves. To relate these vectors to those of the incident wave, we apply the boundary conditions defined in Section 3.2. These stipulate that the components of $\mathbf{E}$ and $\mathbf{H}$ parallel to the interface must be continuous across the interface.

The equality of the fields in medium 1 and medium 2 at the interface requires that the time dependence $\exp(j\omega t)$ be the same for all components and that the transmission and reflection parameters be independent of z. The latter condition requires that the z components of the propagation vectors all be equal.

$$k_{iz} = k_{rz} = k_{tz}$$

$$k_i \sin\theta_i = k_r \sin\theta_r = k_t \sin\theta_t \qquad (3.5.2)$$

The magnitudes of the propagation vectors are:

$$k_i = k_r = k_1 = \omega(\mu_0\epsilon_1)^{\frac{1}{2}} \quad \text{and} \quad k_t = k_2 = \omega(\mu_0\epsilon_2)^{\frac{1}{2}} \qquad (3.5.3)$$

From the first two parts of Equation (3.5.2) we can deduce that

$$k_1 \sin\theta_i = k_1 \sin\theta_r \quad \text{and} \quad \theta_i = \theta_r \qquad (3.5.4)$$

The angle of reflection is equal to the angle of incidence. From the first and last parts of Equation (3.5.2),

$$k_1 \sin\theta_i = k_2 \sin\theta_t$$

or

$$n_1 \sin\theta_i = n_2 \sin\theta_t \qquad (3.5.5)$$

This is Snell's law.

Since the total field at the interface in medium 1 is the sum of the incident and reflected fields, the equality of the tangential fields on the two sides of the interface gives

$$E_{yi} + E_{yr} = E_{yt} \qquad (3.5.6a)$$

and

$$H_{zi} + H_{zr} = H_{zt} \tag{3.5.6b}$$

Equation (3.5.6b) can be expressed in terms of the **E** field by using the intrinsic impedance defined in Equation (3.3.13). Since the **H** field found in this way is that in the plane of the wavefront, and H_z is the z component of the **H** field, we can write

$$H_{zi} = \left[\frac{\epsilon_1}{\mu_0}\right]^{\frac{1}{2}} \cos \theta_i E_{yi} \tag{3.5.7}$$

and similar equations for the z components of the reflected and transmitted waves. Thus, the second equation becomes

$$\left[\frac{\epsilon_1}{\mu_0}\right]^{\frac{1}{2}} [E_{yi} \cos \theta_i - E_{yr} \cos \theta_r] = \left[\frac{\epsilon_2}{\mu_0}\right]^{\frac{1}{2}} E_{yt} \cos \theta_t \tag{3.5.8}$$

We can again cancel the $\mu^{\frac{1}{2}}$ factors and replace $\epsilon^{\frac{1}{2}}$ with the index of refraction n. Since $\theta_i = \theta_r$, this can be reduced to

$$n_1 \cos \theta_i [E_{yi} - E_{yr}] = n_2 \cos \theta_t E_{yt} \tag{3.5.9}$$

Since **E** has only a y component, we can drop the y subscript and refer to the magnitudes of the three **E** fields as E_i, E_r, and E_t.

$$E_i - E_r = \frac{n_2 \cos \theta_t}{n_1 \cos \theta_i} E_t \tag{3.5.10}$$

By using this with Equation (3.5.6a), we can get equations giving the strengths of the transmitted and reflected fields in terms of the incident field strength.

$$E_t = \frac{2n_1 \cos \theta_i}{n_1 \cos \theta_i + n_2 \cos \theta_t} E_i \tag{3.5.11a}$$

and

$$E_r = \frac{n_1 \cos \theta_i - n_2 \cos \theta_t}{n_1 \cos \theta_i + n_2 \cos \theta_t} E_i \tag{3.5.11b}$$

These are the Fresnel equations for a *TE* wave.

A reflection coefficient r_E, defined as E_r/E_i, is found from Equation (3.5.11b)

$$r_E = \frac{n_1 \cos \theta_i - n_2 \cos \theta_t}{n_1 \cos \theta_i + n_2 \cos \theta_t} \tag{3.5.12}$$

Since $k = k_0 n$, $k_{1x} = k_1 \cos \theta_1$, and $k_{2x} = k_2 \cos \theta_t$,

$$r_E = \frac{k_{1x} - k_{2x}}{k_{1x} + k_{2x}} \tag{3.5.13a}$$

Similarly, the transmission coefficient is

$$t_E = \frac{2k_{1x}}{k_{1x} + k_{2x}} \tag{3.5.13b}$$

The corresponding equations for the *TM* wave are

$$r_H = \frac{n_1{}^2 k_{2x} - n_2{}^2 k_{1x}}{n_1{}^2 k_{2x} + n_2{}^2 k_{1x}} \tag{3.5.14a}$$

$$t_H = \frac{2n_2{}^2 k_{1x}}{n_1{}^2 k_{2x} + n_2{}^2 k_{1x}} \tag{3.5.14b}$$

Two additional coefficients relate the reflected and transmitted *power* to that of the incident wave. These coefficients, R and T, called the reflectivity and transmissivity, are given by

$$R = r_E{}^2 \quad \text{or} \quad r_H{}^2 \tag{3.5.15}$$

and

$$T = 1 - R \tag{3.5.16}$$

The reflection and transmission coefficients derived above are used to find the **E** fields. When we know the magnitude and direction of the **E** vector and the intrinsic impedance of the dielectric material, we can find the magnitude and direction of the **H** field vector.

The total field in medium 1 is the sum of the incident and reflected waves. The incident wave is

$$E_i(t,x,z) = A \exp\left[j(\omega t - k_{1x}x - k_{1z}z) \right] \tag{3.5.17}$$

and the reflected wave is

$$E_r(t,x,z) = r_E A \exp\left[j(\omega t + k_{1x}x - k_{1z}z) \right] \tag{3.5.18}$$

The amplitude A will, for the present, be set equal to unity.

Since we have required in writing Equation (3.5.2) that the z components of the propagation vectors all be equal, and since the ωt terms will all be the same, we can drop these terms from the equations; they can be restored if there is need for them. The total E field in medium 1 thus becomes

$$E_1(x) = \exp(-jk_{1x}x) + r_E \exp(+jk_{1x}x)$$
$$= (1 + r_E) \cos(k_{1x}x) - j(1 - r_E) \sin(k_{1x}x) \tag{3.5.19}$$

The magnitude of $E_1(x)$ is

$$|E_1(x)|^2 = [(1 + r_E) \cos(k_{1x}x)]^2 + [(1 - r_E) \sin(k_{1x}x)]^2$$

$$|E_1(x)| = [(1 + r_E^2) + 2r_E \cos(2k_{1x}x)]^{\frac{1}{2}} \tag{3.5.20}$$

The phase shift associated with this $E_1(x)$ is

$$\phi(x) = \tan^{-1}\left[\frac{-(1 - r_E) \sin(k_{1x}x)}{(1 + r_E) \cos(k_{1x}x)}\right] \tag{3.5.21}$$

This is the phase angle of the carrier due to the position in the x direction. Note that when $r_E = 0$, then $\phi(x) = -k_{1x}x$, the phase variation in the x direction of the incident wave.

EXAMPLE 3.2

A plane TE wave is incident on the interface between two dielectric materials at an angle of 75° from the normal to the interface. The indices of refraction are 1.45 in the material through which the incident wave approaches the interface and 1.43 in the other material. The wavelength is 1.3 μm. Calculate the electric field intensities in the two materials in terms of the intensity of the incident wave. Calculate the power flow across the interface.

Solution

The angle of the transmitted (refracted) wave is

$$\theta_t = \sin^{-1}\frac{n_1 \sin\theta_i}{n_2}$$

$$= \sin^{-1}\frac{1.45 \sin 75°}{1.43} = 78.36°$$

The propagation constants are

$$k_1 = \frac{2\pi n_1}{\lambda} = 2\pi \cdot \frac{1.45}{1.3} = 7.0082 \cdot 10^6 \ (\text{m}^{-1})$$

$$k_2 = 2\pi \cdot \frac{1.43}{1.3} = 6.9115 \cdot 10^6$$

$$k_{1x} = k_1 \cos 75° = 1.8139 \cdot 10^6$$
$$k_{1z} = k_1 \sin 75° = 6.7694 \cdot 10^6$$
$$k_{2x} = k_2 \cos 78.36° = 1.3945 \cdot 10^6$$
$$k_{2z} = k_2 \sin 78.36° = 6.7694 \cdot 10^6$$

The reflection and transmission coefficients are

$$r_E = \frac{1.8139 - 1.3945}{1.8139 + 1.3945} = \frac{0.4194}{3.2084} = 0.1307$$

and

$$t_E = \frac{2 \cdot 1.8139}{3.2084} = 1.1307$$

The incident wave is

$$E_i(x) = \exp[-j\,k_{1x}x]$$

The reflected wave is

$$E_r(x) = r_E E_i = 0.1307 \exp[jk_{1x}x]$$

The magnitude of the field in medium 1 is

$$E_1(x) = [(1 + 0.1307^2) + 2 \cdot 0.1307 \cos(2 \cdot 1.814 \cdot x)]^{\frac{1}{2}}$$
$$= [1.0128 + 0.2614 \cos(3.618x)]^{\frac{1}{2}}$$

The transmitted wave is

$$E_t(x) = E_2(x) = 1.1307 \exp[-jk_{2x}x]$$

These fields are plotted in Figure 3.5.2. Note that both the magnitudes and the first derivatives of the fields in the two regions are equal at the boundary.

The power flow across the boundary is found by calculating the x components of the Poynting vector. Let $|E_i| = A$.

In medium 2, $|E_2| = 1.1307\,A$.

$$S_2 = \tfrac{1}{2}\frac{E_2{}^2}{\eta_2} \quad \text{where} \quad \eta_2 = \frac{\eta_0}{n_2}$$

$$= \frac{(1.1307\,A)^2 \cdot n_2}{2\,\eta_0} = 1.8284\left[\frac{A^2}{2\eta_0}\right]$$

$$S_{2x} = S_2 \cos\theta_t = 0.3689\left[\frac{A^2}{2\eta_0}\right] \quad (\text{W/m}^2)$$

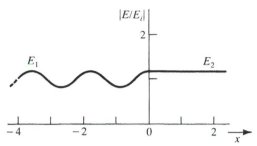

Figure 3.5.2 *Electric field intensities on both sides of a dielectric interface when $\theta_i < \theta_c$. See Example 3.2.*

Similarly,

$$S_{ix} = 0.3753 \, \frac{A^2}{2\eta_0} \quad \text{and} \quad S_{rx} = 0.0064 \, \frac{A^2}{2\eta_0}$$

The sum of the reflected and transmitted power densities is equal to the incident power density.

3.5.2 Total Internal Reflection

Snell's law, Equation (3.5.5), can be written in the form

$$\sin \theta_t = \frac{n_1}{n_2} \sin \theta_i \tag{3.5.22}$$

When $n_2 < n_1$, $\theta_t > \theta_i$; as the wave moves from medium 1 into medium 2, its direction of propagation is bent away from the normal. This bending of the wave as it passes from medium 1 into medium 2 is called refraction.

When $\theta_i = \sin^{-1} n_2/n_1$, $\theta_t = \pi/2$. At this angle, called the critical angle, the wave in medium 2 is propagating parallel to the interface. There is no net propagation of energy in the x direction in medium 2 and no propagation across the interface. The incident wave is totally reflected; this condition is called total internal reflection. This important property is the basis for dielectric waveguides and optical fibers.

When the incident wave approaches the interface at the critical angle, $\theta_t = \pi/2$ and $\sin \theta_t = 1$. For angles of incidence greater than the critical angle, $\sin \theta_t$ is greater than unity. This condition cannot be satisfied with any real θ_t, but it can be satisfied with a complex θ_t. Let $\theta_t = \theta_1 + j\theta_2$, and use the identities

$$\sin \theta_t = \sin (\theta_1 + j\theta_2) = \sin \theta_1 \cosh \theta_2 + j \cos \theta_1 \sinh \theta_2$$

$$\cos \theta_t = \cos (\theta_1 + j\theta_2) = \cos \theta_1 \cosh \theta_2 - j \sin \theta_1 \sinh \theta_2$$

For $\sin \theta_t$ real and greater than unity, $\theta_t = \pi/2 + j\theta_2$. Then

$$\sin \theta_t = \cosh \theta_2 \quad \text{and} \quad \cos \theta_t = -j \sinh \theta_2 \tag{3.5.23}$$

The nature of the fields in medium 2 for the case of total internal reflection can now be considered. We will use the subscript 2, rather than t, to designate the fields in medium 2 for the case of total internal reflection. The x and z components of the propagation vector are

$$k_{2z} = k_2 \sin \theta_t \tag{3.5.24a}$$

$$k_{2x} = k_2 \cos \theta_t = -jk_2(\sin^2\theta_t - 1)^{\frac{1}{2}} \tag{3.5.24b}$$

The propagation constants in medium 2 have surprising values; k_{2z} is greater than k_2, and k_{2x} is imaginary! Note, however, that

$$k_{2z}{}^2 + k_{2x}{}^2 = k_2{}^2 \tag{3.5.25}$$

The imaginary propagation constant k_{2x} represents attenuation in the x direction.

$$E_2(x) = E_2(0) \exp\left[-jk_{2x}x\right] \tag{3.5.26a}$$
$$= E_2(0) \exp\left[-k_2\sinh \theta_2 x\right]$$
$$= E_2(0) \exp\left[-\alpha_2 x\right] \tag{3.5.26b}$$

Thus, the strength of the field in medium 2 will decrease exponentially as the distance from the interface increases. The decay constant is

$$\alpha_2 = k_2\left[\sin^2\theta_t - 1\right]^{\frac{1}{2}} \tag{3.5.27}$$

The real propagation constant k_{2z} represents propagation in the z direction without attenuation.

$$E_2(z) = A \exp\left[-jk_{2z}z\right] = A \exp\left[-jk_2 \sin \theta_t z\right] \tag{3.5.28}$$

We will now examine the nature of the transmitted and reflected waves. The reflection coefficient, Equation (3.5.13a), can be written

$$r_E = \frac{k_1\cos \theta_i + jk_2\left[\sin^2\theta_t - 1\right]^{\frac{1}{2}}}{k_1\cos \theta_i - jk_2\left[\sin^2\theta_t - 1\right]^{\frac{1}{2}}} \tag{3.5.29}$$

$$= \frac{k_{1x} + j\alpha_2}{k_{1x} - j\alpha_2} \tag{3.5.30}$$

This ratio has magnitude equal to unity and a phase, 2ϕ, given by

$$\phi = \tan^{-1}\frac{\alpha_2}{k_{1x}} \tag{3.5.31}$$

This is a phase shift in the reflected wave that is associated with total internal reflection.

The transmission coefficient is

$$t_E = \frac{2k_{1x}}{k_{1x} - j\alpha_2} \tag{3.5.32}$$

$$= \frac{2k_{1x}\exp{(j\phi)}}{[k_{1x}{}^2 + \alpha^2]^{\frac{1}{2}}}$$

$$= 2\cos\phi\exp{(j\phi)} \tag{3.5.33}$$

If the incident wave is

$$E_i(x) = \exp{[-j(k_{1x}x)]} \tag{3.5.34}$$

the reflected wave will be

$$E_r(x) = \exp{[jk_{1x}x]}\exp{[j2\phi]} \tag{3.5.35}$$

and the total wave in medium 1 will be

$$E_1(x) = E_i(x) + E_r(x)$$

$$E_1(x) = \exp{[-jk_{1x}x]} + \exp{[+jk_{1x}x]}\exp{[j2\phi]}$$
$$= 2\exp{[j\phi]}\cos{[k_{1x}x + \phi]} \tag{3.5.36}$$

The wave in medium 2 can be found by using the transmission coefficient t_E.

$$E_2(x) = t_E E_i(0)\exp{(-\alpha_2 x)}$$

$$= 2\cos{(\phi)}\exp{(j\phi)}\exp{(-\alpha_2 x)} \tag{3.5.37}$$

If we now restore the $(\omega t - kz)$ terms and recognize that $k_{1z} = k_{2z} = k_z$, we have complete expressions for the E fields.

$$E_1(t,x,z) = 2\cos{[k_{1x}x + \phi]}\exp{[j(\omega t - k_z z + \phi)]} \tag{3.5.38}$$

$$E_2(t,x,z) = 2\cos\phi\exp{[-\alpha_2 x]}\exp{[j(\omega t - k_z z + \phi)]} \tag{3.5.39}$$

These fields, for a specific numerical example, are plotted in Figure 3.5.3.

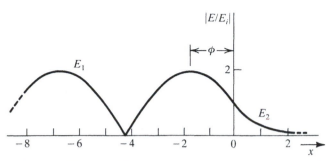

Figure 3.5.3 *Electric field intensities on both sides of a dielectric interface when $\theta_i > \theta_c$. See Example 3.3.*

The effect of the phase angle ϕ on the magnitude of $E(x)$ is to shift the point at which the field has its maximum magnitude by a distance $\Delta x = \phi/k_{1x}$ away from the interface. A phase shift ϕ is also associated with propagation in the z direction. As a result of the complex reflection coefficient, the phase of the propagating wave is advanced by the angle ϕ. A point of constant phase on a wavefront approaching the interface will be shifted forward, discontinuously, by a distance $\Delta z = \phi/k_z$ upon incidence at the interface. The shift, and sometimes the corresponding Δx or other manifestations of the phase shift on reflection, is called the Goos-Haenchen shift.

Equation (3.5.39) indicates that the magnitude of the **E** field, hence those of the **H** field and the Poynting vector, decrease exponentially as the distance from the interface increases. For large α_2, the fields have significant magnitudes only in the region close to the interface; for small α_2, they extend farther into the n_2 region. The depth of penetration of the fields into this region is defined to be $1/\alpha_2$.

The **H** fields in medium 2 can be calculated from Equation (3.2.3).

$$H_{2x} = -\frac{E_2}{\eta_2}\sin\theta_t \tag{3.5.40a}$$

$$H_{2z} = -j\frac{E_2}{\eta_2}[\sin^2\theta_t - 1]^{\frac{1}{2}} \tag{3.5.40b}$$

Real values for E_y and H_x indicate power flow in the z direction, as expected from Snell's law. The magnitude of the Poynting vector is

$$S_{2z} = (1/2)E_{2y} \cdot H_{2x} = \frac{E_{2y}^2}{2\eta_2}\sin\theta_t \tag{3.5.41}$$

When the conditions for total internal reflection are satisfied, the average Poynting vector in medium 2 has only a z component. Although, in the steady-state all of the energy incident on the interface in medium 1 is reflected, there is a wave of constant energy density propagating in medium 2 parallel to the interface. This is an *evanescent* wave. Dielectric waveguides will have evanescent waves that propagate outside the guide.

Since H_z is imaginary, the x component of the Poynting vector is imaginary. An imaginary **H** with a real **E** can be interpreted as **E** and **H** fields with a $\pi/2$ phase angle between them. The instantaneous power density, given by **E** × **H**, is positive, then negative, and so forth, during alternate quarter-cycles. The power flow is away from the interface during part of the cycle, then back toward the interface, and so forth, with average power flow equal to zero. An imaginary Poynting vector is analogous to the instantaneous power flow in an imaginary impedance, that is, a reactance, in an electric circuit; energy flows into the reactance during part of the cycle and out during the next part. The net, or average, power delivered or received during a full cycle is zero.

EXAMPLE 3.3

A plane *TE* wave is incident on the interface between two dielectric materials at an angle of 85° from the normal to the interface. The indices of refraction are 1.45 in the material through which the incident wave is propagating and 1.43 in the other material. The wavelength is 1.3 μm. Calculate the electric field intensities in the two materials in terms of the intensity of the incident wave. (The data for this example are the same as were used in Example 3.2 except that the angle of incidence is greater than the critical angle.)

Solution

The propagation constants k_1 and k_2 are the same as those calculated in Example 3.2. The other propagation constants in medium 1 are

$$k_{1x} = k_1 \cos \theta_i = 7.008 \cdot 10^6 \cos 85°$$
$$= 0.6108 \cdot 10^6 \ (\text{m}^{-1})$$
$$k_{1z} = k_1 \sin \theta_i = 6.9815 \cdot 10^6$$

In medium 2,

$$\sin \theta_t = \frac{n_1}{n_2} \sin \theta_i = 1.0101$$

$$k_{2z} = k_2 \sin \theta_t = 6.9815 \cdot 10^6$$

Note that this is equal to k_{1z}, as it must be; the equality of the propagation constants in the direction parallel to the interface is the basis for Snell's Law.

$$k_{2x} = -jk_2[\sin^2\theta_t - 1]^{\frac{1}{2}} = -j\,0.9861 \cdot 10^6$$

Since

$$k_{2x} = -j\alpha_2 \quad \text{then} \quad \alpha_2 = 0.9861 \cdot 10^6$$

The reflection coefficient is

$$r_E = \frac{k_{1x} + j\alpha_2}{k_{1x} - j\alpha_2} = \exp{(j2\phi)}$$

$$\phi = \tan^{-1}\frac{\alpha_2}{k_{1x}} = 58.23° = 1.016 \, \text{radians}$$

The transmission coefficient is

$$t_E = \frac{2k_{1x}}{k_{1x} - j\alpha_2} = 1.0531 \exp{(j\phi)}$$

As in Example 3.2, we will not write the t and z expressions since they are the same for all equations. The incident wave is the same as in Example 3.2.

The total wave in medium 1 is

$$E_1(x) = E_i + E_r$$
$$= 2\exp{(j\phi)}\cos{(k_{1x}x + \phi)}$$

and the wave in medium 2 is

$$E_2(x) = 1.053\exp{(-\alpha_2 x)}\exp{(j\phi)}$$

The magnitudes of the **E** fields as functions of x are plotted in Figure 3.5.3. Note that both the magnitude and its first derivative are equal at the interface.

SUMMARY

Light waves are electromagnetic waves. Maxwell's equations, wave propagation phenomena, and the laws of reflection and refraction are applicable to the study of light waves. For light waves in dielectric materials, the index of refraction, $n = \epsilon_r^{\frac{1}{2}}$, is an important parameter in characterizing the effects of the material on light wave propagation. The phase propagation constant, β, is equal to $\omega n/c$ or $2\pi n/\lambda$.

The velocity of propagation of a light wave in an infinite medium is $v = c/n$; this velocity is the phase velocity. When β is a nonlinear function of frequency, as, for example, when n is a function of frequency, then the group velocity,

$v_g = d\omega/d\beta$, is a useful concept; the group velocity is the velocity at which any spectrum associated with the wave will propagate. Energy and information move at the group velocity.

When the second derivative of β with frequency is not zero, then different parts of a spectrum move at different group velocities. The result is that the signal at the output of the transmission medium is spread in time. This spreading or stretching of the received signal is called dispersion; dispersion is a form of distortion that limits the information capacity or bandwidth of the communication system. It can arise from material properties or from other characteristics of the transmission medium.

Material dispersion is present in essentially all optical fiber materials. The index of refraction increases with frequency or decreases with wavelength. The dispersion characteristics of materials for optical fibers can be described by graphical or tabulated data that make it possible to determine the index of refraction and the dispersion at various wavelengths.

The reflection and refraction of electromagnetic waves at a dielectric interface can be studied by applying Snell's law and the Fresnel equations. Of special interest is total internal reflection, in which all of the incident energy is reflected. Total internal reflection is the basis for dielectric waveguides, including optical fibers. The analysis of total internal reflection at a dielectric interface shows that although no energy is crossing the interface, there is some energy propagating beyond the interface and moving parallel to it; the external wave is called the evanescent wave.

PROBLEMS

3.1 A dielectric material has propagation constant $\gamma = (0.001 + j\,1.4) \cdot 10^4$ m^{-1} for an electromagnetic wave having wavelength $\lambda = 1$ mm. Find the complex index of refraction of this material.

3.2 The complex dielectric constant of a material is $\epsilon_r = 2.20 - j\,0.25$. A light wave with $\lambda = 1.3$ μm is propagating in this material. Find the attenuation constant α and the phase velocity of this wave.

3.3 Silicon has an $\epsilon_r = 12$, $\mu_r = 1$, and $\rho = 10^5$ ohm-cm. Find the propagation constant γ for light waves having wavelengths of (a) 0.85 μm, (b) 1.3 μm, and (c) 1.55 μm.

3.4 A light wave with wavelength of 1.3 μm has a propagation constant $\mathbf{k} = 3.5 \cdot 10^6\,\mathbf{a}_x + 6.5 \cdot 10^6\,\mathbf{a}_y$. What are the phase velocities of this wave in the x and y directions? What is the phase velocity in the z direction?

3.5 What is the index of refraction of the material in Problem 3.4?

3.6 In a dielectric material having index of refraction $n = 1.225$ the electric field vector is

$$\mathbf{E} = 1.2\,\mathbf{a}_x + 0.7\,\mathbf{a}_y \quad \text{kV/m}$$

Propagation is in the z direction. Calculate the magnetic field vector $\mathbf{H}$.

3.7 Develop an equation for the intrinsic impedance η of a dielectric material in which the conductance σ is not zero. Calculate η at $f = 3 \cdot 10^{14}$ for a material in which $\mu_r = 1$, $\epsilon_r = 1.75$, and $\sigma = 2000$ S/m.

3.8 The spectral width of a signal can be expressed in terms of either the Δf between the edges of the spectrum or the equivalent $\Delta \lambda$. Show that for $\Delta f \ll f$, $\Delta \lambda / \lambda = -\Delta f / f$.

3.9 Show that Equations (3.3.38), (3.3.44), and (3.3.45) are dimensionally consistent.

3.10 By analogy with the equation $v = c/n$, where n is the index of refraction, the *group index* N is defined by $v_g = c/N$. Show that $N = n - \lambda dn/d\lambda$.

3.11 Use the Sellmeier dispersion formula to calculate the group index N and its derivative $dN/d\lambda$ for material A of Table 3.4.1 at $\lambda = 1.3$ μm.

3.12 A plane interface separates two dielectric materials having indices of refraction of 1.4 and 1.5. A TE wave is incident on this interface through the $n = 1.4$ material in a direction inclined 60° from the normal to the interface. Calculate the relative magnitudes and directions of the electric field intensities of reflected and transmitted waves.

3.13 Calculate the critical angle for a dielectric interface having $n_1 = 1.51$ and $n_2 = 1.50$. Calculate the reflection and transmission coefficients, r_E and t_E, for a angle of incidence of 88°.

3.14 (a) Use the data from Example 3.3 and an incident electric field intensity of 1 kV/m to find the power density in the evanescent wave at a point one wavelength ($x = \lambda_2$) from the interface. (b) For what x is the power density in the evanescent wave reduced to 1 percent of that at the interface?

CHAPTER 4

Dielectric Waveguides and Optical Fibers

In Chapter 3, the fundamentals of electromagnetic wave propagation necessary for the study of optical fibers were summarized. In this chapter, these fundamentals will be used to develop the light wave transmission characteristics for various types of fibers. We will introduce the principles of dielectric waveguides and consider the characteristics of the two principal classes of optical fibers.

Dielectric waveguides are based on the principle of total internal reflection. The characteristics of the waveguide can be developed from Maxwell's equations or from the reflections at the sides of the guiding structure. We will use both approaches for one of the waveguides to be considered, but most of our analysis will be based on Maxwell's equations.

We will first analyze the slab waveguide. It has the advantage of a simple geometrical configuration, with correspondingly simple mathematical relationships. It can best be used to illustrate the principles of dielectric waveguides and to establish many of the characteristics that are applicable with little or no change to the optical fiber. However, it is an unrealizable waveguide because it has an infinite dimension.

The two optical fibers to be studied are the step-index fiber, with a sharp cylindrical interface between two regions of different index of refraction, and the graded-index fiber, with a continuously varying index. Both are practical configurations and each has important applications.

Because the optical fiber has a cylindrical configuration, its analysis uses Maxwell's equations in cylindrical coordinates. The mathematical details are substantially more tedious than is the case with the simpler slab waveguide. The derivations are correspondingly more time consuming, and the results are not as easily interpreted. For these reasons, our approach to the study of these dielectric waveguides will be to study the characteristics of the slab waveguide in some detail, then discuss the corresponding results for the step-index and graded-index optical fibers without rigorous derivations of most of these results. For readers who want a more complete treatment of the optical fiber, references [2.6] to [2.10] are suggested.

4.1 The Dielectric Slab Waveguide

The dielectric slab waveguide consists of a slab of dielectric material of index n_1 embedded in a material of index n_2. The slab has width $2d$ and has infinite depth. All fields are assumed to be uniform in the depth dimension. The waveguide structure to be analyzed is shown in Figure 4.1.1.

From previous work, we expect that the requirement for lossless propagation in the guide is that the conditions for total internal reflection are satisfied. In Figure 4.1.1, this is that $n_2 < n_1$ and that the angle of incidence on the side of the guide be greater than the critical angle.

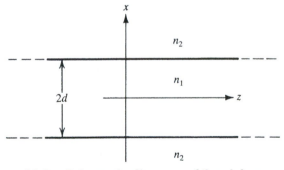

Figure 4.1.1 *Schematic diagram of the slab waveguide. Propagation is in the z direction. The y axis is into and out of the page; the depth of the waveguide in the y direction is infinite. This geometry is used to study propagation, without reference to end effects.*

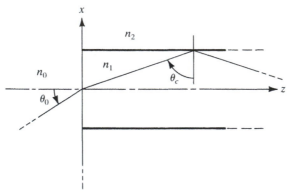

Figure 4.1.2 *Numerical aperture of a dielectric waveguide. The external angle, θ_0, is the maximum angle for which the condition for total internal reflection inside the waveguide can be satisfied. The numerical aperture is $NA = \sin \theta_0$.*

4.1.1 Numerical Aperture

If we assume for the moment that the conditions for total internal reflection at the boundaries of the waveguide are requirements for propagation, we can consider the manner in which a light wave must enter the guide in order to satisfy these conditions. The geometrical relations are illustrated in Figure 4.1.2. By applying Snell's law to the wave incident on the end of the waveguide, it can be shown that total internal reflection can take place inside the guide if, and only if,

$$n_0 \sin \theta_0 \le n_1 \cos \theta_c \tag{4.1.1}$$

The maximum value of $\sin \theta_0$ that can satisfy this equation is called the numerical aperture, NA, of the waveguide. Specifying the NA is the standard way of specifying the range of angles of incidence over which the waveguide will accept input waves. For input angles outside this range, the wave can enter the waveguide but energy will escape through the sides; the energy propagating inside the guide will be rapidly attenuated. This definition of numerical aperture will be applicable to the step-index and graded-index fibers as well as the slab waveguide.

If the outside medium is air, then $n_0 = 1$ and

$$NA = n_1 \cos \theta_c = [n_1{}^2 - n_2{}^2]^{\frac{1}{2}} \tag{4.1.2}$$

4.1.2 Maxwell's Equations in the Slab Waveguide

To derive the fields in the waveguide and the conditions under which useful propagation is possible, we begin with Maxwell's equations [Equation (3.2.3)].

The coordinate system to be used is shown in Figure 4.1.1. We assume a TE mode, with propagation in the z direction; $E_x = 0$ and $E_z = 0$. Since propagation in the z direction is represented by $\exp(-j\beta z)$, then $d\mathbf{E}/dz = -j\beta \mathbf{E}$, with similar relationships for the $\mathbf{H}$ field. Since the fields are uniform in the y direction, $\partial \mathbf{E}/\partial y = 0$. We assume a perfect dielectric with conductivity equal to zero. With these constraints, Maxwell's equations reduce to

$$j\beta E_y = -j\omega\mu\, H_x \tag{4.1.3a}$$

$$0 = -j\omega\mu\, H_y \tag{4.1.3b}$$

$$\frac{dE_y}{dx} = -j\omega\mu\, H_z \tag{4.1.3c}$$

and

$$-j\beta\, H_x - \frac{dH_z}{dx} = j\omega\epsilon\, E_y \tag{4.1.3d}$$

By substituting H_x from Equation (4.1.3a) and H_z from (4.1.3c) into Equation (4.1.3d) we get

$$\frac{d^2 E_y}{dx^2} = \left[\beta^2 - \omega^2\mu\epsilon\right] E_y \tag{4.1.4}$$

This equation is valid in both the slab and the outer material.

Solutions to this equation can be expressed as exponentials or as sine and cosine functions. Assume solutions inside the slab in the form

$$E_y(x) = A \cos k_{1x} x \tag{4.1.5a}$$

or

$$E_y(x) = B \sin k_{1x} x \tag{4.1.5b}$$

By substituting these into (4.1.4), it can be shown that

$$k_{1x}^2 = \omega^2\mu\epsilon_1 - \beta^2 = k_1^2 - \beta^2 \tag{4.1.6}$$

When the $E_y(x)$ is known, the $\mathbf{H}$ fields can be found from Equations (4.1.3a) and (4.1.3c).

To find the fields outside the slab, assume solutions of the form

$$E_y(x) = C \exp\left[-\alpha_2 x\right] \qquad x > d \tag{4.1.7a}$$

and

$$E_y(x) = C \exp\left[\alpha_2 x\right] \qquad x < -d \tag{4.1.7b}$$

By substituting these into Equation (4.1.4), we find

$$\alpha_2{}^2 = \beta^2 - \omega^2 \mu \epsilon_2 = \beta^2 - k_2{}^2 \tag{4.1.8}$$

4.1.3 Propagation Modes

To find the conditions under which propagation is possible, we apply the boundary condition that the tangential fields are continuous across the boundary. First, consider the cosine solution, Equation (4.1.5a). At $x = d$

$$A \cos k_{1x}d = C \exp\left[-\alpha_2 d\right]$$

Then

$$C = A \cos k_{1x}d \exp\left[\alpha_2 d\right]$$

and

$$E_y(x) = A \cos k_{1x}d \exp\left[-\alpha_2(x - d)\right] \qquad x > d \tag{4.1.9}$$

The coefficient A must be evaluated in terms of the power, the field amplitudes, or some other constraint on the magnitude. The fields for $x < -d$ will have equivalent, symmetrical values.

To complete the solution, we apply the boundary condition $H_{1z} = H_{2z}$. These H_z fields can be found by using Equation (4.1.3c).

$$-\frac{jk_{1x}}{\omega\mu} A \sin k_{1x}d = -\frac{j\alpha_2}{\omega\mu} A \cos k_{1x}d$$

which can be reduced to

$$\tan k_{1x}d = \frac{\alpha_2}{k_{1x}} \tag{4.1.10}$$

By using the $\sin k_{1x}d$ solution, Equation (4.1.5b), for the field inside the slab, we can derive a second equation

$$\tan k_{1x}d = -\frac{k_{1x}}{\alpha_2} \tag{4.1.11}$$

Equations (4.1.10) and (4.1.11) give conditions that must be satisfied for the solutions, Equations (4.1.5a) and (4.1.5b), to exist. They restrict the relationship, and hence the values, that k_{1x} and α_2 can have.

It is possible to express α_2 as a function of k_{1x}. Since

$$k_{1x}^2 = k_1^2 - \beta^2 \quad \text{and} \quad \alpha_2^2 = \beta^2 - k_2^2$$

each equation can be solved for β and the two results equated to give

$$\alpha_2^2 = k_0^2 \left[n_1^2 - n_2^2 \right] - k_{1x}^2 \tag{4.1.12}$$

It is convenient to use a normalized frequency variable, V, defined as

$$V = [n_1^2 - n_2^2]^{\frac{1}{2}} k_0 d \tag{4.1.13a}$$

where $k_0 = \omega/c = 2\pi/\lambda$. In cases where $n_1 - n_2$ is small, it is convenient to use $\Delta n = n_1 - n_2$, and $n = (n_1 + n_2)/2$. Then

$$V = [2n\Delta n]^{\frac{1}{2}} k_0 d \tag{4.1.13b}$$

With Equation (4.1.12), V can be expressed in the form

$$V^2 = (\alpha_2 d)^2 + (k_{1x} d)^2 \tag{4.1.14}$$

With the substitution of α_2 from (4.1.14), Equations (4.1.10) and (4.1.11) have only k_{1x} as unknown. They can, in principle, be solved to find values of k_{1x} for which Maxwell's equations are satisfied. These values are called eigenvalues and Equations (4.1.10) and (4.1.11) are the eigenvalue equations.

EXERCISE 4.1

A slab waveguide has $n_1 = 1.49$, $n_2 = 1.48$, and width $2d = 100$ μm. For $\lambda = 0.85$ μm, what is the value of the normalized frequency variable V? If $k_{1x} = 0.900 \cdot 10^6$ m^{-1} is an eigenvalue, find the corresponding α_2 and β.

Answers
63.7, $0.902 \cdot 10^6$ m^{-1}, $10.98 \cdot 10^6$ m^{-1}.

The eigenvalue equations are transcendental equations with no direct, explicit solution. Two methods can be used to find the eigenvalues. The first is a graphical/numerical method. A graph of each side of the equation is plotted. The intersections of the two graphs identify solutions, that is, eigenvalues. With these approximate solutions available from the graph, numerical techniques can be used to find the eigenvalues to any degree of accuracy. This method is illustrated in Example 4.1 and Figure 4.1.3.

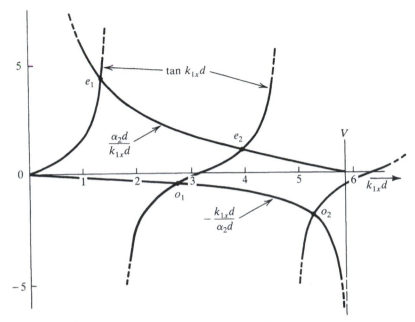

Figure 4.1.3   *Graphs of the eigenvalue equations for a slab waveguide having V = 5.8. There are four eigenvalues, two even solutions designated e1 and e2, and two odd solutions, o1 and o2. See Example 4.1.*

EXAMPLE 4.1

A slab waveguide of width 10 μm has indices of refraction $n_1 = 1.505$ and $n_2 = 1.495$. Find the phase propagation constants for each mode that will propagate in this waveguide when the frequency is $\omega = 2 \cdot 10^{15}$ radians/second.

Solution

$$n_1{}^2 - n_2{}^2 = 0.03$$

$$k_0 = \frac{\omega}{c} = 6.671 \cdot 10^6 \text{ m}^{-1}$$

$$V = 0.03^{\frac{1}{2}} \cdot 6.671 \cdot 10^6 \cdot 5 \cdot 10^{-6}$$
$$= 5.777$$

This V is the cutoff frequency; $k_{1x}d < 5.777$.

To find the eigenvalues, plot Equations (4.1.10) and (4.1.11). These are shown in Figure 4.1.3. From this graph, and some further calculations to improve the accuracy, we can find

$k_{1x}d$ = 1.337, 2.662, 3.957, and 5.173

and

k_{1x} = 0.2674, 0.5325, 0.7914, and 1.0345 · 10^6 m^{-1}

These are the eigenvalues. There are four, and only four, possible modes of propagation in this waveguide at this frequency.

Graphs of the magnitudes of the E fields for the two higher-order modes are shown in Figure 4.1.4.

The phase propagation constants, β, are given by

$$\beta = \left[k_1^2 - k_{1x}^2\right]^{\frac{1}{2}}$$

where $k_1 = n_1 k_0 = 10.04$ · 10^6 m^{-1}. The four values for β are

β = 10.036, 10.026, 10.009, 9.987 · 10^6 m^{-1}

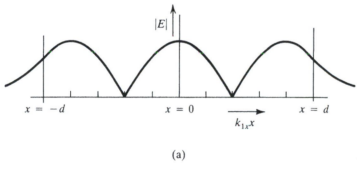

(a)

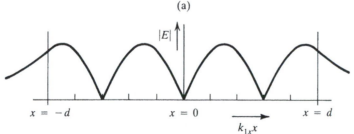

(b)

Figure 4.1.4 *Standing waves in the slab waveguide. (a) The even (cosine) solution, given by Equation (4.1.5a), with $k_{1x}d = 3.9572$. (b) The odd (sine) solution, Equation (4.1.5b), with $k_{1x}d = 5.1727$. These graphs are for the two higher-order modes from Example 4.1.*

Another method for finding the eigenvalues uses approximations for the eigenvalue equations. It is possible to find a simpler equation that is a good approximation to the exact equation over a limited range of the variables; other approximate equations can be used for other ranges of variables. The approximate equation(s) can be used to find the eigenvalues within the applicable range(s) and to find other parameters of interest.

An alternative derivation of the eigenvalue equations is based on the principles of total internal reflection developed in Chapter 3. We assume that a standing wave, such as that calculated in Example 3.3, will exist in the slab, and we will assume symmetry of the fields in the slab. The even (cosine) solution will then have a maximum at the center of the slab and the odd (sine) solution will have a null there. These solutions are illustrated in Figure 4.1.4.

Equation (3.5.36), which represents the standing wave as a cosine function, is based on the origin of the x-axis being at the interface. If we shift the origin to the center of the slab, we can use either a sine or cosine function, depending on the nature of the field at the center. With a maximum at the center, we have the cosine (even) solution. For this case, Figure 4.1.4a shows that the total phase change between the center and the side is given by $k_{1x}d = (m\pi + \phi)$, where m is a positive integer. We have, from Equation (3.5.31), $\tan \phi = \alpha_2/k_{1x}$. We then can write

$$\tan \phi = \tan (k_{1x}d - m\pi) = \frac{\alpha_2}{k_{1x}}$$

and

$$\tan k_{1x}d = \frac{\alpha_2}{k_{1x}}$$

This is Equation (4.1.10), the eigenvalue equation for the cosine solution.

If we assume the sine solution, we can see from Figure 4.1.4b that $k_{1x}d = (m\pi/2 + \phi)$, with m an odd integer. This leads to

$$\tan k_{1x}d = -\frac{k_{1x}}{\alpha_2}$$

which is (4.1.11), the second eigenvalue equation.

The cosine solution could also be derived in terms of $k_{1x}d = (m\pi/2 + \phi)$, with m an even integer. Thus, both use the same expression, but with different constraints on m. The cosine solutions are called even and the sine solutions odd.

4.1.4 *Modal dispersion*

Each eigenvalue defines one mode that will, if excited, be sustained in the waveguide. There can be many different k_{1x} and, therefore, many modes. Each mode has a propagation constant β, phase velocity ω/β, and group velocity $d\omega/d\beta$. Since the propagation constants have different values, the propagation delay times will be different. The receiver in a system with many modes will receive many different replicas of the transmitted signal, spread over a time period corresponding to the fastest and slowest of the group velocities of the various modes.

This multimode propagation is another form of dispersion. It is called intermodal dispersion, or simply modal dispersion, and has effects similar to those discussed in Sections 3.3.5 and 3.4. It limits the maximum rate at which data can be transmitted through the system.

The number of modes that can propagate can be calculated by reference to Figure 4.1.3 and Equation (4.1.14). Since both k_{1x} and α_2 must be real, we can see from Equation (4.1.14) that $k_{1x}d$ cannot be larger than V. Frequencies higher than V cannot propagate; V is the cutoff frequency. From Figure 4.1.3 we see that one and only one eigenvalue is found in each interval of $\pi/2$ on the $k_{1x}d$ scale. The total number of eigenvalues is $2V/\pi$, with any fractional part rounded to the next higher integer.

Modal dispersion can be calculated when the propagation constants are known. It is the difference between the longest and shortest signal propagation times. Because signals propagate at the group velocity, modal dispersion can be expressed as

$$\frac{\Delta\tau}{L} = \frac{1}{v_{g\,min}} - \frac{1}{v_{g\,max}} \quad \text{(s/m)} \tag{4.1.15}$$

The magnitude of this dispersion can be estimated by recognizing that the propagation constant β is approximately that of the external material for modes close to cutoff, and is close to that of the slab material for the low-order modes. This is related to the fact that for modes close to cutoff, most of the energy in the wave propagates in the evanescent wave, that is, in material 2, and for low-order modes most of the energy propagates in material 1. On this basis, we conclude that the extreme values for group velocities are also those of the materials. We then can express the v_g as c/N and the modal dispersion as

$$\frac{\Delta\tau}{L} = \frac{N_1 - N_2}{c} \tag{4.1.16}$$

This can be expressed in terms of Δn by the following approximation:

$$N_1 - N_2 = \frac{N_1}{n_1}\left[n_1 - \frac{n_1}{N_1}N_2\right]$$

We assume that $N_2/n_2 \approx N_1/n_1 \approx N/n$. Then

$$N_1 - N_2 \approx \frac{N}{n}[n_1 - n_2]$$

The modal dispersion can then be written as

$$\frac{\Delta\tau}{L} = \frac{N}{n}\frac{\Delta n}{c} \tag{4.1.17}$$

This is a good approximation for the modal dispersion in a weakly guiding ($\Delta n \ll n$) multimode fiber. Since it is based entirely on material parameters and does not include any waveguide parameters, its accuracy would decrease as the number of modes becomes small.

The number of modes, and therefore the modal dispersion, is reduced by making V small. Equations (4.1.13b) and (4.1.17) indicate that both V and the modal dispersion are reduced by making Δn smaller.

EXERCISE 4.2

The indices of refraction for the materials in a slab waveguide are $n_1 = 1.49$, $N_1 = 1.51$, $n_2 = 1.48$, and $N_2 = 1.50$. The wavelength is $\lambda = 1.3$ μm and $d = 100$ μm. Calculate the number of modes and the modal dispersion in this waveguide.

Answers
54 modes, 33.3 ns/km.

4.2 Step-Index Fiber

The step-index fiber consists of two concentric cylinders. The inner cylinder has index of refraction n_1 and radius a. The outer cylinder has index n_2, with $n_1 > n_2$, and a large outer radius. We will treat the outer radius as if it were infinite; in practical fibers, it is large enough that the evanescent wave reaches negligibly small magnitudes within the outer radius. The inner cylinder is called the core and the outer layer the cladding. A schematic diagram of the step-index fiber is shown in Figure 4.2.1.

To study the propagation of light waves in the step-index fiber, we will consider the solution of Maxwell's equations in cylindrical coordinates for the structure shown in Figure 4.2.1. We will not undertake a thorough and complete solution of the cylindrical boundary-value problem; that involves many details of secondary interest. The approach followed in this section, and following sections, is to outline the procedure for such solutions, to summarize the principal results of these solutions, and to interpret these results for insight concerning the propagation characteristics of light waves in this class of optical fiber.

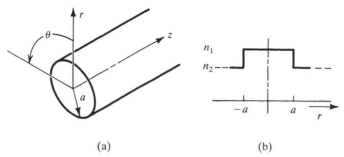

(a)   (b)

Figure 4.2.1 *Schematic diagram of the step-index fiber. (a) Cylindrical coordinates to be used in the analysis of the step-index fiber. (b) Index profile. The radius, b, of the cladding is assumed to be infinite. In a practical fiber, b must be large enough so that the evanescent wave has very small magnitudes at r = b.*

4.2.1 *The Wave Equation and Boundary Conditions*

Maxwell's equations consist of six equations that relate the values of the r, θ, and z components of the **E** and **H** field vectors. We assume that $\exp\left[j(\omega t - \beta z)\right]$ describes propagation in the z direction. Therefore $dE/dz = -j\beta E$ and $dE/dt = j\omega E$, with similar relationships for **H**. We assume the conductivity to be zero. With these substitutions, we can express E_r, E_θ, H_r, and H_θ in terms of E_z and H_z.

$$E_r = -\frac{j}{\kappa^2}\left[\beta\frac{dE_z}{dr} + \omega\mu\frac{1}{r}\frac{dH_z}{d\theta}\right] \tag{4.2.1a}$$

$$E_\theta = -\frac{j}{\kappa^2}\left[\beta\frac{1}{r}\frac{dE_z}{d\theta} - \omega\mu\frac{dH_z}{dr}\right] \tag{4.2.1b}$$

$$H_r = -\frac{j}{\kappa^2}\left[\beta\frac{dH_z}{dr} - \omega\epsilon\frac{1}{r}\frac{dE_z}{d\theta}\right] \tag{4.2.1c}$$

$$H_\theta = -\frac{j}{\kappa^2}\left[\beta\frac{1}{r}\frac{dH_z}{d\theta} + \omega\epsilon\frac{dE_z}{dr}\right] \tag{4.2.1d}$$

where

$$\kappa^2 = \omega^2\mu\epsilon - \beta^2 = k^2 - \beta^2 \tag{4.2.2}$$

Here we have expressed four of the six field vectors in terms of the other two. We can solve the wave equation and the boundary value problem to find the E_z and H_z components.

The wave equation for $E_z(r,\theta)$, in cylindrical coordinates, is

$$\frac{d^2 E_z}{dr^2} + \frac{1}{r}\frac{dE_z}{dr} + \frac{1}{r^2}\frac{d^2 E_z}{d\theta^2} + \kappa^2 E_z = 0 \tag{4.2.3}$$

A similar equation, with the same κ, applies to H_z. These wave equations are applicable in both the core and the cladding.

We have already established that the t and z variations of the fields are given by exp $[j(\omega t - \beta z)]$. The solution of (4.2.3) will have the form

$$E_z(t,r,\theta,z) = Ag(r)h(\theta) \exp [j(\omega t - \beta z)] \tag{4.2.4}$$

The $g(r)$ and $h(\theta)$ are to be determined.

The θ variation must be periodic in θ, with period $2\pi/v$, where v is an integer. Either an exponential or a sinusoid can satisfy this requirement; we will use exp $(jv\theta)$. For $v = 0$, E_z is independent of θ.

If we substitute $h(\theta) = \exp(jv\theta)$ into the wave equation [Equation (4.2.3)], it can be reduced to

$$\frac{d^2 E_z}{dr^2} + \frac{1}{r}\frac{dE_z}{dr} + \left[\kappa^2 - \frac{v^2}{r^2}\right]E_z = 0 \tag{4.2.5}$$

This is Bessel's differential equation. Its solutions are Bessel functions. Note that v, the azimuthal index, is a parameter in this equation.

The constraints we must place on $g(r)$ are that it is finite for $r < a$ and that it must approach zero for $r \gg a$. Bessel functions that satisfy these constraints are the $J_v(\kappa r)$ Bessel functions of the first kind, for $r < a$, and the $K_v(\gamma r)$ modified Bessel functions of the second kind, for $r > a$. The forms of the solutions are then

$$E_z(r,\theta) = AJ_v(\kappa r) \exp(jv\theta) \qquad r < a \tag{4.2.6a}$$

$$H_z(r,\theta) = BJ_v(\kappa r) \exp(jv\theta) \qquad r < a \tag{4.2.6b}$$

for the fields in the core, and

$$E_z(r,\theta) = CK_v(\gamma r) \exp(jv\theta) \qquad r > a \tag{4.2.7a}$$

$$H_z(r,\theta) = DK_v(\gamma r) \exp(jv\theta) \qquad r > a \tag{4.2.7b}$$

in the cladding. The parameters κ and γ are given by

$$\kappa^2 = k_1{}^2 - \beta^2 \tag{4.2.8a}$$

and

$$\gamma^2 = \beta^2 - k_2{}^2 \tag{4.2.8b}$$

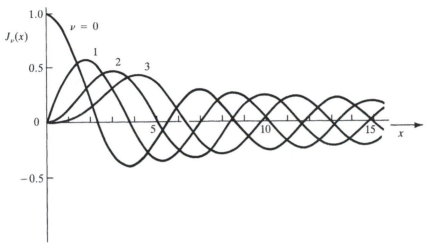

Figure 4.2.2 *Graphs of $J_\nu(x)$ for $\nu = 0, 1, 2,$ and 3.*

Graphs of several $J_\nu(x)$ functions are shown in Figure 4.2.2. The roots of $J_\nu(x) = 0$ are the zeros of the Bessel function; the zeros of Bessel functions will be useful in determining which modes can propagate in the fiber. Table 4.2.1 gives some zeros of low-order Bessel functions. Figure 4.2.3 gives approximate values for the first zero of higher-order functions.

The $K_\nu(x)$ functions are positive for all x; they are infinite for $x = 0$; and they approach zero as x increases.

By eliminating β between Equations (4.2.8a) and (4.2.8b) we have

$$\kappa^2 + \gamma^2 = k_1{}^2 - k_2{}^2 \qquad (4.2.9)$$

Table 4.2.1 Zeros of Bessel Functions

$J_0(x)$	$J_1(x)$	$J_2(x)$	$J_3(x)$	$J_4(x)$	$J_5(x)$
2.405					
	3.832				
5.520		5.136			
	7.016		6.380		
8.634		8.417		7.588	
	10.173		9.761		8.772
11.792		11.620		11.065	
	13.324		13.015		12.339
14.931		14.796		14.372	
	16.471		16.223		15.700
18.071		17.960		17.616	
	19.616		19.409		18.980
21.212		21.117		20.827	
	22.760		22.583		22.218

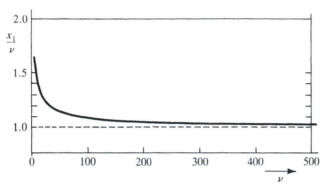

Figure 4.2.3 *The first root, x_1, of $J_\nu(x) = 0$. x_1 is always larger than ν.*

If V is defined to be

$$V = [k_1{}^2 - k_2{}^2]^{\frac{1}{2}}a = 2\pi \frac{a}{\lambda} [n_1{}^2 - n_2{}^2]^{\frac{1}{2}} \tag{4.2.10a}$$

then

$$V^2 = (\kappa a)^2 + (\gamma a)^2 \tag{4.2.10b}$$

This V is the normalized frequency, comparable to that used in considering the slab waveguide in the previous section.

The boundary conditions for the step-index optical fiber are applied by equating the z and θ (tangential) components of the **E** and **H** fields in the core and cladding at $r = a$. As was the case with the dielectric slab, this gives relationships between the amplitudes of the various components, and it gives an eigenvalue equation. The eigenvalue equation establishes relationships between the propagation parameters in the core and cladding. In general, there will be many different solutions, that is, eigenvalues, to the eigenvalue equation. Each eigenvalue defines a set of propagation parameters that represent one possible mode of propagation. Only modes that correspond to an eigenvalue can propagate in the fiber.

The eigenvalue equation can be expressed as a fourth-order determinant that involves the $J_\nu(\kappa r)$, $K_\nu(\gamma r)$, their derivatives, the wavelength, and properties of the materials in the core and cladding. Because it is difficult to solve, and not necessary in most cases of practical interest, we will not consider this equation or its solution. The case that represents most optical fibers, especially those with the better transmission characteristics, is that for which $(n_1 - n_2)$ is small. In this case, the eigenvalue equation reduces to a simpler form. It is considered in the following section.

4.2.2 Weakly Guiding Fibers

An important class of optical fibers is designed with the restriction that $(n_1 - n_2)/n_2 \ll 1$. The reason for imposing this restriction is that it leads to desirable properties in optical fibers. Optical fibers that satisfy this constraint are called weakly guiding fibers. A fortuitous result of the restriction is that the eigenvalue equation is simplified. The resulting equation is

$$\frac{J_v(\kappa a)}{J_{v-1}(\kappa a)} = -\frac{\kappa}{\gamma} \frac{K_v(\gamma a)}{K_{v-1}(\gamma a)} \tag{4.2.11}$$

For $v = 0$, and using the identities $J_{-1}(\kappa a) = -J_1(\kappa a)$ and $K_{-1}(\gamma a) = K_1(\gamma a)$, the eigenvalue equation becomes

$$\frac{J_1(\kappa a)}{J_0(\kappa a)} = \frac{\gamma}{\kappa} \frac{K_1(\gamma a)}{K_0(\gamma a)} \tag{4.2.12}$$

The equation can be solved for the $v = 0$ eigenvalues by the graphical technique shown in Figure 4.2.4(a); this procedure is similar to that used for the slab waveguide in Section 4.1. Intersections of the two graphs identify the eigenvalues. If the accuracy achievable from the graph is not adequate, greater accuracy can be had by using tables of Bessel functions and numerical techniques.

For $v = 1$, the eigenvalue equation is

$$\frac{J_1(\kappa a)}{J_0(\kappa a)} = -\frac{\kappa}{\gamma} \frac{K_1(\gamma a)}{K_0(\gamma a)} \tag{4.2.13}$$

The graphical representation of this equation is also shown in Figure 4.2.4(a).

For all $v > 1$, the equation has the same form as Equation (4.2.11). The J_v/J_{v-1} have the general features of the tangent function. The right side of the equation is always negative, with small magnitudes for low κa and approaching an infinite asymptote as κa approaches V. The graphs for $v = 3$ are plotted in Figure 4.2.4(b) as an example.

One important aspect of the J_v/J_{v-1} function is that as v increases, the first infinite discontinuity moves to the right. This means that for v higher than some v_c there are no solutions to the eigenvalue equation. The maximum azimuthal index, v_c, is determined by V, the cutoff frequency, and $J_{v-1}(\kappa a) = 0$, the first zero for the $v - 1$ order Bessel function. The first zero of $J_{v-1}(\kappa a)$ is the frequency at which J_v/J_{v-1} has its first infinite discontinuity. There can be no eigenvalue until κa is larger than the frequency at which this discontinuity occurs. Thus, if the first zero of J_{v-1} is greater than V, all modes of azimuthal order v are cut off.

Figure 4.2.4(b) shows the eigenvalues for $v = 3$. The first zero of $J_2(\kappa a)$ is at $\kappa a = 5.136$. If V were 5.0, there would be no eigenvalues for $v = 3$.

The first few zeros of $J_v(x)$ for $v = 0$ through 5 are listed in Table 4.2.1. The graph of Figure 4.2.3 gives approximate values of the first zero for v to 500.

The eigenvalues calculated from Equation (4.2.11) are given two subscripts; the first is v and the second identifies the various eigenvalues for that v. Thus, κ_{25} represents the fifth eigenvalue for $v = 2$.

When the eigenvalues have been found, the propagation constants and the phase and group velocities can be calculated. The propagation constant β_{vm}

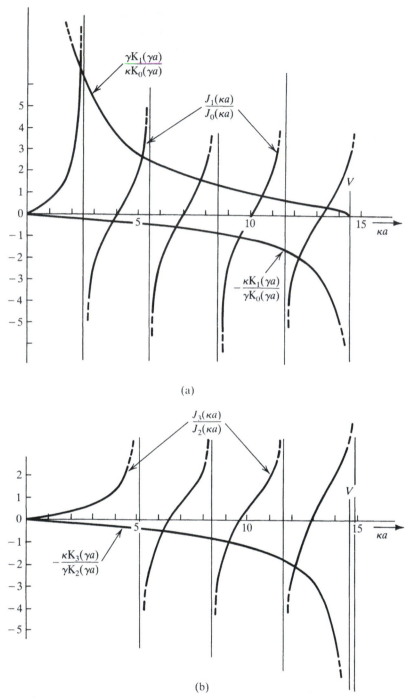

(a)

(b)

Figure 4.2.4   *Graphs of the eigenvalue equation for (a) $v = 0$, and 1; (b) $v = 3$. See Example 4.2. Graphical solutions for $v = 0$ lie in the upper half-plane. Solutions for all $v > 0$ lie in the lower half-plane.*

equals $(k_1{}^2 - \kappa_{vm}{}^2)^{\frac{1}{2}}$, where k_1 is a property of the material and κ_{vm} is the vmth eigenvalue. Both k_1 and κ_{vm} are functions of frequency.

EXAMPLE 4.2

A step-index optical fiber for use at 1.3 μm has radius 25 μm; the indices of refraction are 1.465 and 1.460 for the core and cladding, respectively. Plot the eigenvalue equations for $v = 0$, 1, and 3. Find the propagation constants for the lowest and highest order $v = 0$ modes and for the lowest order $v = 3$ mode.

Solution

The normalized cutoff frequency is

$$V = 2\pi \frac{a}{\lambda} [n_1{}^2 - n_2{}^2]^{\frac{1}{2}} = 14.61$$

By using a table of Bessel functions, graphs of the eigenvalue equations are plotted; see Figures 4.2.4 (a) and (b). The eigenvalues are found to be

$$\kappa_{01} a = 2.24 \qquad \kappa_{05} a = 13.71 \qquad \kappa_{31} a = 5.96$$

and

$$\kappa_{01} = 8.96 \cdot 10^4 \text{ m}^{-1} \qquad \kappa_{05} = 5.48 \cdot 10^5 \text{ m}^{-1} \qquad \kappa_{31} = 2.38 \cdot 10^5 \text{ m}^{-1}$$

The propagation constants are

$$\beta = [k_1{}^2 - \kappa_{vm}{}^2]^{\frac{1}{2}}$$

$$\beta_{01} = 7.080 \cdot 10^6 \text{ m}^{-1} \qquad \beta_{05} = 7.059 \cdot 10^6 \text{ m}^{-1} \qquad \beta_{31} = 7.077 \cdot 10^6 \text{ m}^{-1}$$

The γa for these three modes are

$$\gamma_{01} a = 14.44 \qquad \gamma_{05} a = 5.05 \qquad \gamma_{31} a = 13.34$$

The E_z equations for the 01 and 31 modes are

$$E_z = A_{01} J_0 \left(2.24 \frac{r}{a} \right) \exp \left[j(\omega t - \beta_{01} z) \right] \qquad r < a$$

$$E_z = C_{01} K_0 \left(14.44 \frac{r}{a} \right) \exp \left[j(\omega t - \beta_{01} z) \right] \qquad r > a$$

for the 01 modes, and

$$E_z = A_{31}J_3\left(5.96\,\frac{r}{a}\right)\exp\,(j3\theta)\,\exp\,[j(\omega t - \beta_{31}z)] \qquad r < a$$

$$E_z = C_{31}K_3\left(13.34\,\frac{r}{a}\right)\exp\,(j3\theta)\,\exp\,[j(\omega t - \beta_{31}z)] \qquad r > a$$

for the 31 modes.

EXERCISE 4.3

In a step-index fiber, the cutoff frequency, V, is 8.5. For each azimuthal index, v, how many modes can propagate?

Answer
There can be three modes for $v = 0$; two modes each for $v = 1, 2,$ and 3; one mode each for $v = 4$ and 5; and no modes for $v > 5$.

4.2.3 Modal Dispersion

There are three types of dispersion in optical fibers: material dispersion, waveguide dispersion, and modal dispersion. Modal, or intermodal, dispersion is the dominant effect when it is present. Each mode in a multimode fiber will have a characteristic group velocity and corresponding propagation delay. The modal dispersion is the difference between the longest and shortest propagation delays.

For the analysis of dispersion it is useful to define a normalized propagation constant b_{vm}:

$$b_{vm} = \frac{\beta_{vm}^2 - k_2^2}{k_1^2 - k_2^2} \tag{4.2.14}$$

$$= 1 - \frac{\kappa_{vm}^2 a^2}{V^2} = \frac{\gamma_{vm}^2 a^2}{V^2} \tag{4.2.15}$$

Since $k_2 < \beta < k_1$, then $0 < b < 1$.
Equation 4.2.14 can be solved for β_{vm}:

$$\beta_{vm} = \frac{\omega}{c}\left[n_2^2 + (n_1^2 - n_2^2)b_{vm}\right]^{\frac{1}{2}} \tag{4.2.16}$$

$$\approx \frac{\omega n_2}{c}\left[1 + 2\Delta\,b_{vm}\right]^{\frac{1}{2}} \tag{4.2.17}$$

where Δ is defined[1] as the ratio $\Delta n/n_2$

$$\Delta = \frac{n_1 - n_2}{n_2} \tag{4.2.18}$$

For the weakly guiding fiber $\Delta \ll 1$.

$$\beta_{vm} = \frac{\omega n_2}{c}\left[1 + \Delta \cdot b_{vm}\right] \tag{4.2.19}$$

The group delay in a fiber of length L is $\tau = L/v_g$; it can be found by differentiating β. We will drop the vm subscripts, but recognize that β and b can represent any one of many modes. In differentiating Equation (4.2.19), a term proportional to $d\Delta/d\omega$ has been dropped.

$$\frac{\tau}{L} = \frac{d\beta}{d\omega} = \frac{N_2}{c}\left[1 + \Delta \cdot b\right] \tag{4.2.20}$$

Modal dispersion is the difference between the maximum and minimum group delays.

$$\frac{\Delta\tau}{L} = \frac{N_2}{c}\left[(1 + \Delta \cdot b_1) - (1 + \Delta \cdot b_2)\right]$$

$$= \frac{N_2 \Delta}{c}\left[b_1 - b_2\right] \tag{4.2.21}$$

For fibers with many modes, $b_1 \approx 1$, $b_2 \approx 0$, and $(b_1 - b_2) \approx 1$.

$$\frac{\Delta\tau}{L} \approx \frac{N_2 \Delta}{c} \quad (s/m) \tag{4.2.22}$$

For fibers with small Δ and many modes, this would give a reasonable approximate value for modal dispersion.

The modal dispersion can be reduced by making Δn small, which is the weakly guiding condition. The effect of making Δn small is to reduce V, thus reducing the number of modes. The longest propagation delay is reduced with essentially no change in the shortest.

---

[1] This use of Δ can cause some confusion since the same symbol is used to indicate small increments of other variables, for example $\Delta\tau$ and Δn. However, because it is widely used as defined here, we will conform to this practice. The reader must determine from the context which interpretation is the proper one.

EXERCISE 4.4

A step-index fiber for $\lambda = 0.85 \ \mu m$ has radius $a = 50 \ \mu m$. The indices of refraction are $n_2 = 1.442$, $N_2 = 1.457$, and $\Delta n = \Delta N = 0.005$. Calculate the modal dispersion.

Answer

$1.68 \cdot 10^{-11}$ s/m $= 16.8$ ns/km

4.2.4 Dispersion in Single-Mode Fibers

Modal dispersion can be effectively eliminated by reducing the number of modes that can propagate to one. The single mode fiber carries only the lowest-order, κ_{01}, mode. It can be seen in Figure 4.2.4(a) and Table 4.2.1 that this will require that $V < 2.405$. To make V small, both Δn and the fiber radius must be small. This makes the fiber more difficult to manufacture. It also makes the NA small and coupling light waves into the fiber more difficult. It has been found that single-mode fibers can be built and that they can be used effectively in communication systems.

In single mode fibers the two other types of dispersion, that is, intramodal dispersion, become significant. Material dispersion results when the dielectric constant and index of refraction vary with frequency. Waveguide dispersion is the result of the effects of frequency on the propagation parameters in the waveguide. These two effects will interact. As a result, the analysis becomes cumbersome and the interpretation of the results difficult. We will take advantage of simplifying approximations based on the weakly guiding condition. For a more thorough analysis, refer to [2.7], [2.8], and [5.1].

To study intramodal dispersion, return to Equation (4.2.19). This equation can be differentiated twice to give an expression for the total intramodal dispersion, including both the material and the waveguide dispersion. The resulting equation can, by using approximations based on small Δ, be reduced to [2.9]:

$$\frac{\Delta \tau}{L \cdot \Delta \lambda} = \frac{1}{c} \cdot \frac{dN_2}{d\lambda} - \frac{N_1 - N_2}{\lambda c} \left[V \cdot \frac{d^2(Vb)}{dV^2} \right] \qquad \textbf{(4.2.23)}$$

material   waveguide
dispersion   dispersion

The first term represents the material dispersion and the second the waveguide dispersion. In the derivation of Equation (4.2.23), there are other terms that represent interaction between material and waveguide effects. For the weakly guiding fiber, Gloge [2.9] has shown that the interaction terms are small and the total intramodal dispersion is approximately the sum of the material and waveguide dispersions.

Intramodal dispersion, given by Equation (4.2.23) or equivalent, is a per-unit-spectral-width parameter. We must multiply by the spectral width to find dispersion. This spectral width can arise from two sources. We have already described the bandwidth requirement for transmitting a modulated carrier through the transmission channel. The spectral width, that is, bandwidth, due to modulation defines a minimum spectral width required to handle a modulated signal.

Another significant factor in determining the spectral width of a light-wave signal is the inherent spectral width of the source. The best sources available are not pure sine waves of constant frequency; they can be thought of as having a carrier frequency that wanders randomly throughout a range of frequencies about the nominal carrier frequency. This instability can best be described by defining the nominal frequency and the rms variation about this nominal center frequency. The variation about the center frequency is equivalent to a spectral width.

The total spectrum is a combination of the effects of the modulation and the unstable carrier. The two effects are not simply additive because both are random in nature. They are normally combined by calculating the dispersion for each independently, then adding the squares of the dispersions to find the square of the total dispersion.

The last term in Equation (4.2.23), representing waveguide dispersion, can be reduced to a more easily interpreted expression by recognizing that b is a function of V and that, for the single-mode fiber, V is restricted to a small range of allowed values. Figure 4.2.5 [2.9] shows the derivative part of the waveguide dispersion equation plotted versus V. Because we are interested in a specific,

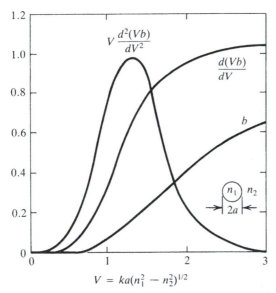

$$V = ka(n_1^2 - n_2^2)^{1/2}$$

Figure 4.2.5 *Normalized propagation constant b and two derivatives,* $\dfrac{d(Vb)}{dV}$ *and* $V\dfrac{d^2(Vb)}{dV^2}$, *as functions of the normalized frequency V. From Gloge,* Applied Optics, *vol. 10, pp. 2442-2445, 1971 [2.9].*

and small, range of values, the curve can be approximated by a reasonably simple equation. For V in the range $1.5 < V < 2.4$,

$$V\frac{d^2(Vb)}{dV^2} \approx \frac{1.984}{V^2} \tag{4.2.24}$$

With this approximation, the waveguide dispersion becomes

$$D_w = \frac{\Delta\tau}{L \cdot \Delta\lambda} = -\frac{N_1 - N_2}{\lambda c} \cdot \frac{1.984}{V^2} \tag{4.2.25}$$

By substituting for V from Equation (4.2.10) and using the approximate relationship

$$\frac{N_1 - N_2}{N_2} \approx \frac{n_1 - n_2}{n_2} \tag{4.2.26}$$

the waveguide dispersion can be expressed

$$D_w = -\frac{1.984\, N_2}{(2\pi a)^2 2cn_2{}^2} \lambda \;\; (s/m^2) \tag{4.2.27}$$

EXERCISE 4.5

A single-mode step-index fiber for 1.3 μm light waves has $a = 3.5\ \mu$m and the following indices: $n_1 = 1.447$, $n_2 = 1.442$, $N_1 = 1.462$, and $N_2 = 1.457$. (a) Confirm that this is a single-mode fiber. For what range of wavelengths would it be single-mode? (b) Calculate the waveguide dispersion.

Answers
(a) Single-mode for $\lambda > 1.099\ \mu$m. (b) $D_w = -6.23$ ps/(km $\cdot$ nm)

It can be seen that in the design of the fiber some control of the waveguide dispersion is possible. To the extent that it is practical to control the fiber radius a, D_w can be set to a desired value at a specific wavelength.

Material dispersion was discussed in Section 3.4. Data such as may be needed for system design are given in Figures 3.4.2 and 3.4.3. Both the index n and the group index N can be estimated from Figure 3.4.2. Dispersion for these materials is given in Figure 3.4.3. These data represent the material characteristics for plane waves in an infinite medium. Equation (4.2.23) indicates that, for the weakly guiding fiber, the waveguide parameters have only a small effect on the material dispersion; dispersion data for plane waves in an infinite medium provide a good approximation for material dispersion in the weakly guiding fiber.

In single-mode fibers, the total dispersion is the sum of the material and wavelength dispersion, as in (4.2.23). The waveguide dispersion, D_w, is negative. In the region of $\lambda < 1.3$ μm, the material dispersion, D_m, for silica fibers is also negative. In the neighborhood of $\lambda = 1.3$ μm, D_m passes through zero. There is some wavelength, λ_0, for which waveguide dispersion is the only dispersion present. At wavelengths above the 1.3 μm region, D_w and D_m dispersion have opposite signs.

It is possible to design the fiber so that in the 1.55 μm wavelength region the total intramodal dispersion, as defined above, is zero! When that condition is achieved, second-order effects will give some dispersion, but it can be very small. The second-order effects are the higher derivatives of β with respect to ω or λ. The dispersion coefficients D are defined as proportional to the second derivative of β. If the third and higher derivatives are not zero, and there is no reason to expect that they are, then the second derivative, and therefore D, will not be constant over the spectrum of the modulated light wave. Although the total dispersion can be made zero at one wavelength, it is not zero over the entire spectrum.

The use of negative D_w to balance the positive D_m has the effect of shifting the zero material dispersion wavelength from 1.3 μm to longer wavelengths. The fibers using this compensation are called dispersion-shifted single-mode fibers. Graphs illustrating this technique are shown in Figure 4.2.6. For a specific

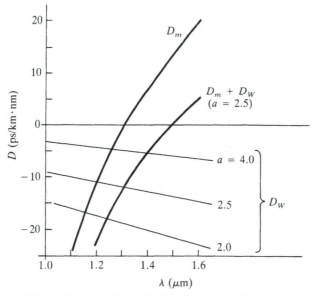

Figure 4.2.6 *Illustration of dispersion-shifted fiber. By selecting an appropriate radius, a, the total dispersion can be made zero at some wavelength in the range 1.3 < λ < 1.6. For smaller λ, both D_m and D_w are negative. For larger λ, the radius, a, required for zero dispersion becomes very small and difficult to achieve.*

wavelength, Equation (4.2.27) can be used to find the radius of the fiber to give zero total dispersion at that wavelength.

It is possible to achieve dispersion of a small fraction of a picosecond per km · nm. By using single-mode fiber and by compensating for material dispersion with the negative waveguide dispersion, the total dispersion in the longer wavelength region can be reduced from tens of nanoseconds per km (modal dispersion) to tens of femtoseconds per km · nm. Thus, very high data rates, or very large bandwidths, become achievable.

EXERCISE 4.6

A step-index fiber has material dispersion $D_m = 20$ ps/km · nm at its operating wavelength $\lambda = 1.55$ μm. The indices of refraction are $n_1 = 1.447$, $N_1 = 1.462$, $n_2 = 1.442$, $N_2 = 1.457$. Find the radius, a, of the fiber that will cause the waveguide dispersion, D_w, to cancel the material dispersion to give zero intramodal dispersion at this wavelength.

Answer
$a = 2.13$ μm.

When a fiber is operated at the wavelength of zero total dispersion, it becomes necessary to recognize higher-order dispersion effects. The zero of dispersion is at a single wavelength, not a band of wavelengths. If the spectrum of the signals being carried is not zero, then parts of the spectrum will lie in frequency ranges where the dispersion is not zero. For the case illustrated in Figure 4.2.6, with the carrier at the wavelength of zero dispersion, part of the spectrum will have positive and part of it negative dispersion. We will usually be concerned with the magnitude of dispersion, without regard for its sign.

A further refinement in the design of the optical fiber has been the use of more complex index profiles. This is illustrated in Figure 4.2.7. The index profile shown there is called quadruple-clad; it is still regarded as a step-index structure. By using such profiles it is possible to achieve relatively low total dispersion over a wide band of wavelengths. There can be two wavelengths of zero total dispersion.

To summarize, dispersion can be reduced in the following ways:

1. Single-mode fibers eliminate modal dispersion.
2. Operation at λ_0, the wavelength of zero material dispersion, eliminates material dispersion at that single wavelength but not over the complete spectrum.
3. For wavelengths longer than λ_0, the wavelength of zero dispersion can be shifted to the longer wavelength by matching the negative waveguide dispersion against the positive material dispersion at that wavelength.
4. By using a more complex index profile, it is possible to achieve a low total dispersion over a wide range of wavelengths.

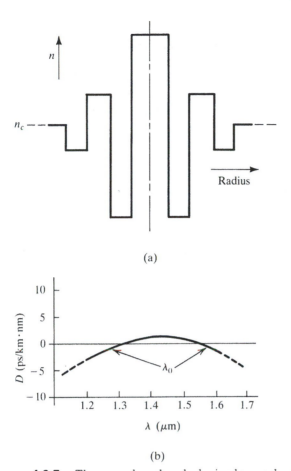

(a)

(b)

Figure 4.2.7 *The quadruple-clad single-mode fiber.* (a) *Index profile.* (b) *Total dispersion versus wavelength. The fiber can be designed to have two wavelengths of zero dispersion, with small dispersion at intermediate wavelengths.*

Modal dispersion, which dominates when it is present, is of the order of tens of nanoseconds per kilometer. A single-mode quadruple-clad fiber can have total dispersion of tens of femtoseconds per km · nm.

4.3 Graded-Index Fiber

The third class of dielectric waveguide that we will consider is the graded-index fiber. It differs from the step-index in that the index of refraction is a continuously variable function of the radial distance from the axis of the fiber. In the simplest configuration, the index is maximum on the axis, decreases with r for $r < a$, and is constant for $r > a$.

The graded-index fiber can provide substantially less modal dispersion than that in step-index fibers. The manner in which the graded-index fiber guides the light wave is not reflection at a dielectric discontinuity, as is the case for the slab and step-index waveguides, but is a continuous bending of the propagation vector **k** as it moves through the region of continuously changing index of refraction. As the wave moves down the fiber, it will, in general, have a component of **k** in the radial direction; this component was the k_{1x} in the slab waveguide and κ in the step-index fiber. As the light wave moves away from the axis into regions of lower index, its velocity will increase and it will be bent back toward the fiber axis. It will eventually be redirected so that it is moving toward the fiber axis into a region of higher index and lower velocity. As it moves toward the fiber axis, it will be bent outward and will again be moving away from the axis. The ray representing the propagating wave will follow a path that spirals around the fiber axis, never reaching either the axis or the cladding. The reduction of dispersion is attributed to the fact that rays having the longer paths will travel through regions of lower index and therefore have higher average velocities than will waves that follow paths that remain closer to the axis of the fiber.

In this section, we will outline the steps in the derivation of the propagation constants and dispersion, summarize the results of such analyses, and interpret these results. More complete and detailed derivations and analyses are available in a number of books and papers, including references [5.1] and [2.12].

4.3.1 *The Index Profile*

The dielectric constant of the graded-index fiber can be represented by

$$\epsilon_r = \epsilon_1[1 - 2\Delta \cdot f(r)] \tag{4.3.1}$$

for $r < a$. The $f(r)$ is usually $(r/a)^\alpha$. $\Delta = (n_1 - n_2)/n_1$, as was defined in Equation $(4.2.18)^2$; as before, $\Delta \ll 1$. The index of refraction is then

$$n(r) = n_1\left[1 - 2\Delta\left(\frac{r}{a}\right)^\alpha\right]^{\frac{1}{2}} \quad r < a \tag{4.3.2a}$$

$$= n_1[1 - 2\Delta]^{\frac{1}{2}} = n_2 \quad r > a \tag{4.3.2b}$$

Here n_1 and n_2 are the maximum index in the core and the index in the cladding, respectively; these definitions are analogous to those used for the same symbols in previous sections. Several index profiles for the graded-index fiber are illustrated in Figure 4.3.1

The value for α can be adjusted to minimize the modal dispersion at a selected wavelength. It often has a value near 2; when $\alpha = 2$, the index profile

---

^{2} The definition of Δ for the graded-index fiber differs from the Δ for the step-index fiber by the use of n_1 rather than n_2 in the denominator.

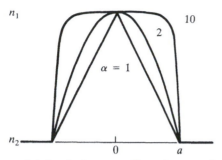

Figure 4.3.1 *Index profiles for graded-index fiber showing the α = 1 (triangular), α = 2 (parabolic), and α = 10 profiles. In many cases the optimum profile is with α slightly smaller than 2. Very large α can represent the step-index fiber.*

is called parabolic. A triangular profile, corresponding to $\alpha = 1$, also has useful properties.

4.3.2 *The Wave Equation and its Solution*

The method of solution outlined in the following is similar to that given in greater detail in reference [2.12]. An alternative derivation that leads to equivalent results in given by Cheo [2.7].

In formulating the wave equation, it was assumed that the material is homogeneous, that is, the index of refraction is the same throughout the material. This condition is not satisfied in the graded-index fiber. The graded-index fibers of practical interest will use the weakly guiding constraints, equivalent to those for the step-index fiber in the previous section. The index is a *slowly varying* function of *r*, and $\Delta \ll 1$; slowly varying means that the change in index within the distance of one wavelength is negligible. In this case, the wave equation can be used with little change; it will use $n(r)$ for the index, but will otherwise have the same form as was used in previous sections.

The components of the electric field intensity will have the form

$$E_z(t,r,\theta,z) = AE(r) \exp(jv\theta) \exp[j(\omega t - \beta z)] \tag{4.3.3}$$

The radial variation, $E(r)$, must satisfy the equation

$$\frac{d^2E}{dr^2} + \frac{1}{r}\cdot\frac{dE}{dr} + \left[\frac{\omega^2}{c^2}n^2(r) - \beta^2 - \frac{v^2}{r^2}\right]E = 0 \tag{4.3.4}$$

With a change of variable $U = r^{\frac{1}{2}}E$, this can be reduced to

$$\frac{d^2U}{dr^2} + u^2(r)\,U = 0 \tag{4.3.5}$$

where

$$u^2(r) = \left[\frac{\omega^2 n^2(r)}{c^2} - \beta^2 - \frac{(v^2 - \frac{1}{4})}{r^2} \right] \qquad \textbf{(4.3.6)}$$

Equation (4.3.5) will have solutions of the form $U = A \exp(\gamma r)$, which requires that $\gamma^2 + u^2(r) = 0$. Then γ will be imaginary when $u^2(r)$ is positive and real when $u^2(r)$ is negative. An imaginary value for the exponent γ corresponds to a wave propagating without loss, and a real value to an exponentially decaying amplitude. The condition for propagation is therefore

$$\frac{\omega^2 n^2(r)}{c^2} > \beta^2 + \frac{(v^2 - \frac{1}{4})}{r^2} \qquad \textbf{(4.3.7)}$$

The nature of this constraint can be seen in Figure 4.3.2. Each side of the inequality is plotted versus r. For $r_1 < r < r_2$, the condition for wave propagation is satisfied; outside this range it is not. By analogy with the step-index fiber, we can expect that there should be standing waves due to propagation in the $r_1 < r < r_2$ range, and evanescent waves for both $r < r_1$ and $r > r_2$.

For the slab and step-index waveguides, the wave equation that results from the application of boundary conditions led to transcendental equations from which the propagation constants could be determined. For the general α-profile graded-index fiber, such relatively straightforward eigenvalue equations are not available. The eigenvalue equation can be determined by using the fact that the total radial variation in phase must be an integral multiple of π radians. Since the radial propagation constant is a function of r, the total phase is found by

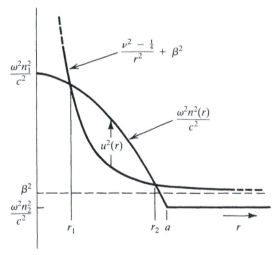

Figure 4.3.2 *Graph of Equation (4.3.7). For $r_1 < r < r_2$, the condition for the existence of a propagating wave is satisfied.*

intergrating $u(r)$ over the range of r in which propagation takes place:

$$m\pi = \int_{r_1}^{r_2} u(r)dr = \int_{r_1}^{r_2} \left[k^2 n^2(r) - \beta^2 - \frac{v^2}{r^2} \right]^{\frac{1}{2}} dr \qquad (4.3.8)$$

where m is an integer. This equation can be solved in closed form for β in only a few special cases.

4.3.3 *Propagation Constants*

The solution of (4.3.8) for the propagation constants for the graded-index fiber will not be undertaken here. Outlines of the method of solution can be found in [2.12] and [5.1]. We will restrict our attention to interpretation of the results of such solutions.

For the weakly-guiding case, many of the individual modes, which have distinct identities in the general case, will degenerate into groups of modes. Each degenerate mode group consists of several modes that, in the limit as Δn becomes very small, all have the same propagation constant. The number of degenerate mode groups is M,

$$M \approx a^2 k_1^2 \left[\frac{\alpha \cdot \Delta}{\alpha + 2} \right] \qquad (4.3.9)$$

Each mode group is identified with a group index m. In the mode group m, there are $2m$ degenerate modes, each having the same β_m.

$$\beta_m \approx k_1 \left[1 - 2\Delta \left(\frac{m}{M} \right)^g \right]^{\frac{1}{2}} \qquad (4.3.10)$$

where

$$g = \frac{\alpha}{\alpha + 2}$$

4.3.4 *Dispersion*

Modal dispersion is defined as the difference between the longest and shortest propagation delays. The propagation delay for each mode, in seconds per meter, is the reciprocal of the group velocity.

$$\frac{\tau}{L} = \frac{1}{v_g} = \frac{d\beta}{d\omega} \qquad (4.3.11)$$

To differentiate β, we put Equation (4.3.10) in the form

$$\beta_m = \frac{\omega n_1}{c} \left[1 - 2\Delta \cdot f_m \right]^{\frac{1}{2}} \qquad (4.3.12)$$

where

$$f_m = \left[\frac{m}{M} \right]^g = K_m [\omega]^{-2g} \tag{4.3.13}$$

This expresses f_m in terms of K_m, which is independent of frequency, and ω. The derivative of (4.3.12) can be reduced to

$$\frac{\tau_m c}{Ln_1} = [1 - 2\Delta \cdot f_m]^{\frac{1}{2}} + 2\Delta \cdot f_m \cdot \frac{\alpha}{\alpha + 2} [1 - 2\Delta \cdot f_m]^{-\frac{1}{2}} \tag{4.3.14}$$

and further to

$$\frac{\tau_m c}{Ln_1} = [1 - 2\Delta \cdot f_m]^{-\frac{1}{2}} \left[1 - \frac{4\Delta \cdot f_m}{\alpha + 2} \right] \tag{4.3.15}$$

By expanding the square root into a power series and retaining only the first three terms, we find

$$\frac{\tau_m c}{Ln_1} = [1 + \Delta \cdot f + \tfrac{3}{2}(\Delta \cdot f)^2 + \ldots] \cdot \left[1 - \frac{4}{\alpha + 2} \Delta \cdot f \right] \tag{4.3.16a}$$

$$= 1 + \frac{\alpha - 2}{\alpha + 2} \cdot \Delta \cdot f + \frac{3\alpha - 2}{\alpha + 2} \cdot \frac{(\Delta \cdot f)^2}{2} + \ldots \tag{4.3.16b}$$

Here f is written without the subscript m.

Our first objective is to find an optimum value for α. This optimum will be defined as the α that makes the propagation delays for $m = 0$ and $m = M$ equal; this eliminates the dispersion between the extreme modes. For $m = 0$, $f = 0$ and

$$\frac{\tau_0 c}{Ln_1} = 1 \tag{4.3.17}$$

For $m = M$, $f = 1$ and

$$\frac{\tau_M c}{Ln_1} = 1 + \frac{\alpha - 2}{\alpha + 2} \cdot \Delta + \frac{3\alpha - 2}{\alpha + 2} \cdot \frac{\Delta^2}{2} \tag{4.3.18}$$

The difference between the propagation delays of Equations (4.3.17) and (4.3.18) should be set equal to zero and this equation solved to find α.

$$\frac{\tau_M - \tau_0}{L} = \frac{n_1}{c} \left[\frac{\alpha - 2}{\alpha + 2} \cdot \Delta + \frac{3\alpha - 2}{\alpha + 2} \cdot \frac{\Delta^2}{2} \right] = 0 \tag{4.3.19}$$

With the approximation $(1 + x)^{-1} \approx (1 - x)$ for small x, we find that the optimum value for α is

$$\alpha_{\text{opt}} = 2(1 - \Delta) \tag{4.3.20}$$

For $0 < m < M$, the dispersion will not be zero. It can be calculated by substituting $\alpha = 2(1 - \Delta)$ into Equation (4.3.16b). Thus, we find

$$\frac{\tau_m - \tau_0}{L} = \frac{n_1 \Delta^2}{2c} f(1 - f) \tag{4.3.21}$$

This has its maximum value when $f = \frac{1}{2}$. The modal dispersion is given by this maximum value of $\Delta\tau$.

$$\frac{\Delta\tau}{L} = \frac{n_1 \Delta^2}{8c} \tag{4.3.22a}$$

When the effects of material dispersion are included in the calculation of modal dispersion, the result is to replace the index n by the group index N.

$$\frac{\Delta\tau}{L} = \frac{N_1 \Delta^2}{8c} \tag{4.3.22b}$$

Note that the modal dispersion given by Equations (4.3.22) is $\Delta/8$ times that found for the step-index fiber. Since Δ can be of the order of 0.01, $\Delta/8$ is of the order of 0.001; the graded-index fiber can be made to have its modal dispersion three orders of magnitude smaller than that of the step-index fiber!

Equation (4.3.22) describes the minimum dispersion available by adjusting α. For other values for α, dispersion is determined by the difference $\tau_0 - \tau_M$ and is given by

$$\frac{\Delta\tau}{L} = \frac{n_1 \Delta}{c} \left[\frac{\alpha - 2}{\alpha + 2} + \frac{3\alpha - 2}{\alpha + 2} \cdot \frac{\Delta}{2} \right] \tag{4.3.23}$$

A graph of modal dispersion versus α plotted from these two equations is shown in Figure 4.3.3. The dispersion increases rapidly as α departs from its optimum value.

In the graded-index fiber material dispersion can be larger than the minimum modal dispersion. When both are significant, intermodal and intramodal dispersions are added on a mean-squared basis. For example, if the material dispersion for the fiber represented in Figure 4.3.3 were found to be 150 ps/km, then the minimum at $\alpha \approx 2$ is increased from 0.03 ns/km to 0.153 ns/km. For α well away from this point, the total dispersion is not changed significantly. A more thorough analysis indicates that when material dispersion is considered, the optimum α is shifted to a slightly larger value.

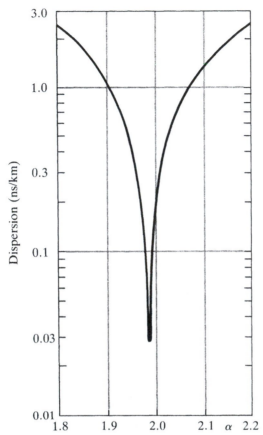

Figure 4.3.3 *Modal dispersion in a graded-index fiber.*

EXERCISE 4.7

A graded-index fiber for 1.3 μm light waves has $a = 50$ μm. The indices of refraction are $n_1 = 1.65$, $N_1 = 1.66$, $n_2 = 1.64$, and $N_2 = 1.65$. Calculate: (a) $\alpha_{optimum}$. (b) The number of degenerate modes. (c) β_m for $m = 1$ and $m = M$. (d) The modal dispersion for $\alpha = \alpha_{opt}$.

Answers:
(a) 1.988. (b) 483. (c) 7.97 $\cdot$ 10^6 m^{-1}, 7.92 $\cdot$ 10^6 m^{-1}. (d) 25.6 ps/km

---

4.4 Effects of Dispersion

We have examined the causes of dispersion in optical fibers, methods for estimating its magnitude, and fiber design techniques for minimizing it. We will now take the point of view that with the dispersion parameters of the fiber given, its effects on communication system performance must be evaluated.

Dispersion is an undesirable, but unavoidable, characteristic of the fiber. Its effect is to place an upper limit on the information capacity of the communication system. The effects of dispersion can be examined in either the time or the frequency domain. The definition and development of the results describing dispersion in optical fibers were set in the time domain. The design of digital communication systems, such as the pulse-code modulation (PCM) system defined in Chapter 2, might be done with a time domain point of view. Continuous wave (CW) systems are more naturally addressed in the frequency domain.

The effects of dispersion in an optical communication system must be examined from two complementary points of view:

1. What characteristics of the communication signals affect dispersion in the receiver signals?
2. In what way(s) does this dispersion limit the performance of the communication system?

4.4.1   *The Calculation of Dispersion*

Modal dispersion was defined in Sections 4.1, 4.2, and 4.3 as the difference in propagation times of the modes with the slowest and fastest velocities. The total modal dispersion was found by multiplying the dispersion per unit length by the total length of the fiber. It is found in practice that when the fiber is longer than a critical length, the total dispersion increases as the square root, rather than the first power, of the fiber length. The reason for this reduction in total dispersion is mode conversion.

Mode conversion is the exchange of energy among the various modes propagating in a multimode fiber. Any departure from the ideal dielectric cylinder will cause some of the energy propagating in the fiber to be scattered in random directions; these scatterers might be irregularities in the radius of the fiber, bends in the fiber, irregularities in the dielectric constant, scratches or cracks in the surface of the fiber, and many other effects. Scattering has the effect of removing energy from the mode in which the scattered energy was propagating. Some of the scattered energy leaves the waveguide and contributes to attenuation; some of it enters other modes and continues to propagate in the fiber.

In a long fiber, little of the transmitted energy propagates the entire length in a single mode. Because the output energy has propagated fractions of the total length in each of several, or many, modes, the concept of a slowest and a fastest mode is no longer a useful or meaningful concept.

A graph illustrating modal dispersion versus fiber length is shown in Figure 4.4.1. For lengths shorter than L_c, the coupling length, modal dispersion is proportional to L. For lengths longer than L_c, further increase in dispersion is proportional to $\sqrt{L}$.

Modal dispersion is largely independent of the characteristics of the transmitted signal. It is a multipath propagation phenomenon. The spectral width of

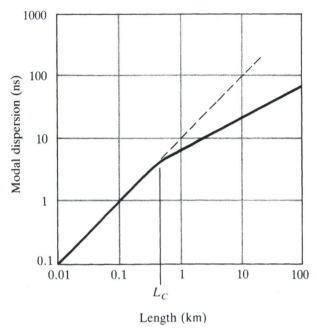

Figure 4.4.1 *Modal dispersion versus fiber length; the numerical scales correspond to an arbitrarily selected fiber. Dispersion is proportional to L for $L < L_c$ and as $[L]^{\frac{1}{2}}$ for $L > L_c$.*

the transmitted signal affects the dispersion in each propagating mode, but since modal dispersion is usually much larger than intramodal dispersion, the spectral width has a small, second-order effect on the total dispersion. An exception to this generality is the graded-index fiber in which the modal dispersion can be made very small; in this case, the intramodal dispersion can be a significant part of the total dispersion (see Exercise 4.8).

Intramodal dispersion in an optical fiber results from the fiber having velocities of propagation that are functions of frequency. When the signal being transmitted must be represented in the frequency domain by a spectrum of width greater than zero, then different parts of this spectrum travel through the fiber at different velocities and arrive at the receiver with different propagation delays. If the transmitted signal had a spectral width equal to zero, there would be no intramodal dispersion. There are two reasons that the spectral width is greater than zero: (1) the carrier frequency may vary about its nominal value due to instabilities in the light wave source, and (2) the modulated carrier has a bandwidth, that is, spectral width, proportional to the bandwidth, at baseband, of the modulating signal.

Thus far, we have represented the carrier by a sinusoid of frequency f_c. It was assumed to be a pure sinusoid of constant, stable frequency. Although this may be a reasonable approximation at lower radio and microwave frequencies, it is not true of the optical sources that are practical for use in current communications systems. The light wave carrier has a spectrum of width greater than zero. It could be represented as a noise-modulated carrier that has a well-defined center frequency but with both amplitude and phase that vary in a random manner; the random variations of amplitude and phase result in an associated spectral width. We could characterize this carrier spectrum with statistical parameters, such as an rms spectral width, but not in a deterministic way that defines the instantaneous carrier amplitude, frequency, and phase as specific functions of time and other parameters. The spectral width of some light wave sources range from tens of nm for a light-emitting diode operating at room temperature to a few Ångstroms for a stabilized diode laser.

We have examined in Chapter 2 the spectrum of a modulated signal. The information-bearing signal has a Fourier transform $X(f)$ of spectral width B. When the light wave carrier is modulated with this $x(t)$, the bandwidth required to pass the modulated signal without distortion is related to B and to the type of modulation. Dispersion in the optical fiber will result in a reduced bandwidth of the fiber and thus limits the bandwidth of the signals that can be transmitted through it. The spectral width due to modulation is usually small compared to that due to the source; its effect is usually neglected.

The total dispersion is the sum of modal, material, and waveguide dispersion. Material and waveguide dispersion are two types of intramodal dispersion and are added linearly to find the total intramodal dispersion. The total intramodal dispersion per unit length is proportional to the spectral width of the light wave signal. Modal dispersion is due to multipath propagation and is independent of spectral width of the signal. Because these two types of dispersion are independent, they do not combine linearly. It has been shown [2.13] that the total rms dispersion σ, in seconds, is

$$\sigma_{\text{total}}^2 = \sigma_{\text{intramodal}}^2 + \sigma_{\text{intermodal}}^2 \tag{4.4.1}$$

This σ is equivalent to the $\Delta\tau$ used in earlier sections. The new symbol is used to designate the rms total dispersion as distinct from the values calculated for specific components of the total dispersion.

EXERCISE 4.8

(a) A step-index fiber has $N_1 = 1.75$ and $\Delta = 0.01$. The intramodal dispersion is 10 ps/km · nm. The light wave source is a light-emitting diode with spectral width $\Delta\lambda = 50$ nm. Find the modal and the total rms dispersion. (b) A graded-index fiber has the same parameters as the step-index fiber in (a) and has $\alpha = 2(1 - \Delta)$. Find the modal dispersion and the total rms dispersion in this fiber.

Answers:
(a) 58.3 ns/km, 58.3 ns/km. (b) 72.9 ps/km, 505.3 ps/km.

4.4.2 *Dispersion In The Time Domain*

Dispersion is defined as the spread in time of the received signal beyond that due to the duration of the signal itself. In deriving equations that could be used to calculate the magnitude of the dispersion, we used the time delay in propagating the signal from the transmitter to the receiver as the basis of the derivation. The units of the dispersion coefficients are s/m^2 for the D parameters and s^2/m for $d^2\beta/d\omega^2$. With either of these, we calculate the dispersion in seconds.

In some studies, the fiber is characaterized by its impulse response. The received signal can be calculated as the convolution of the transmitted signal and the impulse response. The duration of the impulse response provides a reasonable estimate of the pulse stretching due to dispersion. Figures 2.2.4, 2.2.5, and 2.2.6 in Chapter 2 illustrate this point.

The repetition period of pulses in a PCM system must be long enough that successive received pulses do not overlap in time; if they do, reliable decoding would not be possible. It is tempting to use the rms dispersion, in seconds, as the minimum pulse repetition period. However, the duration of the impulse response is often longer than the nominal dispersion. The shape of the impulse response can be very different for different fibers and different operating conditions. It is shown in Section 4.4.3 below, from bandwidth considerations, that $R \approx \frac{1}{4\sigma}$ bits per second is an achievable data rate in a PCM system that uses an optical fiber with rms dispersion of σ seconds.

Dispersion is usually calculated by finding the dispersion per unit length, a property of the fiber, and multiplying this by the total length of the fiber. It can, for some purposes, be left in s/m units and the rms dispersion found in these units. In that way we would calculate the RL (data rate-distance) product, in bit · m/s. For example, for a multimode step-index fiber, $RL = c/4N_2\Delta$; in a silica fiber with $\Delta = 0.01$, $RL \approx 5$ Mb · km/s. Since dispersion is a property of the fiber and transmitter light source, it can be considered constant for a specific system.

The product RL is fixed for a given system, indicating that the designer can trade distance for bit rate but cannot exceed the maximum product RL. With different system parameters, that is, a low-dispersion fiber and a light source with a narrow spectral width, a higher RL figure might be achieved. The RL product is thus a figure of merit for the system.

4.4.3 *Bandwidth*

Continuous-wave signals are more naturally treated in the frequency rather than time domain. Digital signals are also often handled in terms of the bandwidth required to pass them with acceptably low distortion. The bandwidth of the fiber limits the width of the spectrum of the modulated light wave that can be successfully transmitted through the fiber. Manufacturers' data describing fiber characteristics will sometimes be given in terms of bandwidth rather than dispersion.

A common method for calculating the bandwidth of an optical fiber is to assume or derive the shape of the impulse response, then take its Fourier transform to find its spectrum. The bandwidth is then defined as the half-power bandwidth of this spectrum.

The definition of bandwidth in terms of half-power points involves an ambiguity that must be recognized and examined carefully. The optical signal in the fiber is intensity modulated; it is the power of the light wave that represents the modulating signal. The demodulated signal in the receiver is an electrical signal that has the same waveshape as the optical signal in the fiber. The half-power points of these two corresponding signals will be at different frequencies and the half-power bandwidths will be different. The two bandwidths are called the optical bandwidth and the electrical bandwidth. For a specific fiber, the optical bandwidth is greater than the electrical bandwidth. In communication system design, the electrical bandwidth is of primary importance; "bandwidth" will mean electrical bandwidth unless it is specifically designated otherwise. For the purposes of estimating bandwidth, the ratio of optical to electrical bandwidth is approximately $\sqrt{2}$.

The shape of the impulse response, $h(t)$, cannot be calculated with the background available to us at this point in our study of light wave propagation in optical fibers. Shapes that have been calculated and measured indicate that the Gaussian and exponential shapes are reasonable approximations for pulses found in some fibers; other shapes that cannot be represented by simple equations are also found [2.13].

If the impulse response is assumed to be Gaussian with rms width σ, then

$$h(t) = \frac{1}{(2\pi)^{\frac{1}{2}}\sigma} \exp -\frac{t^2}{2\sigma^2} \tag{4.4.2}$$

and

$$H(\omega) = \exp \frac{-\omega^2\sigma^2}{2} \tag{4.4.3}$$

The 3-dB bandwidth is the frequency $\omega_{3\,\text{dB}}$ for which $H(\omega)$ is reduced to $1/\sqrt{2}$ of its value when $\omega = 0$. Thus, $\omega_{3\,\text{dB}} = 0.83/\sigma$. The optical bandwidth for the Gaussian pulse is defined as the frequency for which $H(\omega)$ is reduced to $\frac{1}{2}H(0); \omega_{\text{opt}} = 1.17/\sigma$.

For the exponential impulse response,

$$h(t) = \frac{1}{\sigma} \exp\left[-\frac{t}{\sigma}\right] \tag{4.4.4}$$

and

$$H(\omega) = \frac{1}{1 + j\omega\sigma} \tag{4.4.5}$$

From (4.4.5) we find that the bandwidth is $\omega_{3\,\text{dB}} = 1/\sigma$.

Other approximate treatments of the impulse response lead to $\omega_{3\,dB} = 0.7/\sigma$. We will use the Gaussian approximation, recognizing that the interpretations we derive from it are, at best, rules of thumb for estimating bandwidth and data rate.

With $\omega_{3\,dB} = 0.83/\sigma$, $f_{3\,dB} = 0.13/\sigma$. We found in Chapter 2 that the maximum data rate for PCM is $2f_{3\,dB}$. Then $R = 0.26/\sigma$ or

$$R \approx \tfrac{1}{4\sigma} \tag{4.4.6}$$

and the electrical bandwidth, $B = f_{3\,dB}$, is

$$B \approx \tfrac{1}{8\sigma} \tag{4.4.7}$$

EXERCISE 4.9

An optical fiber has modal dispersion of 20 ps/km and total intramodal dispersion of 10 ps/(km · nm). The light wave source is a junction-diode laser with spectral width $\Delta\lambda = 1$ nm. (a) What are the approximate maximum values for the data rate and the bandwidth of this fiber? (b) What is the maximum length of this fiber over which data can be transmitted at the rate 500 Mbit/s?

Answers

(a) 11.2 Gbit · km/s, 5.6 GHz · km. (b) 22.4 km.

4.5 Attenuation

The two primary factors that limit the length of the optical fiber path are dispersion and attenuation. Each is expressed as a per-km unit; the longer the fiber, the larger the total dispersion and the larger the total attenuation. The maximum path length is determined by one of these per-unit-length quantities.

The total attenuation is the sum of several different kinds of losses between the transmitter and the receiver. The distributed losses in the fiber is one of these. Other sources of power loss are the connectors and splices needed to assemble several lengths of fiber and to couple the transmitter and the receiver to the fiber.

Attenuation can limit the system performance because the receiver has a minimum useful power input and the transmitter a maximum practical power output. The ratio of the transmitter output and the receiver input is the total acceptable attenuation. The attenuation is normally expressed in dB. When the maximum total attenuation, in dB, has been established, then the fixed connector and splice losses can be subtracted to determine the maximum fiber attenuation, in dB. This fiber loss divided by the fiber attenuation, in dB per km, gives the maximum path length.

4.5.1 *Fiber Attenuation*

There are two fundamental physical phenomena that establish a lower limit to the fiber attenuation. For the range of wavelengths of interest for silica-based optical fibers, these are illustrated in Figure 4.5.1.

In the short-wavelength end of the range, the limit is due to Rayleigh scattering. No matter how perfect the glass, there are small irregularities in the material index of refraction that will scatter electromagnetic waves. Propagation through the material is largely undisturbed as long as the wavelength of the electromagnetic wave is large compared to the size of the irregularities. Scattering losses increase as the wavelength approaches the physical dimensions of the irregularities. The Rayleigh-scattering losses decrease as the fourth power of the wavelength. Figure 4.5.1 shows the Rayleigh limit for attenuation at wavelengths in the 0.8- to 1.6-μm range. The Rayleigh-scattering loss increases with doping, as is used to control the index of refraction; the graph could be higher or lower, by a fraction of a dB, for different optical fiber materials.

The lower limit on attenuation at the longer wavelengths is determined by atomic absorption, such as that implied by the atomic model used in Section 3.4.2 to show how the dielectric constant varies with frequency. In the silica materials, there is an infrared absorption peak at longer wavelengths that be-

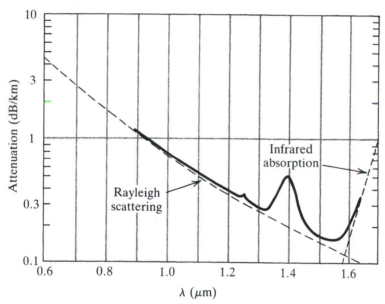

Figure 4.5.1 *Attenuation in silica-based optical fiber materials. For $\lambda < 1.5$ μm, the theoretical minimum attenuation is determined by Rayleigh scattering. For $\lambda > 1.6\,\mu$m, the minimum attenuation is determined by infrared absorption. Except for the OH^- absorption near $\lambda = 1.4\,\mu$m, practical fibers can come close to the theoretical limits. The relative minima at $\lambda = 1.3\,\mu$m and $\lambda = 1.55\,\mu$m make these wavelengths attractive for long-distance optical fiber communication systems. By coincidence, the dispersion in silica fibers also has a minimum at wavelengths near 1.3 μm.*

comes the dominant loss mechanism at wavelengths longer than 1.6 μm. This infrared absorption is also shown in Figure 4.5.1.

The combination of these two fundamental limits gives a total attenuation that can decrease with increasing wavelength until the infrared absorption curve crosses the Rayleigh curve. The intersection of the two curves establishes an absolute minimum for attenuation in this class of optical fibers. The limit is less than 0.2 dB per km. Fibers are available with attenuation of 0.2 dB per km in the 1.55-μm region.

The attenuation of current optical fiber materials approaches this fundamental lower limit very closely. The solid graph in Figure 4.5.1 illustrates the state of the art in optical fiber materials. It is evident from this graph why the 1.3- and 1.55-μm wavelength regions are attractive for optical fiber communication systems.

The principal departure from the fundamental lower limit is in the 1.4-μm region. This loss peak corresponds to water absorption. It is difficult to remove moisture completely from the materials in manufacture. Absorption by OH^- ions peaks at 1.4 μm and appears in high-quality fiber materials as the only significant departure from the minimum attenuation curves. Some practical fibers show other absorption peaks superimposed on the solid curve of Figure 4.5.1; by manufacturing the fiber materials with greater care, and hence greater expense, all but the 1.4-μm peak can be effectively eliminated as significant factors in the fiber attenuation characteristic.

It is evident from Figure 4.5.1 that if the infrared absorption could be eliminated, or shifted to longer wavelengths, then following the Rayleigh curve to longer wavelengths would reduce the minimum attenuation. Current research in new optical fiber materials offers promise that this will be achieved with new fluoride materials. It is believed that the fluoride fibers can have attenuation three orders of magnitude lower than is possible with the current silica fibers; the current lower limit of 0.2 dB per km may be reduced to less than 0.001 dB per km.

In addition to the fundamental limits on scattering and absorption losses, there are the practical factors that can introduce additional losses. The curves in Figure 4.5.1 indicate that the net effect of such practical factors is small. They can be kept small, as indicated in the figure, but if due care is not taken in the use of the fiber, unnecessary losses may be introduced. Such losses result from mechanical damage to the fiber; from violating the minimum bending radius, or from coupling power out of the evanescent wave by bringing the core into close proximity to another fiber or other material capable of receiving optical energy from the fields around an exposed fiber. As the fibers are manufactured and sold to systems users, they are protected from such damage and losses. Excess and unnecessary losses are most likely to occur at the ends of the fiber where they are exposed for splices or for the attachment of connectors.

4.5.2   *Connectors And Splices*

In essentially all practical systems, the ends of the optical fiber must be physically connected to terminal equipment or to the ends of other lengths of fiber. These connections are normally made with connectors or with splices.

In principle, connectors resemble the connectors used in many other kinds of electronic and electrical systems. Two mating connectors must hold the two optical fibers, or fiber and some other optical path, firmly in place so that the optical path between them is well aligned and stable. The ends of the fiber must be smooth, so that surface irregularities will not cause excessive scattering losses. The ends must be held close together, but normally not quite touching. The alignment of the two optical axes is critical; either offset or tilt can cause unnecessary losses.

The losses in optical fiber connectors can be considered to be on the order of one dB per connector; an average connector loss of $\frac{1}{2}$ dB per connector would require care both in selecting and using the connectors.

When several lengths of fiber are connected in tandem to form a long fiber path, splices can be used to join the fiber ends. The fusion splice is formed by carefully aligning the ends of the two fibers to be joined, then fusing them by using an electric plasma or other source of energy. Other splicing techniques involve use of a firm, stable holder to maintain good alignment and an index-matching material to provide good optical coupling, with minimum reflection loss, between the two fiber ends. With the proper equipment and skill, splices averaging 0.1 dB per splice can be achieved.

The distinction between a connector and a splice is that the connector will allow disconnect and reconnect, or connection to any of several points, with relative ease; a splice is a fixed, permanent connection. Splices would be used for joining several lengths of fiber to form a long optical path between two terminals. Connectors might be used at the terminals of the long path to allow flexibility in the terminal equipment to be used for transmitting and receiving.

A more thorough discussion of splices and connectors can be found in Miller and Chynoweth, reference [5.21, Chapters 14 and 15]. Each of these chapters provides further references.

EXERCISE 4.10

A fiber path 50 km long consists of sections of fiber, each having a maximum length of 3 km, spliced together to form one continuous fiber. The attenuation of the fiber is 0.5 dB per km; the splices have added losses of 0.3 dB per splice. Connectors at each end, to connect the fiber to the transmitter and the receiver, have added loss of 1 dB per connector. (a) What is the total loss, in dB, between the transmitter and the receiver? (b) What is the power input to the receiver when the power output of the transmitter is 1 mW?

Answers

(a) 31.8 dB; (b) 0.66 μW.

4.5.3   Bending Losses

Our analysis of light wave propagation in dielectric waveguides was based on a model of infinite extent without any change in the direction of propagation.

Light waves are guided in such structures by total internal reflection from the sides of the waveguide. If the waveguide has any bends or irregularities in its sides, the condition for total internal reflection for some modes may no longer be satisfied. Such modes would then lose energy; the attenuation of the fiber is increased by such losses. The sharper the bend, the larger the number of modes that will suffer increased losses due to angles of incidence on the sides that are greater than the critical angle for total reflection.

The bending losses can be represented with an attenuation constant, α_b [2.15]:

$$\alpha_b = M \exp\left[-\frac{R}{R_0}\right] \quad (\mathrm{m}^{-1}) \tag{4.5.1}$$

where M is independent of R, the radius of the bend. The critical radius, R_0, is a function of the wavelength and the depth of penetration of the evanescent wave into the cladding. R_0 increases with the depth of penetration and decreases as the wavelength increases. In optical fibers, the critical radius is normally of the order of a few cm.

It can be seen in Equation (4.5.1) that for $R \gg R_0$ the bending loss is small. The loss increases as R decreases.

The attenuation constant, α_b, is applicable in the region of the bend. It may be more convenient to express the attenuation in per radian rather than per meter units. The unit conversion consists of multiplying Equation (4.5.1) by R, the radius of the bend.

In many fibers, mechanical stress rather than bending loss determines the minimum bending radius. Mechanical strain that results from sharp bends can cause cracking in the fiber. Stress-induced cracks in the fiber will scatter the incident light waves, producing losses that can be greater than the losses produced by bends and irregularities in the surface of the core. Stress-induced cracks are an aging phenomenon. They are minimized by designing the installation to avoid the necessity for sharp bends, selecting a fiber cable that is stiff enough to make unintentional sharp bends unlikely, and installing the cable with care.

4.6 Directional Couplers

The directional coupler is a dielectric waveguide device with several important applications. In its simplest form, it divides the power of an input light wave between two output waves. It can also combine two inputs to produce output waves composed of the sum of the two inputs. A beam splitter, or half-silvered mirror, is an example of an optical directional coupler for unguided waves. In more complex modes of operation, devices derived from the directional coupler can be made to have electrically controlled power-division ratios, making possible applications such as switching and modulation. In this section, we will consider the principle of the coupling between two dielectric waveguides and the application of that principle to make directional couplers and power combiners. Modulators and switches will be discussed in Chapter 6.

The directional coupler is illustrated in Figure 4.6.1. Two dielectric waveguides are brought into close proximity over a fixed distance L. The distance between them must be small enough so that each waveguide lies within the evanescent wave of the other. There are two input ports and two output ports; in some applications, however, only two or three of the four I/O ports are used. The two waveguides can be cylindrical optical fibers, slab waveguides, or waveguides of other cross section. The cylindrical fiber and implanted waveguide structures each have important applications.

When two parallel dielectric waveguides are closely spaced, the evanescent waves of the waveguides can overlap. Some of the energy carried by the evanescent wave of the first waveguide becomes a guided wave in the second. The amount of energy coupled from the first waveguide into the second depends on the distance between them and the length of the closely coupled region. The propagation constants and the index of refraction of the cladding material also affect the degree of coupling between the two waveguides.

For a symmetrical structure that satisfies the weakly guiding condition, the equations relating the input power and the two output power levels are [2.2]

$$P_2 = P_0 \sin^2 \kappa L \qquad\qquad (4.6.1)$$

$$P_1 = P_0 \cos^2 \kappa L \qquad\qquad (4.6.2)$$

where κ is the coupling coefficient.

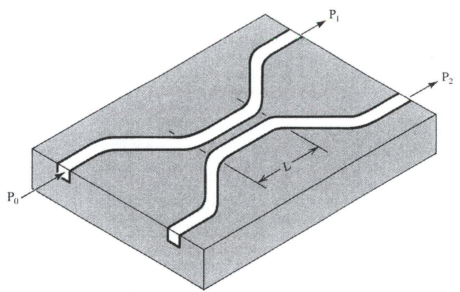

Figure 4.6.1   *Integrated-waveguide directional coupler. Two dielectric waveguides, implanted in a substrate material, are brought into close proximity over a distance L. Optical power into either of the two input ports is divided between the two output ports.*

By choosing κL, it is possible to transfer some or all of the input power to the other waveguide. When used in this way, the coupler is a power divider. Note that for $\kappa L = \pi/2$, all of the power in each waveguide is transferred to the other. For $\kappa L = \pi$, the power in each waveguide is transferred to the other waveguide and back to the original one; there is no net transfer of power. Any power division from zero to 100 percent is possible if κL can be controlled over a range of $\pi/2$ radians.

The equations describing the coupling and the transfer of energy are linear. The principle of superposition is therefore applicable. The coupling from one waveguide to the other can take place in either direction, or it can take place in both directions simultaneously. If we have input light waves at both inputs, each input is divided between the two outputs in the manner described by Equations (4.6.1) and (4.6.2). When used in this mode, the directional coupler could be described as a power combiner. Each output can be made equal to a weighted sum of the two inputs.

EXERCISE 4.11

The coupling coefficient in a directional coupler is $\kappa = 7 \cdot 10^2$. For what interaction length L will the power in the two outputs be equal?

Answer
1.1 mm

One practical coupler, illustrated in Figure 4.6.2, is made from two conventional optical fibers. The cladding of each fiber is first ground (or etched) so that the thickness of the cladding is reduced, thus providing for the close proximity of the cores. The fibers are then twisted together, put under tension, and heated. As the silica materials soften at high temperatures, the fibers fuse to form a region in which the coupling is adequate for a directional coupler. As it cools, the mechanical structure is set into a permanent form. This type of directional coupler is available with standard power-division ratios, in percent, of 10–90, 20–80, ... to 50–50, with insertion loss of the order of 3 dB. This coupler is normally purchased with fiber pigtails or connectors.

Another type of coupler, based on the same principles, is fabricated as an integrated-circuit component by implanting a dielectric waveguide structure

Figure 4.6.2 *Fused-fiber directional coupler.*

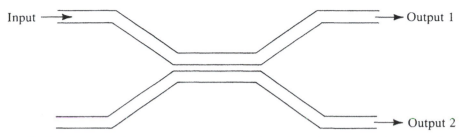

Figure 4.6.3 *Schematic diagram of optical directional coupler. This config-uration is used as the symbol for the optical directional coupler in circuit and system schematic diagrams.*

into a suitable substrate. The implanted-waveguide directional coupler is illustrated in Figures 4.6.1 and 4.6.3. The schematic structure shown in Figure 4.6.3 is used as the symbol for an optical directional coupler in circuit and system schematic diagrams.

Suitable materials for the implanted-waveguide coupler are lithium niobate, $LiNbO_3$, and gallium arsenide, GaAs. Lithium niobate is attractive because it has a large electrooptic coefficient; when the application is an electrooptic modulator or an optical switch, $LiNbO_3$ is the better material. The waveguide is made by diffusing titanium into a $LiNbO_3$ substrate. Gallium arsenide has the advantage of compatibility with the materials with which most other optical integrated-circuit components are made; it is an appropriate choice when the coupler is to have a fixed power-division ratio. Both materials are used in practical devices and systems.

Some applications of directional couplers, described in Chapters 6 and 8, will involve operations with the two outputs that require knowledge of the relative phase between them. Detailed analysis of the directional coupler shows that the phase difference is $\pi/2$ radians, independent of κ and L [2.6]. This $\pi/2$ phase difference can be explained by using conservation of energy. If we neglect insertion losses, the total power is $P_0 = P_1 + P_2$. Since power is proportional to the square of the electrical field strength, we can write $E_0^2 = E_1^2 + E_2^2$. This suggests a right triangle relationship between the E-field phasors, with a $\pi/2$ difference between E_1 and E_2.

4.7 Manufacturing Technologies

Progress in developing the techniques and technologies to make optical fiber communications feasible has been remarkable. Progress in developing the technologies required to mass produce large quantities of optical fiber, with small and critical dimensions, has been equally remarkable. A thorough review of these manufacturing technologies is beyond the scope of this book; reference [2.14] is suggested for interested readers. We will look briefly at the materials and at one of the techniques of current importance for the fabrication of optical fibers.

4.7.1 *Materials For Optical Fibers*

The primary host material for current optical fibers is pure silica, SiO_2. It has an index of refraction in the range of 1.44 to 1.46, depending on wavelength.

By the addition of controlled amounts of selected dopants, the index can be adjusted to higher or lower values. The most commonly used dopant for increasing the index is germanium dioxide, GeO_2. Other materials that would increase the index include oxides of phosphorus and aluminum. Materials that decrease the index include fluorine and boron oxide. Boron or phosphorus are commonly used because they lower the temperatures required for processing; in such cases other dopants may be used as well for index control.

The simplest configuration would use either a germanium-doped core with pure silica cladding or a silica core and fluorine-doped cladding. Either of these gives the required $n_2 < n_1$. In many cases P_2O_5 is added to both core and cladding. Several other combinations of dopants are sometimes used.

4.7.2 *Optical Fiber Fabrication*

The manufacturing process we will describe is the modified chemical vapor deposition (MCVD) process. Other processes include outside vapor deposition (OVD) and vapor-phase axial deposition (VAD). None of these processes is considered to be dominant at present. There is no doubt that progress in process development and improvement will continue.

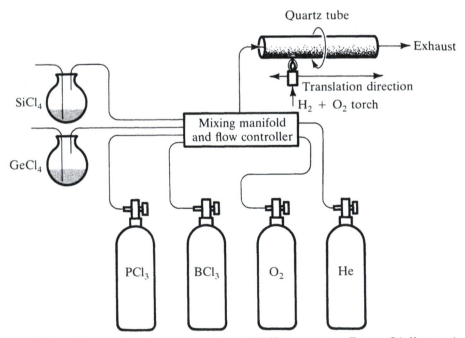

Figure 4.7.1 *Schematic diagram of the MCVD process. From Giallorenzi, Proc. IEEE, Vol. 66, pp. 744–780, 1978; copyright©1978 IEEE [2.23].*

The principle of the MCVD process is the use of a tubular substrate and the deposition of the core and cladding material on the inside of the tube. After deposition of these materials, the tube is collapsed into a solid rod and the fiber is drawn from this rod.

The first stage in the fabrication process is illustrated in Figure 4.7.1. A fused silica tube is mounted in an optical lathe. Chemical vapors are generated, mixed, and injected into one end of the silica tube. A moving hot zone within the tube provides an environment in which glassy particles are formed and deposited. Successive layers, each with a selected type and level of dopants, can be formed.

The second step in the fabrication is the collapse of the tube, now with the core and cladding materials on the inside surface, into a solid rod. This is accomplished with high temperature that allows surface tension to cause the tube to collapse. The solid glass rod is called the preform. Except for the much larger diameter, the preform has the same index profile as the fiber is to have.

The final step is to draw the fiber from the preform. This process is illustrated schematically in Figure 4.7.2. The end of the preform is heated and tension

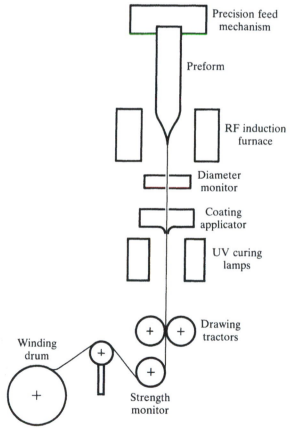

Figure 4.7.2 *Schematic diagram of the fiber-drawing process.*

applied to pull the fiber. Of course, it is necessary to control the temperature and tension very carefully. After drawing, the fiber is coated for protection and wound on a spool for handling and storage.

The substrate tube is approximately one m long and two cm in diameter. The fiber produced from the substrate is typically several km long and 125 μm in diameter.

SUMMARY

The optical fiber is a dielectric waveguide based on the principle of total internal reflection. The application of Maxwell's equations with suitable boundary conditions yields equations that must be satisfied for light waves to propagate in the waveguide. One set of these equations gives relationships between the propagation constants, k_1 and k_2, of the materials and the propagation constant, β, of the waveguide. The other set, the eigenvalue equations, gives relationships that must be satisfied in order that propagation modes can exist. In general, there can be many solutions to the eigenvalue equations. Each solution, an eigenvalue, defines one mode of propagation in the guide.

The step-index fiber consists of a core with an index of refraction n_1 enclosed in a cladding with an index n_2, with $n_2 < n_1$. The wave equation for the optical fiber is Bessel's differential equation; the solutions, which describe the fields both in the core and in the cladding, are Bessel functions.

$$E_z(t,r,\theta,z) = AJ_v(\kappa r) \exp (jv\theta) \exp \left[j(\omega t - \beta z) \right] \quad r < a$$

$$E_z(t,r,\theta,z) = CK_v(\gamma r) \exp (jv\theta) \exp \left[j(\omega t - \beta z) \right] \quad r < a$$

The propagation constants are related by

$$\kappa^2 = k_1{}^2 - \beta^2$$

and

$$\gamma^2 = \beta^2 - k_2{}^2$$

A normalized frequency variable is defined as

$$V = 2\pi \frac{a}{\lambda} [n_1{}^2 - n_2{}^2]^{\frac{1}{2}} = [(\kappa a)^2 + (\gamma a)^2]^{\frac{1}{2}}$$

Since both κa and γa must be real, V represents the highest value that either can reach.

The weakly guiding fiber, for which $(n_1 - n_2) \ll 1$, is the case of primary interest. For this case, the eigenvalue equation is

$$\frac{J_v(\kappa a)}{J_{v-1}(\kappa a)} = -\frac{\kappa}{\gamma} \frac{K_v(\gamma a)}{K_{v-1}(\gamma a)}$$

For each azimuthal index, v, there will be a set of eigenvalues. Each eigenvalue, κ_{vm}, has a corresponding propagation constant, β_{vm}, given by

$$\beta_{vm}{}^2 = k_1{}^2 - \kappa_{vm}{}^2$$

Since the phase and group velocities are functions of β, each mode has its own group velocity. Signals propagating via different modes will have different propagation delays between the transmit and receive ends of the fiber. The received signals are thus dispersed in time, a form of distortion. For the step-index fiber, this modal dispersion is given by

$$\frac{\Delta\tau}{L} \approx \frac{N_2\Delta}{c} \text{ s/m}$$

where N_2 is the group index and $\Delta = (n_1 - n_2)/n_2$. By making Δ and the fiber radius a small, so that $V < 2.405$, the fiber can be made to have only one mode, thus eliminating modal dispersion.

When modal dispersion has been eliminated, intramodal dispersion becomes significant. Material and waveguide dispersion combine linearly to give a total intramodal dispersion. At wavelengths above 1.3 μm, material dispersion is positive and waveguide dispersion negative. The wavelength at which the sum is zero can be controlled by the selection of index and a.

A dispersion parameter D, in ps/(km · nm), is commonly used to express intramodal dispersion. The product of D times the spectral width, in nm, reduces intramodal dispersion to the same units, s/m or ns/km, that are used for modal dispersion.

The graded-index fiber has an index of refraction that is a function of the radial distance from the center of the core. The index variation usually has the form

$$n(r) = n_1\left[1 - 2\Delta\left(\frac{r}{a}\right)^\alpha \right]^{\frac{1}{2}} \quad r < a$$

$$= n_1[1 - 2\Delta]^{\frac{1}{2}} = n_2 \quad r > a$$

The graded-index fiber is a multimode fiber. Its modal dispersion can be made very small by proper choice of $n(r)$. For $\alpha = 2(1 - \Delta)$, the modal dispersion is

$$\frac{\Delta\tau}{L} = \frac{N_1\Delta^2}{8c}$$

This is $\Delta/8$ times the modal dispersion of the step-index fiber; by making Δ small, it is possible to have $\Delta/8 \approx 10^{-3}$. The modal dispersion in the graded-index fiber can be made small enough so that intramodal dispersion is of comparable magnitude.

When both modal and intramodal must be considered, the total rms dispersion, σ, in seconds, is given by

$$\sigma_{total}^2 = \sigma_{intramodal}^2 = \sigma_{intermodal}^2$$

The effects of dispersion are to limit the information capacity, RL(bit-km/s), or the bandwidth, $\omega_{3\ dB}$. As general approximations, $RL \approx \frac{1}{4\sigma}$ and $\omega_{3\ dB} \approx 1/\sigma$.

Attenuation in optical fibers limits the length of fiber that can be used without regeneration (amplification) of the signal. The minimum attenuation at wavelengths up to 1.55 μm is set by Rayleigh scattering; practical limits are approximately 3 dB/km at 0.85 μm, 0.5 dB/km at 1.3 μm, and 0.2 dB/km at 1.55 μm. Above 1.55 μm, the attenuation increases rapidly due to infrared absorption. Silica fibers are not considered useful at longer wavelengths. New fiber materials are expected to decrease the minimum attenuation to 0.001 dB/km. Connectors and splices, necessary at the ends of the fiber and for joining sections to form longer spans, add further losses of the order of 1.0 dB per connector and 0.1 dB per splice.

Attenuation, in dB/km, and total dispersion, in ns/km, can limit the length of fiber that can be used without detection, amplification, and retransmission of the signal. At low data rates, the limit is determined by the attenuation and at high data rates, by dispersion.

PROBLEMS

4.1 A dielectric slab waveguide has indices of refraction of 1.43 and 1.40. The external medium is air.
(a) What is the value of the numerical aperture of this waveguide?
(b) What is the total range of angles of incidence, θ_0, over which input light waves would satisfy the conditions for total internal reflection inside the guide?

4.2 A slab waveguide of width $2d = 5$ μm has $n_1 = 1.53$ and $n_2 = 1.51$. Calculate the normalized frequency variable V when the frequency is $\omega = 2 \cdot 10^{15}$ s^{-1}. How many modes would propagate in this guide?

4.3 Find the eigenvalues for the waveguide of Problem 4.2. Calculate them to an accuracy of three significant figures.

4.4 In a slab waveguide having $n_1 = 1.43$, $n_2 = 1.42$, and wavelength $\lambda = 0.85$ μm, what is the maximum slab thickness, $2d$, for which there would be only a single mode that could propagate in the waveguide? Repeat for $\lambda = 1.55$ μm.

4.5 (a) Use equations (4.1.5a) and (4.1.9) to show that, for the even (cosine) solutions, both $E_y(x)$ and $dE_y(x)/dx$ are equal on the two sides of the interface at $x = d$.
(b) Use Equation (4.1.5b), for the odd (sine) solution, to derive an equation for the field for $x > d$. Show that both $E_y(x)$ and $dE_y(x)/dx$ are equal on the two sides of the interface at $x = d$.

4.6 A dielectric slab waveguide has $n_1 = 1.53$, $n_2 = 1.51$, $d = 5\ \mu m$, and $\omega = 2 \cdot 10^{15}$. The normalized frequency and the eigenvalues are given by $V = 8.220$; $k_{1x}d = 1.400, 2.491, 4.179, 5.543, 6.865$, and 8.054. The maximum power density for each mode in the waveguide is 10^6 W/m^2; for even modes this peak power density occurs at $x = 0$.
(a) Which eigenvalues correspond to even modes and which to odd modes?
(b) Plot the magnitude of $E_y(x)$ for $-10 < x < 10$ for the $k_{1x}d = 1.400$ mode.
(c) Plot the magnitude of $E_y(x)$ for $-10 < x < 10$ for the $k_{1x}d = 8.054$ mode.

4.7 For propagation in the slab waveguide, the condition

$$n_1 k_0 < \beta < n_2 k_0$$

must be satisfied. Interpret the extreme values, $\beta = n_1 k_0$ and $\beta = n_2 k_0$, in terms of the angle θ that the propagation vector, k, makes with the side of the guide.

4.8 In a step-index fiber with $n_1 = 1.51$ and $n_2 = 1.49$, the wavelength of the light wave is $1.3\ \mu m$. If the eigenvalue of the 01 mode is $0.5 \cdot 10^6\ m^{-1}$, what is the propagation constant β?

4.9 A step-index optical fiber has $V = 5.8$. Determine which modes can propagate in this fiber. Identify the modes with the double subscript notation. Calculation of the eigenvalues is not necessary.

4.10 Identify all modes that will propagate in a step-index fiber that has $V = 4.0$.

4.11 For the step-index fiber of Example 4.2, plot the eigenvalue equation for $v = 2$. Use this graph to estimate the eigenvalues. Write the complete equation for $E_z(r,\theta,z,t)$ in both the core and the cladding for each of these modes.

4.12 A step-index fiber has $n = 1.61$ and $\lambda = 1.55\ \mu m$. The radius a and the index difference Δn are to be selected to make this a single-mode fiber. Plot a graph of a versus Δn representing the condition for single-mode propagation, and indicate on the graph the values for a and Δn for which this will be a single-mode fiber.

4.13 A multimode step-index fiber has $n_1 = 1.75$, $N_1 = 1.77$, $n_2 = 1.73$, $N_2 = 1.75$, and $a = 50\ \mu m$. The length of the fiber is 30 km and the wavelength of the light wave is $1.3\ \mu m$.
(a) Calculate the modal dispersion.
(b) Explain the effects of reducing V on the magnitude of the modal dispersion.

4.14 The materials in a step-index fiber are materials A and B in Table 3.4.1 and Figure 3.4.2. Calculate the modal dispersion for this fiber at $\lambda = 1.3\ \mu m$ (see problem 3.11).

4.15 A single-mode step-index fiber has the following indices of refraction:

$$n_1 = 1.460 \qquad n_2 = 1.450$$

$$N_1 = 1.475 \qquad N_2 = 1.464$$

The material dispersion at $\lambda = 1.55\ \mu m$ is $D_m = 21.5$ ps/km $\cdot$ nm. Specify the fiber radius necessary to make the total intramodal dispersion zero at this wavelength.

4.16 In a single-mode step-index fiber for $\lambda = 1.54\ \mu m$, the indices of refraction are $n_1 = 1.550$, $N_1 = 1.565$, $n_2 = 1.540$, and $N_2 = 1.553$, and the material dispersion is 15 ps/km $\cdot$ nm. Use Equation (4.2.23) and Figure 4.2.5 to find a V for which the intramodal dispersion will be zero at this wavelength.

4.17 Calculate the modal dispersion, in ns/km, for

(a) a step-index fiber with the following parameters:

$$n_1 = 1.465 \qquad n_2 = 1.455 \qquad a = 50\ \mu m$$

$$N_1 = 1.480 \qquad N_2 = 1.469 \qquad \lambda = 1.3\ \mu m,\ \text{and}$$

(b) a graded-index α-profile fiber with α optimized to minimize modal dispersion. The graded-index fiber has the same parameters as the step-index fiber of (a), with the index of the core being interpreted as the index at the center of the core.

CHAPTER 5

Electroluminescence and Light-Emitting Diodes

Transmitters for optical communication must have sources of light waves at suitable wavelengths. The power output and spectral characteristics of the source and the availability of a means for modulating the light wave are essential considerations in selecting a light wave source.

There are two classes of sources in common use for optical fiber systems. They are the laser and the light-emitting diode. For optical fiber communication systems, both are made as semiconductor *p-n* junction diodes. For some systems, either of them may provide a satisfactory design, but in many cases one is clearly better than the other for the specific communication system under consideration.

In this chapter, we will begin by reviewing some of the principles of semiconductor devices, especially the behavior of the *p-n* junction. The general principles of light emission from the *p-n* junction and the use of this phenomenon to develop useful light-emitting diodes are presented. Those characteristics of the light-emitting diode that are important for communication system design are given special attention. The semiconductor *p-n* junction diode laser is treated in Chapter 6.

5.1   **Review of Semiconductors**

Before examining the characteristics of light-emitting devices, we will review some of the principles of semiconductor materials and devices. It is assumed that the reader has some prior understanding of semiconductors and that, therefore, the material offered in this section is not entirely new. The objectives are to present a brief review for those who may need it and to collect for easy reference the concepts and relationships that will be needed for the study of light-emitting diodes and lasers.

5.1.1   *Thermal Equilibrium*

By definition, a semiconductor is a material that has a moderate number of charge carriers that are free to move under the influence of an electric field. It is characterized by an energy-level diagram such as that shown in Figure 5.1.1. Electrons are normally held in the semiconductor crystal structure by covalent bonds, the forces that form the crystal and hold it together. However, they are excited by thermal energy and have a range of energies given by the Fermi-Dirac distribution function, Equation (5.1.2). In semiconductors, the band gap energy, $\mathcal{E}_g$, is small enough that, at normal temperatures, the thermal energy of electrons in the semiconductor crystal is sufficient to give some of them energies in the conduction band.

When an electron has an energy level that falls within the band gap, its kinetic energy is not sufficient to free it from the covalent bond that holds it in the crystal. If an electron acquires energy greater than $\mathcal{E}_c$, it can become free of these bonds and thus becomes a conduction electron. The vacancy created by

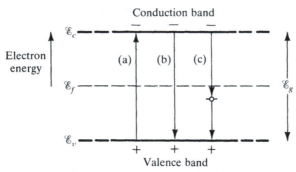

Figure 5.1.1   *Electron energy-level diagram. $\mathcal{E}_c$ and $\mathcal{E}_v$ represent the edges of the conduction band and valence band, respectively. $\mathcal{E}_f$ is the Fermi level. Transition (a) is the excitation of an electron from a bound state in the valence band into the conduction band by the acquisition of energy greater than the band gap energy. Transition (b) is the recombination of an electron in the conduction band with a hole in the valence band. Transition (c) represents recombination via a recombination center within the band gap.*

freeing a bound electron, that is, a hole, is also a charge carrier and contributes to conduction. The conductivity of the material is proportional to the number of electrons in the conduction band and the number of holes in the valence band.

The exclusion principle of quantum mechanics indicates that only certain energy levels are available for occupancy by electrons and that each available energy level can be occupied by only one electron. It is thus possible that an electron might have sufficient energy to escape the crystal bonds but that it cannot do so because no suitable energy states in the conduction band are available. The available states are represented by $N(\mathscr{E})$, the density of available energy states at energy $\mathscr{E}$. In the conduction band,

$$N(\mathscr{E}) = \frac{4\pi}{h^3} \left[2m_e\right]^{\frac{3}{2}} \left[\mathscr{E} - \mathscr{E}_c\right]^{\frac{1}{2}} \quad (\mathrm{J^{-1}m^{-3}}) \tag{5.1.1a}$$

and in the valence band

$$N(\mathscr{E}) = \frac{4\pi}{h^3} \left[2m_h\right]^{\frac{3}{2}} \left[\mathscr{E}_v - \mathscr{E}\right]^{\frac{1}{2}} \quad (\mathrm{J^{-1}m^{-3}}) \tag{5.1.1b}$$

where m_e and m_h are the effective masses of conduction electrons and holes, respectively. This is the number of states per cubic meter per joule. The integral of $N(\mathscr{E})d\mathscr{E}$ over some range of energies gives the number of available states per cubic meter having energies within this range.

The Fermi-Dirac distribution gives the probability that an available state at energy level $\mathscr{E}$ is occupied by an electron.

$$f(\mathscr{E}) = \frac{1}{\exp \dfrac{\mathscr{E} - \mathscr{E}_f}{kT} + 1} \tag{5.1.2}$$

In this equation, $\mathscr{E}_f$ is the Fermi level. The probability that an available state at the Fermi level is occupied by an electron is 0.5. The probability that any specific level higher than the Fermi level will be occupied is less than 0.5.

The density of electrons, commonly called the carrier concentration, in the conduction band can be determined by integrating the product of the density of states and the Fermi-Dirac distribution of electron energies.

$$n = \int_{\mathscr{E}_c}^{\infty} f(\mathscr{E}) \cdot N(\mathscr{E}) \, d\mathscr{E}$$

$$= N_c \exp \frac{-(\mathscr{E}_c - \mathscr{E}_f)}{kT} \quad (\mathrm{m^{-3}}) \tag{5.1.3}$$

where

$$N_c = 2 \left[\frac{2\pi m_e kT}{h^2}\right]^{\frac{3}{2}}$$

Similarly, the density of holes, that is, hole concentration, in the valence band is the integral of the product of the density of states and the probability that a state is *not* occupied by an electron.

$$p = \int_{-\infty}^{\mathscr{E}_v} [1 - f(\mathscr{E})] \cdot N(\mathscr{E}) \, d\mathscr{E}$$

$$= N_v \exp \frac{-(\mathscr{E}_f - \mathscr{E}_v)}{kT} \quad (\mathrm{m}^{-3}) \tag{5.1.4}$$

where

$$N_v = 2 \left[\frac{2\pi m_h kT}{h^2} \right]^{\frac{3}{2}}$$

$N(\mathscr{E})$, $f(\mathscr{E})$, and the carrier concentrations are illustrated on an electron-energy level diagram in Figure 5.1.2

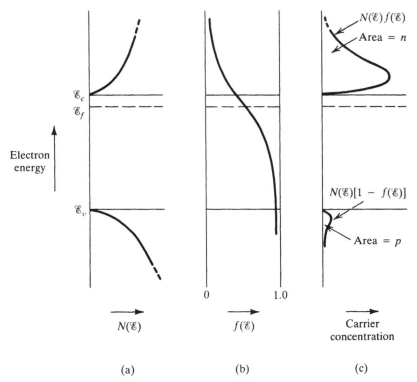

Figure 5.1.2 *Electron and hole concentrations in an n-type semiconductor. This is a schematic diagram illustrating the method for calculating the carrier densities for a semiconductor in thermal equilibrium; it is not to scale. (a) The density of states, $N(\mathscr{E})$, in the conduction and valence bands. (b) The Fermi-Dirac distribution function, $f(\mathscr{E})$. (c) Carrier concentrations in the conduction and valence bands.*

The *np* product defines the intrinsic carrier density, n_i, an important parameter.

$$np = n_i{}^2 = N_c N_v \exp \frac{-\mathcal{E}_g}{kT} \tag{5.1.5}$$

where $\mathcal{E}_g = (\mathcal{E}_c - \mathcal{E}_v)$. This can also be expressed as

$$n_i{}^2 = K_1 T^3 \exp \frac{-\mathcal{E}_g}{kT} \tag{5.1.6}$$

where K_1 is a constant.

In a pure semiconductor material, one with no impurity atoms, $n = p = n_i$. Such materials are called intrinsic materials.

The relationships (5.1.3) through (5.1.6) apply when the material is in thermal equilibrium. Semiconductor devices are based on the behavior of the materials when this equilibrium condition is disturbed. The nonequilibrium behavior of semiconductor materials is reviewed in Section 5.1.4 and optical devices based on this behavior are described in Chapters 5 to 7.

5.1.2   *Extrinsic Semiconductors*

Most semiconductor devices, including optical sources and detectors, use extrinsic semiconductor materials, that is, material in which some impurity atoms have been introduced. Certain impurity atoms, when introduced into the semiconductor crystal in small concentrations, have Fermi levels very close to the edge of the conduction or valence band. When the intrinsic material is silicon (valence 4) and the dopant is a material of valence 5 and has Fermi level slightly below the edge of the conduction band, one of the valence electrons from each impurity atom can be excited into the conduction band with very little thermal energy. It is often assumed that at room temperature all of the impurity atoms are ionized. In that case, the electron density in the conduction band is $n = p + N_D$, where N_D is the density of valence 5 dopant atoms. Since each electron in the conduction band is matched by either a hole in the valence band or an ionized impurity atom, the material remains electrically neutral.

It can be shown that Equation (5.1.5), developed for the intrinsic semiconductor, is applicable to extrinsic semiconductor materials as well. Therefore, for an *n*-type material,

$$np = p^2 + pN_D = n_i{}^2 \tag{5.1.7}$$

The electron density n is now larger and p smaller than the intrinsic concentration. The electrons are the majority carriers and the holes the minority carriers.

In most cases of practical interest, the doping level is substantially larger than the intrinsic concentration. In that case, $n = N_D$ and

$$p = \frac{n_i{}^2}{N_D} \tag{5.1.8}$$

This is an *n*-type dopant and the resulting extrinsic material is an *n*-type material. The impurity atoms are donors, hence the subscript *D*. Similarly, for *p*-type semiconductors, the dopants are atoms of valence 3, with Fermi level near the edge of the valence band. The acceptor density is N_A, and the carrier densities are $p = N_A$ and

$$n = \frac{n_i^2}{N_A} \tag{5.1.9}$$

In the *p*-type material, holes are the majority carriers and electrons the minority carriers.

In semiconductors, the band gap energies are of the order of 1 eV and energies required to ionize the dopants are usually less than 0.05 eV. Intrinsic carrier concentrations are of the order of 10^{16} m^{-3} for silicon and 10^{13} m^{-3} for GaAs. Doping levels are typically two or more orders of magnitude greater than n_i.

5.1.3 The p-n Junction

A *p-n* junction is formed when adjacent regions in a single crystal of semiconductor material are doped with donor and acceptor impurities to form contiguous *n* and *p* regions. The interface between the two regions, assumed here to be a plane, is the *p-n* junction. In this section, we will examine the electrical characteristics of such a junction. We will later study the junction diode as a source or a detector of light waves.

An *n*- or a *p*-type material, in isolation from other effects, will be electrically neutral. For each electron excited into the conduction band by thermal ionization, for example, there will be a corresponding hole in the valence band. For each conduction electron produced by ionization of a donor impurity atom, there will be a matching positive charge in the nucleus of the donor atom; this charge is bound in the crystal structure and is not free to contribute to conduction. Similarly, for each hole produced by an acceptor impurity atom, there must be a matching bound negative charge. In the *n* region, for example, there will be bound positive charges consisting of ionized donor atoms, a slightly larger number of conduction electrons, and a few holes [see Equation (5.1.7)].

When the *n* and *p* regions are contiguous in a single semiconductor crystal, with no electric fields present, the free majority carriers, electrons in the *n* region and holes in the *p* region, will each tend to diffuse into the other region. In the vicinity of the junction, where both types of charge carriers exist together, there will be some recombination of holes and electrons, reducing the number of carriers. As carriers are thus consumed, the bound positive charges in the *n* region and the bound negative charges in the *p* region will no longer be fully compensated by free electrons and holes, respectively. The *n* region will have a net positive charge and the *p* region a net negative charge. The overall *p-n* junction device, consisting of both the *n* and the *p* regions, will remain electrically neutral; the bound positive and negative charges will be equal and will exactly compensate each other. The uncompensated bound charges will be in the regions

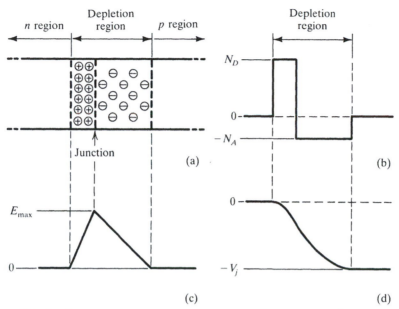

Figure 5.1.3 *The p-n junction in thermal equilibrium. (a) Net charge density. The n and p regions are electrically neutral. The depletion region, taken as a whole, is electrically neutral. There will be an electric field, E, in the depletion region, directed from the p region toward the n region. (b) The charge density in the depletion region. (c) The electric field intensity, E, in the depletion region. The electric field intensity in the n and p regions, outside the depletion region, is zero. (d) The electrostatic potential, V, in the depletion region, relative to the potential in the n region.*

immediately adjacent to the junction. The resulting charge distributions are illustrated in Figure 5.1.3(a) and (b); the corresponding electric field intensity, E, and the potential, V, are shown in (c) and (d).

It is evident from Figure 5.1.3 that an electric field will be established across the junction by the uncompensated bound charges. This electric field is in the direction that opposes majority carrier diffusion. As diffusion causes the numbers of uncompensated bound charges to increase, the strength of the electric field increases.[1] Diffusion of majority carriers across the junction will continue until the tendency to diffuse is exactly balanced by the tendency of the electric field to oppose diffusion. At equilibrium, the voltage across the junction is the diffusion potential, given by

$$V_d = -V_T \ln \frac{N_D N_A}{n_i^2} \quad \text{(V)} \qquad\qquad \textbf{(5.1.10)}$$

---

[1] In Section 7.1.4, we will examine the electric field strength and the corresponding terminal voltage quantitatively. For the present purposes, these results are not needed.

where

$$V_T = \frac{kT}{q}, \quad \text{the thermal potential}$$

If forward bias of the *p-n* junction is defined as positive, then the diffusion potential is negative. The diffusion potential can also be calculated by dividing the difference between the Fermi levels of the *p* and *n* regions, in electron volts, by *q*, the charge on an electron.

EXERCISE 5.1

A GaAs *p-n* junction diode has doping levels $N_A = 10^{23}$ m^{-3} and $N_D = 10^{22}$ m^{-3}. Calculate (a) the majority and minority carrier concentrations in the *p* and *n* regions, and (b) the diffusion potential.

Answers

10^{23} m^{-3}; $3 \cdot 10^{1}$ m^{-3}; 10^{22} m^{-3}; $3 \cdot 10^{2}$ m^{-3}; -1.22 V.

Moving an electron across the depletion region, from the *n*-region into the *p*-region, will require work to overcome the force that the electric field exerts on the electron. The potential difference through which the electron moves times the charge of the electron is the work done. The potential, *V*, across the junction is a potential barrier opposing the transfer of electrons across the junction. The product qV is a corresponding energy barrier. When the junction is in thermal equilibrium, the energy barrier is qV_d, where V_d is the diffusion potential, Equation (5.1.10). When the junction has a forward-bias voltage *V*, the energy barrier is $q(V_d - V)$.

An energy level diagram of the *p-n* junction is shown in Figure 5.1.4. The energy barrier to the flow of electrons to the right is represented by the slope and height of the conduction band-edge curve; there is a similar barrier to the flow of holes to the left. The shape of these band-edge energy barriers is the same as the shape of the potential barrier, shown, for the case of thermal equilibrium, in Figure 5.1.3(d).[2]

The diffusion of majority carriers across the depletion region, where they become minority carriers, is referred to as minority carrier injection, or simply, carrier injection. Carrier injection is the diffusion across the depletion region of those majority carriers that have energies greater than the energy barrier. It is evident from Figure 5.1.2(c) that no matter how high the energy barrier, some carriers will have enough energy to overcome it.

---

[2] The curve of Figure 5.1.3(d) must be inverted for use in the energy-level diagram. This potential curve is based on the polarities associated with conventional current, that is, the flow of positive charges. When studying the flow of electrons, we are dealing with negative current. We use negative potential so that energy will have its conventional meaning.

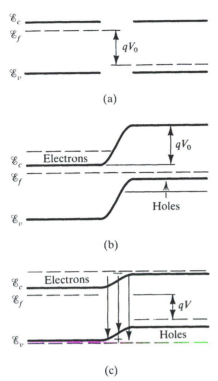

Figure 5.1.4 *Energy level diagram for* (a) *the isolated p- and n-type materials,* (b) *the p-n junction in thermal equilibrium, and* (c) *the p-n junction with an applied forward-bias voltage V.*

When the diode is forward biased, the barrier is lowered, and minority carrier injection will increase. Minority carrier injection, and the subsequent recombination with majority carriers, is the principal mechanism for current flow in the *p-n* junction diode. We expect that for forward voltages of the order of the diffusion potential and larger voltages, substantial current will flow. When the polarity of the applied voltage is reversed, the width of the depletion region is increased and the current is reduced to a small reverse saturation current. These *V-I* characteristics are expressed in the familiar equation

$$I = I_s[e^{V/V_T} - 1] \quad (A) \tag{5.1.11}$$

5.1.4 *Nonequilibrium Behavior*

Any exchange of energy into or out of the semiconductor, other than the steady-state thermal energy of the material, will disturb the thermal equilibrium. If the disturbance is a transient one, thermal equilibrium will be reestablished after the energy transient has been dissipated. If the disturbance is a persistent one, a new condition of equilibrium will be established. Two important classes of disturbances are the absorption of electromagnetic radiation and the flow of current in a forward-biased diode. Each of these disturbances could be either transient or persistent.

The absorption of energy from an electromagnetic wave of power density P can be represented by

$$P(x) = P(0) \exp(-\alpha x) \quad \text{(W)} \tag{5.1.12}$$

where α is the attenuation constant of the material. The energy absorbed from the wave can produce electron-hole pairs by exciting electrons from the valence band into the conduction band or it can be absorbed as heat, increasing the temperature above the equilibrium temperature.

Thermal equilibrium is the condition for which the rate of thermal generation of electron-hole pairs (EHP) is equal to the rate of recombination; the net rate of change of the number of electrons in the conduction band is zero. The rate of thermal generation of conduction electrons, that is, of EHP, is considered to be essentially independent of the numbers of conduction electrons and holes, n and p. The rate of recombination of EHP is proportional to n and to p; the constant of proportionality is B_r. The rate of recombination is then

$$r_r = B_r np \quad (\text{m}^{-3}\text{s}^{-1}) \tag{5.1.13}$$

This relationship holds whether equilibrium exists or not.

For thermal equilibrium, $n_0 p_0 = n_i^2$, where n_0 and p_0 are the equilibrium values of n and p. At equilibrium, the rate of thermal generation of EHP is equal to the rate of recombination.

$$r_{\text{th}} = g_{\text{th}} = B_r n_i^2 \quad (\text{m}^{-3}\text{s}^{-1}) \tag{5.1.14}$$

This relationship gives the rate of thermal generation of EHP.

To examine the effects of a departure from thermal equilibrium, consider that an impulse of energy is absorbed in the material, creating excess electron-hole pairs. Recombination, described by Equation (5.1.13), will cause the number of excess electrons to decrease. Thermal generation, given by (5.1.14), provides additional electrons. Since with an excess of electrons and holes, the rate of recombination will exceed the rate of thermal generation, there will be a negative rate of change of electron density.

$$-\frac{dn}{dt} = B_r np - B_r n_i^2 \quad (\text{m}^{-3}\text{s}^{-1}) \tag{5.1.15}$$

If the nonequilibrium condition is a deficiency of electrons, that is, np is less than n_i^2, then n will increase to restore equilibrium. This rate equation describes the transient behavior of n as it responds to a nonequilibrium condition. Rate equations such as this, and occasionally of higher order, will be important in the study of optoelectronic devices for optical communication systems.

To solve this rate equation, let n and p be expressed as the sum of their equilibrium values and the disturbance.

$$n = n_0 + \Delta n \qquad p = p_0 + \Delta p$$

By substituting these expressions for n and p into the rate equation, we have

$$-\frac{1}{B_r}\frac{dn(t)}{dt} = (n_0 + \Delta n)(p_0 + \Delta p) - n_i^2 \qquad (5.1.16)$$

This equation can be reduced by recognizing that $n_0 p_0 = n_i^2$, $\Delta p = \Delta n$, and $(\Delta n)^2$ is negligibly small.

$$\frac{d(\Delta n)}{dt} = -B_r(p_0 + n_0)\Delta n \quad (\text{m}^{-3}\text{s}^{-1}) \qquad (5.1.17)$$

This recombination rate will be given the symbol r, with subscripts to identify different classes of recombination.

In extrinsic materials, one type of carrier will have much higher concentration than does the other. In a p-type material, for example, $p_0 \gg n_0$. The solution to this rate equation describes an exponential decay of the excess minority carriers.

$$\Delta n(t) = \Delta n(0) \exp\left(-B_r p_0 t\right) = \Delta n(0) \exp\left(-\frac{t}{\tau_r}\right) \qquad (5.1.18)$$

The time constant, $\tau_r = 1/(B_r p_0)$, is called the recombination lifetime, and B_r is the recombination coefficient. In many cases, Δn represents a small fraction of the majority carriers and essentially all of the minority carriers. In these cases, the recombination lifetime is the minority carrier lifetime. The minority carrier lifetime is inversely proportional to the density of majority carriers. The recombination coefficient is a property of the intrinsic material and is applicable for either n- or p-type doping.

When a current flows in the junction diode, the thermal equilibrium conditions are disturbed by the addition of excess electrons into the n region and excess holes into the p region. If the current is constant, that is, a persistent disturbance, then a new condition of equilibrium will be established. The total rate at which carriers are generated is the sum of the externally supplied and thermal generation rates. The rate of recombination is, as before, equal to $B_r np$.

We wish to solve for the equilibrium carrier densities when a constant current is flowing. The units for n and p are m^{-3}. The units for current should be reduced to electrons per second per unit volume. This is normally handled by using current density J, in amperes per square meter, flowing through a recombination region of thickness d. The current density can then be expressed as J/qd (electrons per second per cubic meter). The rate equation thus becomes

$$\frac{d(\Delta n)}{dt} = \frac{J}{qd} - \frac{\Delta n}{\tau_r} \quad (\text{m}^{-3}\text{s}^{-1}) \qquad (5.1.19)$$

The condition for equilibrium is that the derivative be equal to zero. The steady-state excess electron density when a constant current is flowing into the junction is

$$\Delta n = n - n_0 = N_0 = \frac{J}{qd}\,\tau_r \quad (m^{-3}) \tag{5.1.20}$$

The transient response, $\Delta n(t)$, to a step function for J can be found by solving (5.1.19). It can be shown that $\Delta n(t)$ will follow a simple exponential equation, with time constant τ_r, from its initial value to its new steady-state value. If the initial and final values of J are J_1 and J_2, respectively, then Δn is given by

$$\Delta n = \frac{J_2 \tau_r}{qd} - \frac{(J_2 - J_1)\tau_r}{qd} \exp\left(-\frac{t}{\tau_r}\right) \quad (m^{-3}) \tag{5.1.21}$$

EXERCISE 5.1

Find $\Delta n(t)$ when $\tau_r = 10$ ns and J is a step function: $J/qd = 10^{15}\,u(t)$. Calculate the time required for $\Delta n(t)$ to rise from 10 to 90 percent of its final value.

Answers

$\Delta n(t) = 10^7[1 - \exp(-10^8 t)]$; 22 ns.

5.2 Principles of the Light-Emitting Diode

The *p-n* junction diode, with energy supplied by a direct-current source, can be a source of light for optical communication systems. When an electron in the conduction band recombines with a hole in the valence band, energy approximately equal to the band gap energy is released. This energy may be in the form of a photon. The diode can thus produce light. Since we know that the energy of a photon is hf, then if we know the band gap energy, we can find the frequency of the photon or of the light wave produced by the emission of many such photons. The band gap energies of the materials we will consider are of the order of one electron-volt. For photon energy of 1 eV, the wavelength is 1.24 μm and the frequency is $2.42 \cdot 10^{14}$ Hz.

There are two major classes of junction-diode light sources. The first, the light-emitting diode, or LED, is the principal subject of this chapter. The second light-emitting device, the laser, is the subject of Chapter 6.

5.2.1 *LED Power Output*

Equation (5.1.19) indicates that, in the steady state, the number of recombinations per second will be J/qd per unit volume or I/q total. If the fraction of recombinations that produce photons is η, then there will be $\eta I / q$ photons per second. Since each photon is equivalent to hf joules, the optical power produced by the junction diode is

$$P_0 = \eta \frac{I}{q} hf \quad \text{(W)} \tag{5.2.1a}$$

Since, in light wave systems, we will usually describe the spectral characteristics in terms of wavelength rather than frequency, the power output can be expressed

$$P_0 = \frac{\eta hc}{q\lambda} I \quad \text{(W)} \tag{5.2.1b}$$

By substituting the appropriate constant values for h, c, and q, the equation can be written

$$P_0 = \frac{1.24 \eta I}{\lambda} \quad \text{(W)} \tag{5.2.1c}$$

where λ must be in μm.

The quantity η in these equations is the quantum efficiency of the light-emitting diode. Depending on the way in which η is defined, Equations (5.2.1a to c) can be made to represent the optical power produced in the diode, the power output from a specific output port, or the power delivered into an optical fiber.

The quantum efficiency is usually expressed as the product of two components. The first, the internal quantum efficiency, is the ratio of the radiative recombinations to the total recombinations. The second, the external quantum efficiency, is the ratio of the photons collected in the optical output of the device to the total photons produced in the device. It is easy to see that the product of these two terms is the ratio of output photons to total recombinations. It is this product that we would normally use in equations such as (5.2.1).

Recombination of electrons and holes can take any of several different forms. The energy released by the electron is equal to the difference between its initial and final energy levels. If the initial energy is in the conduction band and the final energy is in the valence band, then the energy released is equal to or slightly greater than the band gap. The energy thus released may be either thermal or radiated energy. If thermal, it consists of kinetic energy in crystal lattice vibrations and will cause an increase in temperature of the crystal. If radiated, it is represented as a photon that will propagate at the speed of light in the semiconductor material and in an undetermined direction. In some cases, the initial or final energies may be in the forbidden region (see Figure 5.1.1); these too could be either thermal or radiated energy, but the quantity of energy is smaller, and the radiated wavelength longer, than that corresponding to the full band gap.

When the current flowing in the diode introduces excess electrons into the conduction band, the excess carriers will recombine at rates established in Section 5.1. The recombination rate, given in Equations (5.1.18) and (5.1.19), is characterized by the minority-carrier lifetime, τ_r. Let us define the recombination rate to be r; from (5.1.19), $r = \Delta n / \tau_r$. We now need to distinguish between several different mechanisms for recombination. The mechanism that produces the output desired from the LED is radiative recombination; this will be interpreted to be optical radiation at or near the wavelength of interest. Other mechanisms include decay to the valence band with the energy being transformed into thermal energy and various kinds of transitions to energy levels within the forbidden band; all of these transitions, which do not contribute to useful output, are grouped under the nonradiative recombination classification. The rates associated with radiative and nonradiative recombination are represented by r_{rr} and r_{nr}, respectively.

The internal quantum efficiency can be defined as the ratio of the rate of radiative recombination to the total recombination rate.

$$\eta_{\text{int}} = \frac{r_{rr}}{r_{rr} + r_{nr}} \tag{5.2.2}$$

It is possible to build devices with internal quantum efficiencies as high as 50 percent.

On the other hand, the external quantum efficiency is relatively small. It is defined as the ratio of photons contributing to the output power to the total number of photons produced per unit time. There are four major reasons that photons produced by radiative recombination may not appear in the LED output. These are illustrated in Figure 5.2.1. First, the device designer has no control over the direction in which light will be radiated; we can assume that all directions are equally probable. Thus, half of the light will be radiated into the lower and half into the upper hemisphere. Second, some light will approach the output interface of the device at angles greater than the critical angle and will be totally reflected. Third, light waves not totally reflected will be partially reflected at the interface. Finally, some photons will be absorbed, that is, the light wave attenuated, in the semiconductor material between the active junction region and the output interface. The combination of these losses can give external quantum efficiencies of less than 1 percent.

In the light-emitting diode, the region from which the light will be emitted can be an important consideration. It is clear that light will be emitted from the region(s) where recombination takes place. Most recombination occurs between injected minority carriers and majority carriers within one diffusion length of the edge of the depletion region. In a light-emitting diode, the doping, and hence the majority-carrier concentration, is much higher on one side of the junction than on the other. This causes a higher injected minority-carrier concentration in the region with the lower majority-carrier concentration. In this case, most light will be generated in the region with the high minority-carrier concentration.

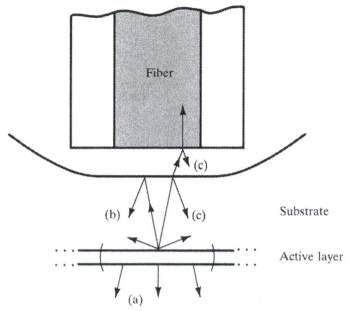

Figure 5.2.1 *External quantum efficiency. Useful output, that is, light propagating in the fiber, is less than the light produced in the active region of the LED because of (a) radiation in directions other than toward the fiber, (b) total internal reflection at the substrate-air interface, (c) partial reflection at the substrate-air and air-fiber interfaces, and (d) absorption in the substrate and other semiconductor materials between the active region and the output interface.*

5.2.2 *Frequency Characteristics of the LED*

The frequency region in which the LED will radiate is determined by the energy of the radiated photons. This energy is approximately equal to the band gap energy. However, reference to Figure 5.1.2 will show that there is a range of possible energies for electrons in the conduction band and for holes in the valence band. In principle, any electron can combine with any hole, thus releasing energy given by the difference in their respective energy levels.

Useful recombination radiation takes place in a region near the junction where there is a significant number of excess minority carriers. An energy level diagram is shown, to approximate scale, with electron and hole concentrations in Figure 5.2.2. The energy band for electrons extends for several kT above the edge of the conduction band. Similarly, the energy band for holes extends for several kT below the edge of the valence band. It is apparent that the maximum intensity of radiation would be at photon energies of the order of kT greater than the band gap, and that the range of photon energies extends from the band gap energy to perhaps $5kT$ higher than the band gap energy. The spectrum of this recombination radiation is illustrated in Figure 5.2.3 for several materials and the corresponding wavelength regions. It can be shown that the spectral

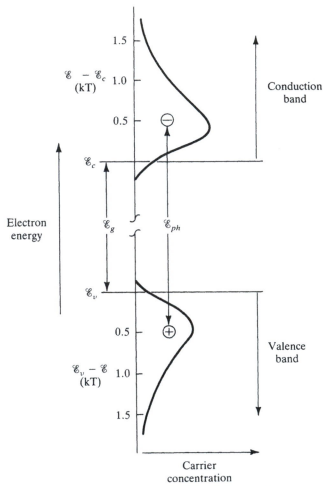

Figure 5.2.2 *Carrier concentrations versus electron energy level. The electron energies are to approximate scale. Free carriers can exist inside the band gap in heavily doped semiconductors; these "band tails" are shown. The center of the radiated spectrum, corresponding to the most probable photon energy, is approximately kT greater than the band gap. The width of the spectrum of recombination radiation is in the range 3kT to 4kT.*

width for the three LEDs of Figure 5.2.3 corresponds to $\Delta\mathscr{E}_{ph} \approx 3.3kT$ (see Problem 5.7).

Recall that the dispersion in single-mode optical fibers is proportional to the spectral width of the source. For the LED transmitter, the spectral width can be estimated to be a few kT. In addition to providing a basis for calculating dispersion, this also indicates that the spectral width is directly proportional to temperature. It will increase if the diode is allowed to increase significantly in temperature and it could be reduced by cooling the LED.

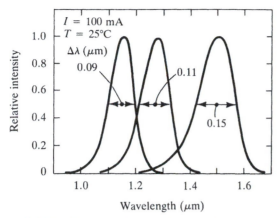

Figure 5.2.3   *The spectrum of recombination radiation for three InGaAsP/InP LEDs of different material composition. From Wada, et al.,* IEEE Journal of Quantum Electronics, *vol. QE-18, pp. 368-373; copyright © 1982, IEEE [3.2].*

EXERCISE 5.3

A GaAs LED operating at room temperature has total quantum efficiency $\eta = 0.2$. When the diode current is 50 mA, find the wavelength; the spectral width, in nm; and the optical power output.

Answers

0.85 μm; 50 nm; 14.6 mW.

EXERCISE 5.4

An InGaAsP LED, with band gap of 0.90 eV at 320 K, is used as the transmitter in a PCM system. The dispersion of the fiber is 30 ps/(km · nm). Find the total dispersion over a 50-km length of fiber. By what factor is the dispersion reduced if the temperature is reduced to 77°K?

Answers

194 ns; 3.9.

The light wave from the LED does not have a well-defined sinusoidal center frequency as is the case with a microwave or other electronic oscillator. It consists of many photons emitted independently of each other, each having a frequency and phase that are independent of those of other photons. In that respect, it is very much like the light from an incandescent lamp. Since the light wave has no well-defined frequency or phase, it cannot be used for frequency or phase modulation. The LED is useful for intensity modulation, including both cw modulation and PCM, but for little else. However, it is a light wave source of major importance for optical fiber communication systems.

5.2.3 *Modulation of the LED*

The means for intensity modulation of the LED are evident in Equation (5.2.1). Since the power output is linearly proportional to the input current, intensity modulation is possible by adding a signal current to the bias current so that the total current, and therefore the total power, can vary linearly over a suitable range.

EXAMPLE 5.1

The input-output relationship for an LED is $P = 0.1I$. The maximum average power is 1 mW and the maximum instantaneous power is 3 mW. The LED is to be amplitude modulated with an $x(t)$ that has a dc component of 0.2 and a periodic component with range of ± 2.28. Find the modulating current, $i(t) = I[1 + mx(t)]$.

Solution

The range of $x(t)$ is $-2.08 < x(t) < 2.48$. The power will have the form $p(t) = P[1 + mx(t)]$, where m and P are to be determined. The average value of $p(t)$ will be

$$\langle p(t) \rangle = P[1 + 0.2m] = 1 \text{ mW}$$

The minimum value of $p(t)$ is

$$p(t)_{min} = [1 - 2.08m] \geq 0$$

from which we find that $m \leq 0.48$. The maximum value of $p(t)$ is

$$p(t)_{max} = P[1 + 2.48m] \leq 3 \text{ mW}$$

from which we find that $m \leq 0.88$. Thus, $m \leq 0.48$.

We now use the equation for average power to find P.

$$P[1 + 0.48 \cdot 0.2] \leq 1 \text{ mW}$$

$$P = \frac{1}{1.096} = 0.91 \text{ mW}$$

Finally,

$$p(t) = 0.91[1 + 0.48x(t)] \text{ mW}$$

and

$$i(t) = 10p(t) = 9.1[1 + 0.48x(t)] \text{ mA}$$

It was shown in Equation (5.1.19) that there is a maximum rate at which the LED can respond to a disturbance in carrier densities; the response was characterized with a time constant, the lifetime of minority carriers. We should expect that there is a corresponding maximum frequency for sinusoidal intensity modulation. The response of the LED to sinusoidal modulation of the input current, and therefore of the output power, can be derived from Equation (5.1.19).

If in Equation (5.1.19) we define J_0 to be the constant bias current and N_0 the steady state value for Δn, the excess carrier density in the steady-state condition, then the equilibrium condition is given by

$$\frac{J_0}{qd} - \frac{N_0}{\tau_r} = 0 \quad (\text{m}^{-3}\text{s}^{-1}) \tag{5.2.3}$$

To consider the response to sinusoidal modulation, let

$$J = J_0[1 + m_J \exp(j\omega t)] \tag{5.2.4a}$$

and

$$\Delta n = N_0[1 + m_N \exp\{j(\omega t - \theta)\}] \tag{5.2.4b}$$

By substituting these into (5.1.19) we can find

$$j\omega m_N N_0 \exp(-j\theta) \exp(j\omega t) = \frac{m_J J_0}{qd} \exp(j\omega t)$$

$$- \frac{m_N N_0 \exp(-j\theta)}{\tau_r} \exp(j\omega t) \tag{5.2.5}$$

If we multiply this by τ_r and use (5.2.3) to relate J_0 to N_0, this can be reduced to

$$j\omega\tau_r m_N \exp(-j\theta) = m_J - m_N \exp(-j\theta) \tag{5.2.6}$$

and further to

$$m_N \exp(-j\theta) = \frac{m_J}{1 + j\omega\tau_r} \tag{5.2.7}$$

By substituting (5.2.7) into (5.2.4b) and recognizing that the power output is proportional to Δn, we can write

$$P(\omega) = P_0 \left[1 + \frac{m_J}{1 + j\omega\tau_r} \exp(j\omega t) \right] \quad \text{(W)} \tag{5.2.8}$$

Thus, the modulation of the output power is seen to differ from that of the input current by the factor $1/(1 + j\omega\tau_r)$. The modulation bandwidth is $\omega_{3\,dB} = 1/\tau_r$.

Pulse code modulation (PCM) signals consist of pulses of light that are turned on and off, with suitable amplitudes, according to the information being transmitted and the pulse code in use. Binary PCM, for which the pulse has only two possible amplitudes, one of which is normally zero, is the dominant system. We can determine the maximum rate for PCM pulses, hence the maximum data rate, from the solutions to the rate equation, [Equation (5.1.19)].

To use an LED as the transmitter for a binary PCM system, the input current would consist of a rectangular pulse of suitable amplitude to represent the binary *1* and no pulse (a pulse of amplitude zero) to represent the binary *0*. The optimum pulse shape from the LED would be one with maximum energy and minimum duration. We have found that the response of the LED to a rectangular current pulse has an exponential rise and decay. The minimum duration of the pulse must therefore be approximately two to three time constants. If the pulse duration is $T = 2.5\tau$, then the maximum data rate is $R = 1/T = 1/(2.5\tau)$.[3]

EXERCISE 5.5

The recombination lifetime of the LED in a PCM transmitter is 15 ns. Binary decisions are made at the end of each pulse period to determine whether a pulse was transmitted or not. Intersymbol interference occurs when the decay transient from one pulse period is present at the end of the following pulse period. For this problem, assume that the intersymbol interference is due entirely to the time constants of the LED. What is the intersymbol interference, expressed as a percent of the peak pulse amplitude, when the PCM data rate is $R = 50$ Mb/s; 25 Mb/s; and 10 Mb/s?

Answers
26.4 percent; 6.9 percent; 0.13 percent.

---

[3] The factor T/τ, here taken to be 2.5, should not be interpreted to be a universal constant. Appropriate values for the ratio T/τ depend on several system design considerations that will lead to different values for different circumstances.

5.3 Heterostructures

The semiconductor junction diodes discussed in previous sections were based on a single semiconductor material, doped to form *p* and *n* regions. Because the intrinsic material is the same throughout, the band gap and index of refraction are essentially the same throughout the device. It would be possible for device designers to create better device characteristics if they could exercise some control over these material properties. For example, quantum efficiency might be increased by using a wider band gap material that is transparent to the radiated light for the material between the active region and the optical output, thus eliminating absorption of the radiated light in this material.

A device structure that uses two or more different materials to form a single-crystal semiconductor device is called a heterostructure. The fact that the materials must still form a single crystal places constraints on the materials that can be used for heterostructure devices. Each material must form natural crystals having essentially the same physical dimensions for the crystal lattice, that is, the same lattice constants, so that the resulting crystal will not have mechanical strains at the interface between the two materials. If this condition can be satisfied, then materials having different electrical and electronic properties can be used to fabricate heterostructures having desirable device characteristics.

A heterojunction is formed when two materials, having similar lattice constants but different electronic properties, are formed into a single crystal. We will assume an abrupt, plane interface between the two materials. Figure 5.3.1 illustrates the formation of an *n-P* heterojunction. The uppercase *P* designates the material with the larger band gap.

The energy-band diagram for the heterojunction is constructed from the diagrams of the isolated materials by (1) aligning the Fermi levels on both sides of the junction, (2) maintaining continuity of the vacuum level, and (3) drawing the band edges parallel to the vacuum level. This produces discontinuities in the band-edge energy levels equal to the differences in the band-edge energy levels, $\Delta \mathscr{E}_c$ and $\Delta \mathscr{E}_v$, in Figure 5.3.1(a). The double spike in the valence band edge at the junction can inhibit the flow of carriers across the junction. The transitions from the energy levels of material 1 to those of material 2 take place within the depletion region. The slopes and shapes of the curves in this transition region are determined by the density of bound charges, that is, the doping levels, in the depletion region.

The *n-P* heterojunction illustrated in Figure 5.3.1(b) has the following properties:

1. Because the band gaps of the two materials are not the same, light generated in material 1, with photon energy approximately equal to $\mathscr{E}_{g1}$, can pass through the *P*-type material without absorption. If $(\mathscr{E}_{g2} - \mathscr{E}_{g1})$ is, for example, greater than a few kT, then material 2 is essentially transparent to recombination radiation generated in material 1.

2. The barrier to the flow of electrons is higher, by $\Delta \mathscr{E}_c$, than that of the *n-p* homojunction.

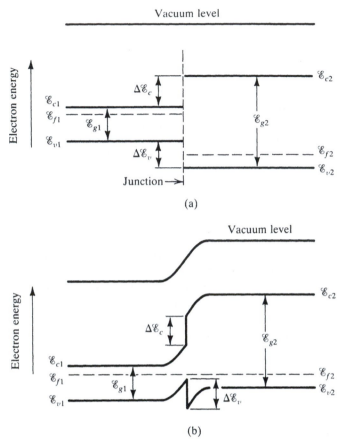

Figure 5.3.1 *An n-p heterojunction made from a narrow-band-gap n-type semiconductor and a wide-band-gap p-type semiconductor. (a) energy level diagram for the isolated semiconductors, with their vacuum energy levels aligned. (b) Energy level diagram for the n-p heterojunction. An important feature of this diagram is that the barrier to the flow of electrons is greater than in the corresponding n-p junction with equal-band-gap n and p materials*

An N-n-P double heterostructure (DH) is illustrated in Figure 5.3.2. This is a p-n junction diode, but with useful properties that are made possible by the use of the double heterostructure. The barriers to the flow of electrons and holes across the junction region are not at the same junction, as is the case for the homojunction diode. In the n region between the two junctions, both types of carriers exist in high concentrations. The minority-carrier concentration, p, can have magnitude comparable to the majority-carrier concentration in the P region. When excess carriers are provided by forward bias, recombination takes place within, and is largely confined to, this central region. This carrier confinement feature can be used to advantage in making LEDs and lasers.

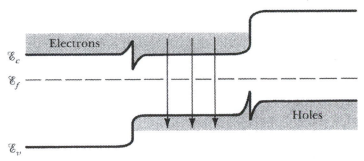

Figure 5.3.2 *An N-n-P double heterojunction. The enhanced barriers to the flow of electrons in one direction and holes in the other are at different heterojunctions. The concentration of minority carriers in the center, narrow-band-gap, n region is substantially greater than in the homo-junction diode [see Figure 5.1.4(c)]. Recombination radiation takes place in a well-defined region that has dimensions selected by the device designer.*

A second advantage stems from the fact that the dielectric constant in the low-band-gap material of the center region is higher than in the high-band-gap material of the outer regions. The center region can therefore function as a dielectric waveguide. Since light can be guided within the region in which it is generated, improvements in external quantum efficiency are possible.

5.4 Light-Emitting Diodes II: Devices

The principles upon which the behavior of light-emitting diodes is based are important for understanding their capabilities and limitations and for interpreting new developments as they are reported. These principles are reviewed in the earlier sections of this chapter. This section describes some specific types of LED devices that have proved to be useful and summarizes some of the design considerations that the communication system designer must face. The topics to be covered include the structure of two important types of LED devices, the semiconductor materials used in them, and some consideration of the lifetime and reliability of the LED.

5.4.1 *Device Structures*

The physical structure of the light-emitting diode must provide means for causing current to flow in the light-producing regions of the diode and for coupling the light produced to the output interface. Of course, there are many other details to be considered. The ultimate device will result from design decisions that will sacrifice some aspects of the performance parameters in order to optimize others.

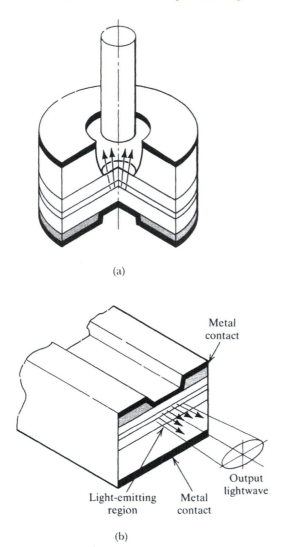

(a)

(b)

Figure 5.4.1 *Two basic LED structures, (a) the surface-emitting, or Burrus, LED and (b) the edge-emitting LED. In both devices, the size of the active region is determined primarily by the pattern of current flow through the active layer. (a) The lower metallic contact is insulated from the semiconductor except in a circular area, approximately the same diameter as the output fiber. (b) The upper metallic contact is insulated from the semiconductor except for a narrow stripe running the length of the device. Adapted from* Semiconductor Devices for Optical Communication, *H. Kressel, editor, Topics in Applied Physics, vol. 39, second edition, Springer-Verlag, Inc., Heidelberg, 1982 [3.4].*

There are two principal classes of light-emitting diodes, based on the way in which the light output is collected from the active region. They are the edge-emitting diode and the surface-emitting diode, illustrated schematically in Figure 5.4.1. The surface-emitting LED takes its output from the light passing through one of the large-area surfaces; it can have a higher external quantum efficiency than does the edge-emitting LED. The edge-emitting LED provides its output from light that propagates parallel to the junction toward the output interface; the intensity of the output is higher and the optical output beam width smaller than for the surface-emitting LED.

We have seen that the light is generated by recombination radiation in the immediate vicinity of the *p-n* junction. The active region then tends to be very thin, that is, of the order of 1 μm, and to have an area parallel to the junction that can provide the required total volume for the active region. Typical dimensions for the edge-emitting LED are $w = 10$ μm and $L = 300$ μm. For the Burrus LED, the diameter of the active region is comparable to that of the fiber to which the output is coupled.

A typical structure for the edge-emitting LED is shown in Figure 5.4.2. The lower metallic contact is in contact with the semiconductor material along a stripe 13-μm wide and running the full length of the structure. This structure is common in both LEDs and lasers and is called the stripe-contact, or simply stripe, LED. The active region is sandwiched between two similar layers having higher band gaps and lower indices of refraction than it does; this is therefore a double-heterostructure device. These layers of lower index constitute a slab

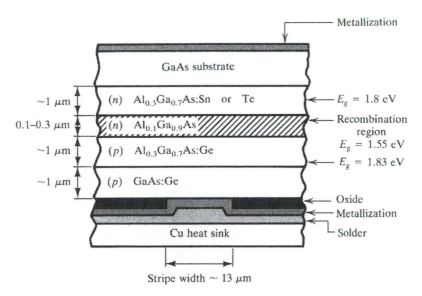

Figure 5.4.2 *Cross section of an N-n-P double heterostructure edge-emitting LED. From* Semiconductor Devices for Optical Communication, *H. Kressel, editor, Topics in Applied Physics, vol. 39, second edition, Springer-Verlag, Inc., Heidelberg, 1982 [3.4].*

waveguide that tends to confine a fraction of the recombination radiation to the active region. The output is taken from one end, as shown schematically in Figure 5.4.1(b).

A schematic diagram of a surface-emitting LED is shown in Figure 5.4.3. The surface-emitter is commonly designed to couple its output light directly into an optical fiber; the fiber pigtail is a part of the device as it comes from the manufacturer. The lower contact has cross section similar to that of the fiber to which it is matched. Although the surface emitter could be designed as a homostructure device, it would usually use a double heterostructure because of the better internal quantum efficiency and bandwidth available with the *DH* configuration. The surface-emitting LED is often called the Burrus LED, in recognition of its developer.

The Burrus LED has the advantage over the edge-emitting LED of higher external quantum efficiency and higher output power. Because its output is taken from a smaller cross section, the edge emitter can have a more intense beam.

The edge emitter has an output beam with width 120° in the direction parallel to the junction and 30° normal to the junction. The beam width of the Burrus diode is 120° in both directions. The narrower width from the edge emitter is attributed to the guiding properties of the double heterostructure.

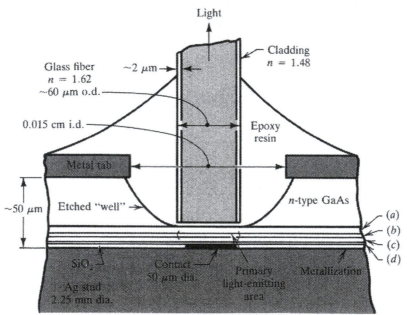

Figure 5.4.3 *A surface-emitting LED. This is an N-p-P double heterostructure AlGaAs diode on a GaAs substrate. (b) is the active, narrow-band-gap region. (a) and (c) are the wide-band-gap regions. (d) is a thin p-type GaAs layer, for contact purposes. From Burrus and Miller,* Optics Communications, *vol. 4, pp. 307-309, Elsevier Science Publishers, Amsterdam, 1971 [3.5].*

5.4.2 *Materials*

The first consideration in selecting the semiconductor material is that its band gap energy must be suitable for producing light of the desired wavelength. Beyond that, the secondary considerations include the index of refraction, the recombination coefficient, the technology for producing material of adequate purity, the ease of fabricating devices with this material, and so forth.

The materials that have become dominant for LED and laser devices are gallium arsenide, GaAs, and other materials from groups III and V of the periodic table of elements. In all such devices there would be at least one element each from groups III and V, and in many cases there will be three or four elements in the crystal. In the ternary and quaternary compound semiconductors, there must still be at least one element from each group. These are called III-V compound semiconductors; they are a very important class of semiconductor material for many types of semiconductor devices.

Among the more useful III-V compound semiconductor materials for LEDs and lasers are GaAs, GaP, InAs, and InP. Each of these materials will form a single crystal, analogous to the single-element crystals formed by silicon and germanium. Each has a characteristic band gap and lattice constant. The band gaps (wavelengths) available from these materials range from 1.35 eV (3.5 μm) for InAs to 2.26 eV (0.55 μm) for GaP. Properties of several semiconductor materials are given in Appendix C.

Ternary compounds are formed by using one element from one group with carefully determined proportions of two elements from the other group. GaAlAs is an example of a ternary compound. GaAs and AlAs have lattice constants that differ by less than 0.15 percent. By mixing appropriate proportions of gallium and aluminum, it is possible to make a material with band gap (wavelength) anywhere in the range from 2.16 eV (0.57 μm) for AlAs to 1.42 eV (0.87 μm) for GaAs. A ternary material is represented as $Ga_xAl_{1-x}As$, where x is the fraction of gallium in the Ga-Al mix.

Quaternary materials can be mixed in a similar manner and written in the form $In_xGa_{1-x}As_yP_{1-y}$, where x and $(1-x)$ give the proportions of the group III elements and y and $(1-y)$ the proportions of the group V elements.

Figure 5.4.4 displays the lattice constants and band gap energies for several III-V semiconductor materials. Charts similar to this are useful in understanding which materials should be used to form a high-quality crystal with the desired band gap energy. The first requirement is that the lattice constants for two crystals must be closely matched. When the lattice match has been assured, then the proportions of the two crystals can be determined. For example, it can be seen that GaAs and AlAs have very nearly the same lattice constant. The proportions of GaAs and AlAs are chosen according to where along the line from GaAs to AlAs the desired operating point is located.

The wavelengths of most importance for silica-based optical fibers are 1.3 and 1.55 μm. GaAlAs materials are not useful in this range. Figure 5.4.4 suggests that In, and especially InAs, should be useful in designing electroluminescent devices for these longer wavelengths. The InGaAsP quaternary compounds use a mixture of indium and gallium from group III and arsenic and phosphorous

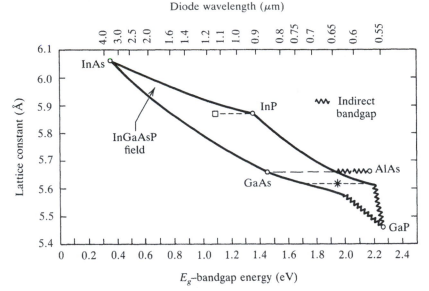

Figure 5.4.4 *Lattice constant versus LED band-gap energy and wavelength. Lattice-matched single-crystal compound semiconductors can be made to cover the spectrum from 0.57-μm to 3.5-μm wavelengths. For the shorter wavelengths, AlGaAs material on a GaAs substrate would be suitable. For the 1.3- and 1.55-wavelength regions where silica fibers have attractive properties, InGaAsP/InP quaternary compound materials are commonly used. The square, lattice-matched to an InP substrate, is $In_{0.8}Ga_{0.2}As_{0.35}P_{0.65}$; an LED made of this material would have bandgap energy of 1.1 eV and wavelength of 1.13 μm. From* Semiconductor Devices for Optical Communication, *H. Kressel, editor,* Topics in Applied Physics, *vol. 39, second edition, Springer-Verlag, Inc., Heidelberg, 1982 [3.4].*

from group V. The area identified as the InGaAsP field shows the range of wavelengths and lattice constants that this compound material encompasses.

The dark rectangle within the InGaAsP field will serve to illustrate the principles of the quaternary compounds. It represents a material having the same lattice constant as InP and a band gap energy of 1.1 eV. It is then possible to use InP as the substrate material and to grow onto that substrate a crystal having the same lattice constant and the desired band gap. The approximate proportions of indium and gallium and the proportions of arsenic and phosphorous can be estimated from this chart by viewing it as resembling a parallelogram and drawing a grid within this parallelogram. The upper side is 100 percent indium and the lower side 100 percent gallium. The left side represents 100 percent arsenic and the right side 100 percent phosphorous. the material represented by the dark square is $In_{0.8}Ga_{0.2}As_{0.35}P_{0.65}$[3.41].

Electroluminescent devices can be and are made of single elements. Such materials have the advantage of relative simplicity but the major disadvantage that they are restricted to a single wavelength. To the extent that this wavelength is useful, such devices will be useful. Some find application as sources of visible light. For optical fiber communications, however, single-element light sources are not available at the wavelengths for which LEDs and lasers are required. Compound semiconductor materials have the overwhelming advantage that they can be made to cover the entire range of interest for current optical fiber applications. As the new fluoride fibers emerge, with the requirement for light wave sources at longer wavelengths, materials will evolve to provide sources at these wavelengths.

5.4.3 *Reliability*

The reliability and lifetime of system components, especially the active devices, are important considerations in system design. Reliability of the LED can be defined in terms of the probability that the LED will perform according to specifications or the probability that it will fail to perform as required. One common unit for expressing reliability is mean time before failure. Failure can be discussed under two types, catastrophic failure and gradual degradation of performance.

Catastrophic failures are due to such factors as physical or mechanical damage or severe overload. The construction and packaging of system components and subsystems can protect them against many of the mechanical and environmental hazards that can be anticipated. To a limited extent, the designer can protect the system from overload failures by the use of current-limiting and voltage-limiting circuits or devices and by providing redundancy in the system so that it can continue to function even though some system components have failed. For example, the repeaters in the transatlantic optical fiber communication system, TAT-8, have four laser transmitters for each channel; when a laser fails, a spare can be switched in to replace it.

Gradual degradation of performance, that is, aging, is a more predictable and manageable mode of failure. The LED is said to have failed when its power output falls below some specified level; the failure level is set at a level where the system will perform satisfactorily but with a margin of safety that is unacceptable. Its lifetime is defined as the time period over which the LED can be expected to maintain power output above this specified level. When it falls below this level, it is said to have failed, even though it may have failed by only a very small amount. The degradation will surely continue, so that once it has failed in this sense, it will no longer satisfy the requirements of the system.

The gradual degradation of LED and laser devices might be expressed in terms of time before a 50 percent drop in output power, for example, or time before some other performance parameter falls below its critical level. Figure 5.4.5 shows an example of such data. The system designer must determine how much degradation is acceptable for the system under consideration. However, the designer cannot determine with absolute certainty that the lifetime of the system, or a subsystem, will satisfy the design objectives. Ultimately, the best one can do is to find an acceptable probability that this objective will be met.

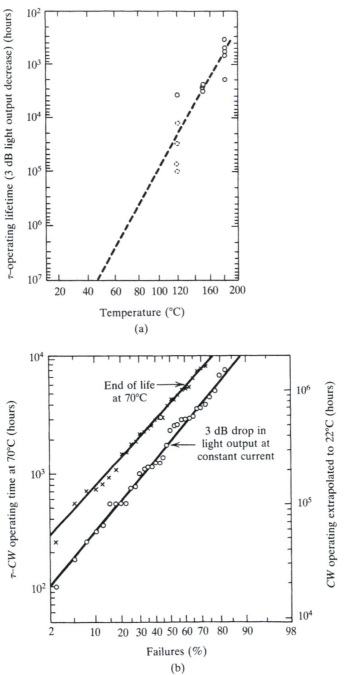

Figure 5.4.5 (a) *Lifetime versus temperature for 1.3-μm InGaAsP/InP edge-emitting LEDs [3.6]. Data taken at high temperatures in accelerated life tests are extrapolated to estimate lifetimes for operation at lower temperatures. From Ettenberg et al.,* IEEE journal of Lightwave Technology,*vol. LT-2, pp. 1016–1023; copyright c 1984 IEEE [3.6]. (b) Distribution of operating time to end-of-life for AlGaAs stripe-contact diodes [3.4]. This can be interpreted as indicating that 50 percent of these diodes should operate for at least $3.5 \cdot 10^5$ hours at 22°C with less than a 3-dB drop in light output. From* Semiconductor Devices for Optical Communication, *H. Kressel, editor, Topics in Applied Physics, vol. 39, second edition, Springer-Verlag, Inc., Heidelberg, 1982 [3.4].*

As with many engineering design decisions, the "acceptable" level will be a compromise between the cost and the value of various alternatives.

The reasons for the aging in LEDs are not completely understood, but many aspects of aging are evident from lifetime test data and from research on aging phenomena. Aging in LEDs is usually characterized by reduction of the internal quantum efficiency. The degradation rate in heterojunction LEDs increases with operating current density as J^m, where $m = 1.5$ to 2. It is evident that for long life and/or high reliability, one should consider operating the LED at current levels below the maximum ratings specified by the manufacturer. It is also recognized that high temperature operation will accelerate aging. Temperature control, by minimizing heat generated within the device and by providing heat sinks to remove unavoidable heat, should be considered when designing for long lifetime.

SUMMARY

Optical power is produced in a forward-biased p-n junction diode by the radiative recombination of holes and electrons. The power is proportional to I, the input current, and hf, the energy of the photon. The constant of proportionality is the quantum efficiency, η.

$$P_0 = \eta \frac{I}{q} hf \quad \text{(W)}$$

The quantum efficiency is the product of η_{int}, an internal quantum efficiency, and η_{ext}, an external quantum efficiency. The internal quantum efficiency is the ratio of photons generated to electron-hole recombinations. External quantum efficiency is the ratio of photons appearing in the output to the photons generated.

The energy of the radiated photons is $\mathscr{E}_{ph} = \mathscr{E}_g + mkT$, where m is of the order of 1. The frequency of the light wave can be found from $\mathscr{E}_{ph} = hf$ (J). The wavelength is $\lambda = 1.24/\mathscr{E}_{ph}$, where $\mathscr{E}_{ph}$ is in eV and λ in μm. The spectral width of the LED light wave is $\Delta\lambda = (\lambda/\mathscr{E}_{ph})\Delta\mathscr{E}_{ph}$, where $\Delta\mathscr{E}_{ph}$ is of the order of $3kT$ (J).

Because the relationship between P and I is linear, the LED can be intensity modulated by modulating the input current. Frequency and phase modulation are not feasible when the light wave source is an LED.

The rate at which the LED can respond to a change in excitation can be represented by the recombination lifetime, τ_r. This is also called the minority-carrier lifetime. It is the time constant of an exponential response to a change in excess carrier concentration. When the modulating signal is a pulse, as with PCM, τ_r is the time constant for the rise and the decay of the output light wave pulse. When the modulation signal is a continuous waveform with a Fourier spectrum $M(\omega)$, the envelope of the intensity-modulated light wave will have a Fourier spectrum $M(\omega)/(1 + j\omega\tau_r)$. The modulation bandwidth is $\omega_{3\text{ dB}} = 1/\tau_r$.

The double heterostructure (DH) LED consists of a narrow-band-gap material, constituting the active region, sandwiched between two wide-band-gap materials. This structure can be used to confine most recombination to the central, narrow-band-gap region. The wavelength of the radiation is determined by the narrow-band-gap energy. The double heterostructure LED has several advantages over the single *p-n* junction diode. The recombination region, from which light is emitted, is well defined by the structure of the device. The internal quantum efficiency can be higher than that of the homojunction diode. Because the wide-band-gap material has lower index of refraction, the double heterostructure constitutes a dielectric waveguide that confines the light wave and results in higher external quantum efficiency. The wide-band-gap material can be made transparent at the wavelength of the light wave, thus reducing attenuation and further increasing the internal quantum efficiency.

LED devices are available in two basic types, edge-emitting and surface-emitting. The edge-emitting LED is usually a stripe-contact device in which the metallic contact that defines the path of current in the active region is a long narrow stripe. The stripe, and therefore the active region, is of the order of 10-μm wide and several hundred μm long. The active region, usually the narrow-band-gap region of a DH structure, is of the order of 1-μm thick. Light generated in the active region propagates parallel to the stripe and is emitted from one end of the active region. Light propagating in other directions is lost. The surface-emitting LED is usually designed to couple light directly into an optical fiber. The surface through which light is emitted is circular, with a diameter similar to that of the fiber. Because the output is taken from a side of larger area, the external quantum efficiency of the surface-emitter can be larger than that of the edge-emitter.

Currently LEDs are made from III-V compound semiconductor materials. To make devices to serve the wavelength regions most useful for optical fiber systems, ternary and quaternary compound semiconductor materials are required. GaAlAs is commonly used for the 0.85-μm region and InGaAsP for the 1.3- and 1.55-μm regions.

The reliability of LEDs and lasers is expressed in terms of aging rates and mean time before failure. High-temperature and high-current-density operation will accelerate aging and shorten useful lifetime. Design for long life and reliable operation would favor operation at temperatures and currents well below the maximum ratings suggested by the manufacturer. System reliability is improved by including redundant circuits and components in the design.

PROBLEMS

5.1 The active region of a 1.55-μm LED is 0.3-μm thick and 50-μm in diameter. The overall quantum efficiency is $\eta = 0.02$. What current density is required to deliver 70 μW to the fiber?

5.2 Show that the optical output of the LED can be expressed as $P_0/I = \eta \mathscr{E}_{ph}$ (W/A) where $\mathscr{E}_{ph}$ is the photon energy in electron-volts.

5.3 A GaAs LED is used as the transmitter for a binary PCM system. It is modulated with a current pulse of 10 mA for a binary *1* and 0 mA for a *0*. The quantum efficiency of the LED is $\eta = 0.1$. What is the power output of the LED when the current is on?

5.4 An LED with $I = 100$ mA at room temperature generates light at $\lambda = 1.3$ μm with total quantum efficiency of 0.02. When reverse biased, the diode has reverse saturation current of 1 pA. Calculate the power efficiency of this diode, that is, the optical power output divided by the electrical power input. Is it surprising to find that the power efficiency is greater than the quantum efficiency?

5.5 The external quantum efficiency of the Burrus LED (see Figure 5.2.1) is degraded by reflections at the dielectric interfaces (a) between the LED and the external medium, and (b) between the external medium and the optical fiber. Assume that the transmissivity T of the LED to the fiber is the product of the Ts at the two interfaces. The indices of refraction are 3.5 for the LED material and 1.5 for the fiber material. Calculate the LED-to-fiber transmissivity, at near normal incidence, when the external medium is (a) air, with $n = 1$, and (b) an index-matching fluid with $n = 2.3$.

5.6 A silicon LED operates at room temperature.
   (a) What are the wavelength and the spectral width of the light wave from this LED?
   (b) This LED is used as the transmitter in a graded-index optical fiber communication system. The fiber has attenuation and dispersion of 1.2 dB/km and -30 ps/km · nm, respectively. What are the total attenuation and dispersion if the length of the fiber is 30 km?

5.7 The wavelengths and spectral widths of the three LEDs of Figure 5.2.3 are, in μm, 1.15 and 0.093, 1.27 and 0.110, and 1.50 and 0.145. Assume that the spectral width is mkT (eV) and find m for each LED.

5.8 (a) Show that, if $\mathscr{E}_{ph} = \mathscr{E}_g + kT$, then

$$\frac{d\lambda}{dT} = -\frac{\lambda^2}{1.24}\left[\frac{d\mathscr{E}_g}{dT} + \frac{k}{q}\right]$$

where $\mathscr{E}_g$, λ, and T are in eV, μm, and °K, respectively.
   (b) The effect of temperature on the band gap is given by [3.1, p. 15]

$$\mathscr{E}_g(T) = \mathscr{E}_g(0) - \left[\frac{\alpha T^2}{T + \beta}\right]$$

For GaAs, $\mathscr{E}_g(0) = 1.519$ eV, $\alpha = 5.405 \cdot 10^{-4}$, and $\beta = 204$. Find $d\lambda/dT$ at $T = 300$ °K.

5.9 An edge-emitting LED has dimensions $w = 10$ μm, $d = 0.5$ μm, and $L = 300$ μm. The recombination lifetime is 0.5 ns. The index of refraction of the material in the active region is 3.7. Output light is taken from one end of the long dimension. When the injected current is turned off, the excess carriers and the photons in the active region will not decay to zero instantaneously. Estimate the time required for the output light to decay to less than 1 percent of its value when the injected current is switched off. What part of the decay time is due to the recombination lifetime and what part to the propagation delay required for photons to exit the active region?

5.10 Plot graphs of $\Delta n(t)$ when $\tau_r = 15$ (arbitrary units) and (a) $J/qd = 10$ for $t > 0$, and (b) $J/qd = 10 + 20 \exp(-t/5)$ for $t > 0$.

5.11 (a) Derive an equation for $\Delta n(t)$ when $J/qd = A + B \exp(-t/\tau)$.
   (b) Show that when $B = 2A$ and $\tau = \tau_r/3$, then $\Delta n(t) = A\tau_r[1 - \exp(-t/\tau)]$.

5.12 A silicon LED with minority-carrier lifetimes of $\tau = 10$ ns is excited by a modulated current of the form $I = I_0 + I_1 \cos \omega_1 t$; $I_0 = 10$ mA, $I_1 = 2$ mA. The output power will have the form $p(t) = P_0 + P_1 \cos(\omega_1 t + \theta)$, where $P_0 = 5$ mW. Find $p(t)$ for $f_1 =$ (a) 1 MHz, (b) 10 MHz, and (c) 100 MHz.

CHAPTER 6

Lasers

Lasers provide a higher-quality light wave source than does the light-emitting diode. For many optical communication system applications, the advantages that the laser can offer are valuable; for some, the LED is adequate.

When it became evident that low-loss optical fibers would make optical fiber communication systems technically and economically feasible, research to develop optical sources and detectors was launched in earnest. The wavelengths that would be optimum for long-distance transmission of light waves through optical fibers were identified and efforts to develop sources at those wavelengths were given high priority. They were successful in developing useful laser and LED sources. Efforts to develop still better lasers and LED sources continue.

The semiconductor junction-diode laser is the dominant laser in optical fiber communication systems. The physical dimensions are measured in tens of microns, as is the diameter of the optical fiber. This general similarity in size makes coupling the light wave from the laser into the optical fiber more efficient than would be possible with lasers of much larger physical dimensions. The spectral and temporal properties of the laser are substantially better than those of the LED, but are still not all that are needed for some applications.

In this chapter, we will summarize the principles on which the laser is based, examine the principles and characteristics of the semiconductor diode laser, and show how this laser is used as the transmitter in an optical fiber communication system.

6.1 Principles of Lasers

The laser is an optical cavity resonator with a gain mechanism that can provide energy to compensate for the losses in the passive cavity. The laser cavity resonator will have many resonant frequencies, each of which can be represented by an impulse response of the form

$$h(t) = A \exp\left(-\frac{\alpha t}{2}\right) \cos \omega t$$

The total impulse response is the sum of several terms, all having this form. The resonant frequencies, or wavelengths, are determined by the dimensions and index of refraction of the cavity. Each resonant frequency corrresponds to a possible mode of oscillation.

The damping constant, α, represents all of the losses, including absorption, scattering, and the output power. There must be some amplification with gain to offset the loss represented by α. The laser will oscillate at each resonant frequency at which the gain is sufficient to overcome the losses.

In this section, we will examine the principles of the laser and develop relationships that can be used to calculate the wavelength(s) and amplitude(s) of its output. In Section 6.2, this general treatment of lasers will be extended to analyze the semiconductor *p-n* junction diode laser.

6.1.1 *The Optical Cavity Resonator*

The optical cavity resonator, shown in Figure 6.1.1, consists of two plane, parallel, reflecting surfaces of infinite extent. This is the model on which our

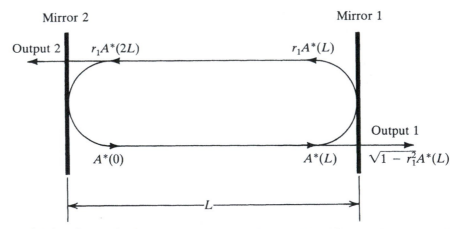

Figure 6.1.1 *An optical cavity resonator. It consists of two plane, parallel mirrors, with reflection coefficients r_1 and r_2, separated by a distance L. This configuration is known in optics as the Fabry-Perot etalon, or Fabry-Perot interferometer. The $A^*(x)$ is the complex amplitude of the electric field vector, defined in Equation (6.1.2).*

analysis of the laser will be based. Some lasers have spherical reflectors. All have some means by which the lateral extent of the optical waves is limited to finite dimensions. However, this simple model is adequate for the purpose of understanding the principles of lasers and it is a good model for studying most aspects of semiconductor diode lasers.

We represent the laser as two plane reflectors (mirrors) with a plane electromagnetic wave propagating along the axis normal to the mirrors. The propagating wave is expressed in terms of the electric field phasor.

$$E(t,x) = A \exp \left[-\frac{\alpha_s x}{2} \right] \exp \left[j(\omega t - \beta x) \right] \quad \text{(V/m)}$$
$$= A^* \exp \left[j\omega t \right]. \tag{6.1.1}$$

where

$$A^* = A \exp \left[-\frac{\alpha_s x}{2} \right] \exp \left[-j\beta x \right] \tag{6.1.2}$$

In these equations, α_s is the *intensity* attenuation constant; in representing the attenuation of the electric field phasor, we must use $\alpha_s/2$.

The gain in most lasers is distributed along the length of the laser cavity. It amplifies the wave as it propagates, and is represented by a term of the form $\exp [gx]$. If this gain is included in Equations (6.1.1) and (6.1.2), it will be associated with α_s in the form $\exp [(g - \alpha_s)x/2]$.

To analyze the characteristics of the cavity, we assume that the mirrors have reflection coefficients r_1 and r_2 and are separated by the distance L, the length of the cavity. Let one of the mirrors be located at $x = 0$ and the other at $x = L$. If a wave of the form of Equation (6.1.1) is launched from $x = 0$, propagates to the second mirror and back to the first, reflecting from each mirror, it will have returned to its starting point. The returned wave can be described

$$E(t,0) = r_1 r_2 A \exp \left[(g - \alpha_s)L \right] \exp \left[j(\omega t - 2\beta L) \right] \quad \text{(V/m)} \tag{6.1.3}$$

with frequency ω and complex amplitude

$$A^* = r_1 r_2 A \exp \left[(g - \alpha_s)L \right] \exp \left[-j2\beta L \right] \tag{6.1.4}$$

The condition for steady-state oscillation is that the complex amplitude, that is, magnitude and phase, of the returned wave must be equal to the original amplitude and phase. This gives two equations

$$r_1 r_2 A \exp \left[(g - \alpha_s)L \right] = A \tag{6.1.5}$$

and

$$\exp\left[-j2\beta L\right] = 1 \tag{6.1.6}$$

The first of these equations will enable us to determine which modes have sufficient gain for oscillation to be sustained and to calculate the amplitudes of these modes. The second equation will enable us to find the resonant frequencies of the Fabry-Perot optical-cavity resonator.

From Equation (6.1.5), we find that one condition for oscillation to begin is

$$g \geq \alpha_t = \alpha_s + \frac{1}{L}\ln\frac{1}{r_1 r_2} \tag{6.1.7a}$$

A more common form of this equation uses the reflectivity, R, rather than the reflection coefficient, r. R is the ratio of the reflected to incident power; r is the ratio of reflected to incident electric field intensity. $R = r^2$.

$$g \geq \alpha_t = \alpha_s + \frac{1}{2L}\ln\frac{1}{R_1 R_2} \tag{6.1.7b}$$

The right-hand side of this equation is the total loss. The first term, α_s, includes all of the distributed losses such as scattering and absorption. The second term represents the losses at the mirrors, a lumped loss, that is averaged over the length $2L$ so that it has the same dimensions and can be treated in the same way as α_s and g.

If the gain is equal to or greater than the total loss, oscillations will be initiated. As the amplitude increases, nonlinear (saturation) effects will reduce the gain. The stable amplitude of oscillation is the amplitude for which the gain has been reduced so that it exactly matches the total loss. We will examine the gain characteristics further in Section 6.2.

EXERCISE 6.1

The attenuation constant, α_s, in a GaAs laser material is 600 m^{-1}. The length of the cavity is 500 μm. What is the minimum gain, g, for which lasing can occur? Note that the reflection coefficient, r, for the air-GaAs interface can be calculated using the dielectric constant of GaAs; see Appendix C.

Answer
2860 m^{-1}.

The condition imposed by Equation (6.1.6) is that

$$2\beta L = 2\pi m$$

where m is any integer. By substituting $\beta = 2\pi n/\lambda$, where n is the index of refraction, this condition can be expressed

$$\lambda = \frac{2Ln}{m} \quad \text{(m)} \tag{6.1.8}$$

Since λ/n is the wavelength in the material inside the cavity, the condition imposed by Equation (6.1.8) is that the length of the cavity must be an integral number of half wavelengths.

In all lasers, the gain, g, is a function of frequency. There will be a band of frequencies for which Equation (6.1.7) is satisfied. Within that band, there will usually be several, or many, frequencies for which Equation (6.1.8) is satisfied. At each of these frequencies both of the conditions for oscillation are satisfied and lasing will occur. Each of these frequencies, with its intensity and other characteristics, is called a *mode* of oscillation of the laser. Some lasers are single mode and some lasers are multimode.

It will be useful to know the frequency, or wavelength, separation between the lasing modes in a multimode laser. This can be determined by solving Equation (6.1.8) for successive values for the integer m. The question is most easily addressed by finding the frequency difference between successive modes. By using Equation (6.1.8) and $c = f\lambda$, we can write

$$m = \frac{2Ln}{\lambda_m} = \frac{2Ln}{c} f_m$$

and

$$m - 1 = \frac{2Ln}{c} f_{m-1}$$

By subtracting these equations, we get

$$1 = \frac{2Ln}{c} (f_m - f_{m-1}) = \frac{2Ln}{c} \Delta f$$

and

$$\Delta f = \frac{c}{2Ln} \quad \text{(Hz)} \tag{6.1.9}$$

By using $\Delta f/f = \Delta\lambda/\lambda$, this can be written in terms of λ.

$$\Delta\lambda = \frac{\lambda^2}{2Ln} \quad \text{(m)} \tag{6.1.10}$$

The resonant wavelengths, Equation (6.1.8), and the mode spacing, Equations (6.1.9) and (6.1.10), are based on the approximation of uniform plane waves propagating between the mirrors of the resonator. These are called longitudinal modes. A more realistic, and more complex, analysis must include the effects of the sides, as well as the ends, of the resonator cavity. For each longitudinal mode, there may be several transverse modes that represent one or more reflections at the sides of the resonator as the wave propagates from one mirror to the other. Thompson [3.8] provides an introduction to the subject of transverse modes in semiconductor lasers.

EXERCISE 6.2

In the GaAs laser of Exercise 6.1, the wavelength of the output light wave is approximately 0.80 μm. The gain, g, exceeds the total loss, α_t, throughout the range $0.75 < \lambda < 0.85$ (μm). How many modes will exist in this laser?

Answer
568

6.1.2 *Stimulated Emission*

The physical basis for the LED is *spontaneous* radiative transitions from elevated electron energy levels in the conduction band to their reference levels in the valence band. The physical basis for the laser is *stimulated* radiative transitions of these electrons. When an electromagnetic wave propagating in a material that has electrons at elevated energy levels is incident on one of the high-energy electrons, it can under the proper circumstances stimulate a downward transition. The stimulated radiative transition can cause a photon to be emitted with the same phase and same direction of propagation as the incident electromagnetic wave. The wave thus gains energy, that is, it is amplified. Laser is an acronym for *l*ight *a*mplification by the *s*timulated *e*mission of *r*adiation.

In most classes of lasers, the high-energy electron states are excited states within the atoms of the laser material. Each electron is uniquely associated with a specific atom. When the electron is excited into a high-energy state, it leaves vacant the lower-energy ground state from which it came. In the semiconductor laser, the electrons in high-energy states, that is, in the conduction band, are not associated with individual atoms. However, in most respects the principles of the semiconductor laser and other lasers are essentially the same.

In this section, we will examine stimulated emission from the classical, that is, atomic, point of view, and give examples of lasers based on this phenomenon. In the following section, the semiconductor junction laser will be treated, and the lasers that have found application as transmitters for optical fiber communication systems will be described.

Let us assume that the valence electrons of an atom can be raised from their ground state energy, $\mathscr{E}_1$, to an excited energy level, $\mathscr{E}_2$, by some means not yet identified. Atoms in the excited state have a finite lifetime in that state. They will eventually decay to their original condition. If there are n_2 electrons per unit volume in the excited state, the rate at which electrons will decay spontaneously from the excited to the ground state is

$$r_{sp} = A_{21}n_2 \quad (\text{m}^{-3}\text{s}^{-1}) \tag{6.1.11}$$

The coefficient A_{21} is a function of frequency and properties of the material; it is often referred to as the Einstein A coefficient.

If there is an incident electromagnetic wave of frequency $f_{21} = (\mathscr{E}_2 - \mathscr{E}_1)/h$, with energy density ρ_{21}, then there will be both absorption and stimulated downward transitions caused by this incident wave. This ρ_{21} is both a volume density and a spectral density; its units are joules per meter cubed per hertz or joule-seconds per meter cubed. The absorption will result in the transfer of electrons from the ground to the excited state at the rate

$$r_a = B_{12}\rho_{21}n_1 \quad (\text{m}^{-3}\text{s}^{-1}) \tag{6.1.12}$$

The stimulated downward transitions from the excited to the ground state will be at the rate

$$r_{st} = B_{21}\rho_{21}n_2 \quad (\text{m}^{-3}\text{s}^{-1}) \tag{6.1.13}$$

B_{12} and B_{21} are the Einstein B coefficients.

Absorption will remove energy from the electromagnetic wave and stimulated downward transitions will increase the energy of the wave. The energy from spontaneous decay is not related to the incident wave and will have only a small effect on it.

For a material in equilibrium, the rates of upward and downward transitions must be equal.

$$B_{12}\rho_{21}n_1 = A_{21}n_2 + B_{21}\rho_{21}n_2 \tag{6.1.14}$$

It can be shown [2.4, Chapter 5], that

$$B_{12} = B_{21} = B \tag{6.1.15}$$

and

$$\frac{A}{B} = \frac{8\pi h f^3}{v^3} \tag{6.1.16}$$

In this equation and some that follow, we use v rather than c/n for the internal velocity to avoid using the same symbol, n, for both the index of refraction and the electron concentration.

The net effect of absorption and stimulated emission can be found by combining Equations (6.1.12) and (6.1.13). For example, the net rate of downward transitions is

$$r_d = (n_2 - n_1)B\rho_{21} \quad (\text{m}^{-3}\text{s}^{-1}) \tag{6.1.17}$$

These transitions will provide a net gain in the energy, or power, of the wave.

When the variation of ρ with frequency must be considered, $\rho(f)$, in J-s/m^3, can be interpreted as $\rho g(f)$, where $g(f)$ is a dimensionless normalized frequency spectrum function and ρ is the total energy density in J/m^3. Then $\int g(f)df = 1$ and $\int \rho(f)df = \rho$ (J/m^3). For a sinusoidal wave of frequency $f = f_0$ and zero spectral width, $g(f) = \delta(f_0)$.

To calculate the gain of the wave as it propagates, we first recognize that this energy, ρ, propagating with velocity v, can be interpreted as a wave of power,

$$P = \rho v \quad (\text{W/m}^2) \tag{6.1.18}$$

The power gained by the wave is

$$\frac{dP}{dx} = r_d hf \quad (\text{W/m}^3) \tag{6.1.19}$$

If we now substitute (6.1.17) and (6.1.18) into (6.1.19) and rearrange terms, we can find

$$\frac{dP}{P} = (n_2 - n_1)\frac{Bhf}{v}dx \tag{6.1.20}$$

and

$$P(x) = P(0)\exp[gx] \quad (\text{W/m}^2) \tag{6.1.21}$$

where

$$g = (n_2 - n_1)\frac{Bhf}{v} \quad (\text{m}^{-1}) \tag{6.1.22}$$

The coefficient B can be eliminated from this equation by using Equation (6.1.16) and $A = 1/\tau_{sp}$.

$$g = (n_2 - n_1)\frac{\lambda^2}{8\pi\epsilon_r\tau_{sp}} \tag{6.1.23}$$

Thus, for amplification, it is necessary that $n_2 > n_1$. This is called a population inversion, in contrast to the normal condition for which the lower energy state has the larger concentration.

The means whereby a population inversion is created is called pumping. Energy is somehow supplied to increase the concentration in the excited state, usually by transferring electrons from the ground state into the excited state. There are a variety of pumping methods for various different lasers. Two of these will be described, in connection with the description of specific lasers, in Section 6.1.3.

Amplification by the stimulated emission of radiation is possible as long as the population inversion is maintained. The population is created by pumping at a fixed average rate. When gain is present, the inequality of Equation (6.1.7) is satisfied, lasing will begin, and the strength of the electromagnetic field will increase exponentially. The rate at which excited electrons are stimulated to decay to the lower energy state, emitting a photon, is proportional to the strength of the incident field. Thus, as the intensity of the lasing field builds up, the rate at which electrons are removed from the excited state increases. Since they are being supplied to the excited state by pumping at a fixed average rate, they cannot be removed from that state at a higher average rate. As the intensity increases, the gain will therefore decrease, until the equality of Equation (6.1.7) is statisfied. This self-adjustment of the gain is a nonlinear saturation phenomenon that is present in some form in all oscillators.

6.1.3 *Practical Lasers*

The attainment of a population inversion is not a trivial problem. It is evident from Equations (6.1.12) through (6.1.15) that pumping with an electromagnetic wave at frequency $f = (\mathscr{E}_2 - \mathscr{E}_1)/h$ at best can achieve equal populations at the two levels. In this section, we will examine two practical lasers as illustrations of practical pumping techniques.

The first laser to be operated successfully was a ruby laser. It used absorption from an incident light wave to pump electrons from the ground state to an excited state, but achieved lasing from a different, intermediate excited level. The principles of this three-level laser system are illustrated in Figure 6.1.2. Pumping with an intense light can achieve approximate equality in the populations of the ground and higher excited states. Very rapid spontaneous decay from the upper to the lower excited state, with very slow spontaneous decay from the lower excited state back to the ground state, allows electrons to be accumulated in state $\mathscr{E}_2$. This is called a metastable state. Because of the different rates of excitation and decay in the metastable state, its population can be greater than that of either of the other two states. A population inversion exists between states 2 and 1. That is the lasing transition.

The ruby laser consists of a ruby rod, with ends polished and perhaps coated to function as the end mirrors of the laser cavity resonator. A flash lamp serves as the pump. During the flash, the population of the metastable state builds up until it exceeds the lasing threshold, as given by Equation (6.1.7). When the threshold is exceeded, lasing begins. The inverted population is quickly depleted and lasing stops. The ruby laser can generate several pulses of light output during each flash of the pump lamp.

A refinement of the three-level system uses four electron energy levels. Pumping is from the ground state to level 4. The spontaneous decay from level

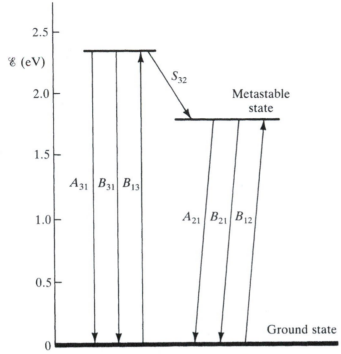

Figure 6.1.2 *Energy level diagram of a three-level laser. Pumping is from the ground state to the higest state, by the absorption of energy from an electromagnetic field. Very rapid decay from the upper state to the metastable state causes a population inversion and lasing transition between the metastable state and ground state.*

4 to 3 and from level 2 to ground are both very fast. Level 3 is the metastable state and the lasing transition is from 3 to 2. This has the advantage over the three-level system that the population of level 2, the lower lasing state, is always small because of the rapid spontaneous decay. Pumping is therefore more efficient and population inversion easier to achieve.

Figure 6.1.3 shows the pumping scheme of the helium-neon (HeNe) laser. The laser structure consists of a glass tube containing a low-pressure mixture of helium and neon. The mirrors are usually outside the tube. The gases are ionized by an electrical discharge. As is indicated in Figure 6.1.3, the helium atoms are excited to level 3 by electron impact. This excitation is transferred to neon atoms by atomic collisions. This state, in neon, is metastable. The lasing transition is from 3 to 2 in neon. The spontaneous decay out of level 2 must be rapid in order to sustain the population inversion.

There are many other lasers that differ in detail, in the materials used, and in the wavelengths and power generated. Some are solid, some liquid, and some gaseous. Most are based on atomic systems using excited electron energy levels, but without removing the electron from the atom. The energy level diagrams are more complex than the illustrations in Figures 6.1.2 and 6.1.3 suggest; the

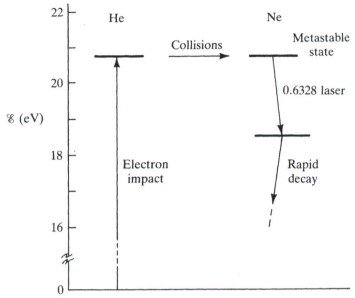

Figure 6.1.3   *Energy level diagram of the helium-neon laser. Pumping is by an electrical discharge and electron impact on helium atoms. The energy of the excited helium atoms is transferred to a metastable state in neon atoms by collisions. The lasing transition is to a lower-energy state that decays very rapidly, making it possible to sustain a population inversion.*

ruby and HeNe lasers represented by those figures actually have several lasing wavelengths, pumping to two or more levels, and decay to and from the lasing levels via intermediate states not shown on these diagrams [2.4, Chapter 7].

The semiconductor junction diode laser is similar to those described above in that the lasing transition provides amplification by the stimulated emission of radiation; it also requires a population inversion before it can provide this amplification. It is very different from most lasers in the way it achieves the requisite population inversion. The population inversion is between the conduction band and the valence band; the valence band represents the ground state and the conduction band is an excited state. By forward-biasing the diode, electrons are provided from the external circuit into the conduction band. Pumping is provided by a direct-current source. In Section 6.2, we will study the physical basis for the semiconductor laser and, in Section 6.3, its characteristics as an optical communication transmitter.

6.1.4   *Coherent Light Waves*

One great advantage of the laser over the LED is the coherence of its output light wave. We will briefly discuss the meaning and significance of coherence; a careful definition and thorough analysis of coherent light waves are beyond the scope and objectives of this book [3.7].

An LED, an incandescent lamp, and a "red-hot" object are sources of incoherent radiation. Such sources emit individual photons in a random, uncorrelated manner. The resulting optical outputs have no consistent temporal structure in the sense that the output from a microwave oscillator does. A sinusoidal wave, $e(t,z) = E \exp j(\omega t - \beta z + \phi)$, with E, ω, and ϕ constant, has a high degree of coherence. The output of an LED cannot be represented in this form; it has a low degree of coherence.

A more realistic characterization of the output of a microwave oscillator, or laser, has the form of the $e(t,z)$ above but recognizes that E, ω, and ϕ are not constant. They will vary in a random manner that can only be described statistically. The reasons for the random variations include internal noise, the spontaneous emission of photons, and the coupling of random external effects into the oscillator resonator. For example, if a laser operating at a stable frequency and phase received a photon from some source other than stimulated emission, the addition of this photon to the oscillation would cause a small discontinuity in both amplitude and phase. The amplitude disturbance would decay due to nonlinear saturation behavior of the gain mechanism. The phase disturbance would be regenerated and sustained; the stable phase, ϕ, after the disturbance would be different from that before the disturbance.

A pure sinusoid of constant amplitude, frequency, and phase has a spectrum of zero width; its Fourier transform is an impulse in the frequency domain. The random disturbance of the amplitude and phase can be viewed as modulating these parameters with random variables. The width of the modulation spectrum is a measure of how rapidly the amplitude and phase can change. After a time period proportional to the reciprocal of the spectral width, the phase of the wave is essentially independent of its initial value.

Coherence is useful for several reasons. First, it makes feasible a greater variety of modulation and detection techniques, thus providing the system designer with more options. Coherent communication systems have a potential for better performance than is available with incoherent systems. For intensity modulation, coherence of the light wave carrier is of little concern. For frequency and phase modulation, a high degree of coherence is essential. Coherent detection techniques, including heterodyne and homodyne detection, require that the phase of the received carrier and that of the local oscillator be well defined and stable; both light waves must have a high degree of coherence.

For most purposes, the optical properties of coherent light waves are better than those of incoherent light. Coherent light waves can be formed into narrower, more highly directional beams. They can be focused more sharply. Many optical components provide better resolution with coherent rather than incoherent light.

6.2 Semiconductor Lasers

The techniques outlined in the previous section are applicable in principle to the semiconductor junction-diode laser. The A and B coefficients for spontaneous and stimulated emission are applicable, and the necessity for a population inversion to have gain remains. One important difference results from the fact that, in the semiconductor, the excited electrons are not associated with specific host atoms and have no unique ground state to which they decay. In the semiconductor, the probability of spontaneous or stimulated transitions depends on the availability of a suitable state at the new energy level as well as on the A and B coefficients and the energy density of the incident light wave.

In the semiconductor, knowing the electron concentration in the conduction band is not sufficient for writing the probabilities of transitions. Both an electron in the conduction band and a hole in the valence band are required. The existence of an electron at an excited level does not assure that the ground state is vacant, as it does in the atomic system. The probabilities for transitions must be written in terms of the Fermi-Dirac distribution functions. The equations corresponding to (6.1.12) or (6.1.13) would be

$$r d\mathscr{E} = B\rho(\mathscr{E}) f_1 (1 - f_2) d\mathscr{E} \qquad \text{(6.2.1)}$$

where f_1 and f_2 are the Fermi-Dirac distribution functions for the initial and final states, respectively; and it is understood that this must be integrated over all $\mathscr{E}$ for which

$$\mathscr{E}_2 - \mathscr{E}_1 = \mathscr{E} = hf \qquad \text{(6.2.2)}$$

We will not pursue this approach. Another starting point, the rate equations, can be related to the transition probabilities and will better serve our present purposes.

6.2.1 *The Rate Equations*

Rate equations can adequately describe the behavior, and some important limitations, of the laser diode as a communication system transmitter. These equations are simple in form and are easily interpreted. Since our primary objective is to use these devices, an approach that provides insight at the expense of the ultimate in rigor is preferred. Those whose interests are in device development or basic research will require a more thorough and detailed study of the laser dynamics than is presented here [2.4, 3.8].

The approach we will take, and the results derived, have been shown to be valid by both theoretical and experimental results. Lau and Yariv [3.9] have analyzed the validity of the rate equations in the form to be used here and the limits on the use of these equations. The equations represent an average behavior of the active medium within the laser cavity. They are not applicable when the time period is short compared to the transit time of the light wave in the laser cavity. A GaAs laser with length $L = 300$ μm has transit time of approximately 4 ps. These simplified rate equations, and results derived from them, should be questioned when the time scales are shorter than about 10 ps or the modulation bandwidth is greater than 100 GHz.

There are two rate equations, one for the electron density, n, and one for the photon density, ϕ. They are

$$\frac{dn}{dt} = \frac{J}{qd} - \frac{n}{\tau_{sp}} - Cn\phi \quad (\mathrm{m}^{-3}\mathrm{s}^{-1}) \tag{6.2.3}$$

and

$$\frac{d\phi}{dt} = Cn\phi + \delta\frac{n}{\tau_{sp}} - \frac{\phi}{\tau_{ph}} \quad (\mathrm{m}^{-3}\mathrm{s}^{-1}) \tag{6.2.4}$$

The terms in these equations can be justified by bookkeeping entries, that is, accounting for all factors that affect the numbers of electrons and photons. The symbols used are identified as the terms in these equations are discussed in the following paragraphs.

In Equation (6.2.3), the first term indicates that the electron concentration in the conduction band is increased by current flow in the diode. It is the same as the term we used in the rate equation for the LED. The second and third terms are the electrons lost from the conduction band by spontaneous and stimulated transitions, respectively. The second term is equivalent to the $A_{21}n_2$ in Equation (6.1.11), where $A_{21} = 1/\tau_{sp}$. In the third, the stimulated-emission term, the C coefficient incorporates the B coefficients and changes in the units in which the equation is written.

The second rate Equation (6.2.4) shows the stimulated emission term, $Cn\phi$, as a source of photons. The second term is the fraction of the photons produced by spontaneous emission that adds to the energy of the lasing mode; δ is small and this term is often neglected. The last term is the decay in the number of photons due to losses in the optical cavity; it is related to the total attenuation, α_t, the right-hand side of Equation (6.1.7).

The photon lifetime, τ_{ph}, can be related to α_t by recognizing that both represent the same losses, one in the form $\exp(-t/\tau_{ph})$ and the other in the form $\exp(-\alpha_t x)$. The x and t are related by the velocity of propagation, c/n. Thus

$$\tau_{ph} = \frac{n}{c\alpha_t} \tag{6.2.5}$$

Rate equations such as (6.2.3) and (6.2.4) are applicable to any type of laser. The dynamic behavior that they can describe is found in most lasers and is similar in most respects to the results found from these simplified equations.

These rate equations can be used to study both the transient and the steady-state behavior of the laser. We will first examine the steady-state conditions, then the response of the laser to pulse modulation, and finally the bandwidth of the laser under sinusoidal modulation.

6.2.2 *The Steady-State Solution to the Rate Equations*

The steady-state solutions to the rate equations are found by defining the steady-state to be characterized by $dn/dt = 0$; $d\phi/dt = 0$; $n \neq 0$; and $\phi \neq 0$. We will also see the conditions required for the fields in the optical cavity, represented by ϕ, to build up from small initial values; this requires that $d\phi/dt$ be positive when ϕ is small.

From Equation (6.2.4), with $\delta = 0$, we can see that whatever the value of ϕ, $d\phi/dt$ can be positive only if

$$Cn - \frac{1}{\tau_{ph}} \geq 0$$

The n that satisfies the equality is the threshold value. For n larger than this threshold value, ϕ can increase; for smaller n, it cannot.

$$n_{th} = \frac{1}{C\tau_{ph}} \qquad (\text{m}^{-3}) \tag{6.2.6}$$

It is also evident from Equation (6.2.4) that the steady-state value for n is n_{th}. For $\phi \neq 0$, $d\phi/dt = 0$ if, and only if, $n = n_{th}$.

We now define a threshold current as the J required to sustain $n = n_{th}$ when $\phi = 0$. From Equation (6.2.3), this is

$$\frac{J_{th}}{qd} = \frac{n_{th}}{\tau_{sp}} \qquad (\text{m}^{-3}\text{s}^{-1}) \tag{6.2.7}$$

This gives the current required to sustain an excess electron concentration, n, when spontaneous emission is the only decay mechanism. It is the same relation found for the LED, given as Equation (5.1.19).

We can now use Equation (6.2.4) to find ϕ_s, the steady-state photon concentration. By substituting Equation (6.2.7) into (6.2.3), we find

$$0 = \frac{(J - J_{th})}{qd} - Cn_{th}\phi_s$$

and

$$\phi_s = \frac{1}{Cn_{th}} \left[\frac{J - J_{th}}{qd} \right] \qquad (\text{m}^{-3}) \tag{6.2.8a}$$

Since $Cn_{th} = 1/\tau_{ph}$, this could be expressed

$$\phi_s = \frac{\tau_{ph}}{qd}(J - J_{th}) \qquad (\text{m}^{-3}) \tag{6.2.8b}$$

Since ϕ, the number of photons per unit volume, is an inherently positive quantity, a negative ϕ is meaningless. The current must exceed its threshold value for ϕ to be greater than zero, and ϕ is proportional to the amount by which J exceeds its threshold value. Power density, in W/m^2, can be calculated from photon density by recognizing that each photon carries energy hf, and that half of the photons are moving in each of two directions, with velocity c/n.

Equation (6.2.8) is plotted as a dashed line in Figure 6.2.1 and a practical laser characteristic as the solid line. The linear power output versus current characteristic given by Equations (6.2.8) is typical of practical laser characteristics in their normal operating ranges. At higher currents, nonlinear behavior is observed that cannot be described with the simplified rate equations. At currents near threshold, output due to spontaneous emission is present. This could be deduced from Equation (6.2.3) but is not included in the ϕ field of the laser rate equations.

The solid graph of Figure 6.2.1 consists of three parts, a linear LED characteristic for low current, the linear laser characteristic for currents above threshold, and an intermediate section between the linear LED and laser parts. In this curved, intermediate section, the diode is basically an LED, the gain is

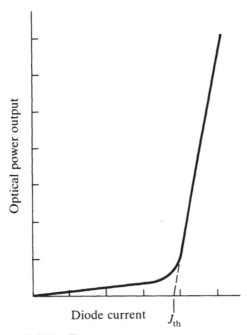

Figure 6.2.1 *The power versus current characteristic of a semiconductor laser. The power and current scales are in arbitrary units. Lasing cannot occur until the current exceeds a threshold value. For currents below threshold, some light from spontaneous emission may be present.*

below its threshold value, but the output due to spontaneous emission is enhanced by some stimulated emission. This intermediate characteristic is used as the basis for some commercial devices, called superluminescent diodes.

EXERCISE 6.3

An InGaAsP laser for 1.3 μm has dimensions $d = 0.7$ μm, $w = 1.3$ μm, and $L = 500$ μm. The distributed attenuation is $\alpha_s = 600$ m^{-1}, $\tau_{sp} = 4$ ns, and $n = 3.5$. (a) Calculate the photon lifetime, τ_{ph}. (b) If $J_{th} = 2 \cdot 10^6$ A/m^2 and $J = 3 \cdot 10^6$ A/m^2, calculate ϕ_s. (c) Calculate the output power density, in W/m^2.

Answers
(a) 3.95 ps. (b) 3.53 $\cdot$ 10^{19} m^{-3}. (c) 1.60 $\cdot$ 10^8 W/m^2.

Modulation of the laser is similar to modulation of the LED. The linear P versus I characteristic provides linear intensity modulation of the laser output power by modulating the input current. For cw modulation, a bias current is set in the linear laser range. A signal current added to the bias current causes the output power to vary about its average value, following the current variations in a linear manner.

For pulsed signals, the bias current can be zero. Current pulses with peak values greater than the threshold current will produce pulses of output optical power. The bias current for pulsed signals is often set at a value near the threshold current because the time required for the output pulse to reach its peak value can be made shorter if the current and the population of electrons in the conduction band do not have to build up from zero.

The dynamic behavior of the laser under sinusoidal modulation and the transient behavior under pulse modulation are treated in Section 6.3.

6.2.3 *Frequency Characteristics of the Semiconductor Laser*

The frequencies (wavelengths) at which the laser will operate are determined by the resonant frequencies of the optical cavity resonator and the frequency range over which stimulated emission can provide sufficient gain. The first of these requirements involves the phase and the second the amplitudes of the waves propagating in the cavity. These relations were stated and discussed briefly in Section 6.1.1.

The phase condition for oscillation is satisfied at or near a resonant frequency. Secondary effects will pull the frequency away from its natural resonant frequency but will not pull it far, except in unusual circumstances. If the phase condition is satisfied at multiple frequencies, then all of these for which there is adequate gain will oscillate. Each such oscillation is called a mode. The optical cavity resonator shown in Figure 6.1.1 has multiple modes. The resonant frequencies derived from Equation (6.1.8) are $f_0 = mc/2Ln$ (Hz), where m is *any* integer. There are an infinite number of resonant frequencies, spaced at intervals $c/2Ln$ (Hz).

The second requirement for oscillation is that the gain at the frequency of oscillation must be greater than the cavity losses at that frequency. In the semiconductor laser, the gain mechanism is stimulated transitions of electrons in the conduction band to available energy levels, that is, holes, in the valence band. Each such transition produces a photon with an energy slightly greater than the band gap and with frequency given by $\mathscr{E}_{ph} = hf$. Since there is a range of energy levels over which these transitions can occur, there is a corresponding range of frequencies at which amplification by stimulated emission is available. This range is equivalent to the spectral line width of the LED. The gain is greatest at frequencies corresponding to energy of the order of kT greater than the band gap and is very small for energies several kT greater than $\mathscr{E}_g$.

For many purposes, the shape of the gain versus wavelength curve can be assumed to be Gaussian.

$$g(\lambda) = g(0) \exp\left[\frac{-(\lambda - \lambda_0)^2}{2\sigma^2} \right] \quad (\text{m}^{-1}) \tag{6.2.9}$$

where λ_0 is the center and σ the width of the gain spectrum. The maximum gain, $g(0)$, is proportional to the population inversion. Within limits imposed by nonlinear phenomena, temperature rise, and so forth, the gain can be increased by increasing the pumping level. In the semiconductor diode laser, the wavelength of maximum gain is determined largely by $\mathscr{E}_g$, the spectral width by temperature, and the gain by the diode current.

If the gain versus frequency curve can be plotted, and the losses are known, then the range of frequencies for which the gain exceeds the loss is readily apparent. The laser will operate at any frequency within this range for which the phase condition is satisfied, that is, at one frequency for each cavity resonance within the range of adequate gain. Figure 6.2.2 shows a typical spectrum of a multimode laser.

EXERCISE 6.4

A GaAs laser for $\lambda = 0.85$ μm has length 350 μm, $\alpha_t = 3200$/m and $n = 3.7$. It has a Gaussian gain characteristic, with peak gain of 5000/m, and spectral width at half maximum gain of 75 nm. What is the value for σ, the width parameter in the Gaussian spectrum function? How many modes will be active in this laser?

Answers
$\sigma = 31.85$ nm; 215 modes.

One simple method for reducing the spectral width and number of modes is to reduce the gain or increase the loss. This would reduce the range over

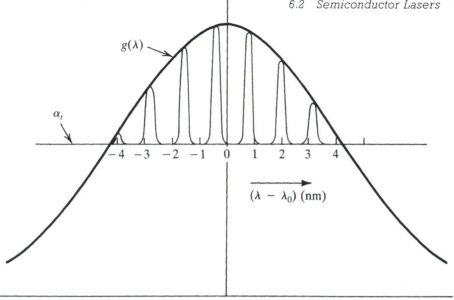

Figure 6.2.2 *Spectrum of the output of a multimode laser. There will be an oscillation at every frequency for which (1) the gain is greater than the losses, and (2) the optical cavity is resonant. The intensity of each mode of oscillation is proportional to the difference between the gain and the cavity losses. In this example, the spacing between resonant modes is 1.2 nm.*

which the gain is adequate for oscillation at the expense of output power. This can reduce the spectral width, but is not a useful way to achieve single-mode operation.

Many lasers, and especially semiconductor lasers, tend naturally to have many modes. For some applications, this is acceptable. However, for many applications a single-mode laser would be preferred. In optical fiber transmission, the light wave from a multimode laser has a greater spectral width, and therefore greater intramodal dispersion, than would a single-mode laser. In coherent communication systems, a single-mode laser is required because the light wave from a multimode laser has inherently poor coherence.

Two types of single-mode lasers have evolved. They are the compound-cavity laser and the distributed-feedback laser. In each, the spacing between modes of the resonator can be made much greater than the $c/2Ln$ (Hz) spacing of the modes of a simple cavity.

The compound-cavity laser, illustrated in Figure 6.2.3(a), uses two cavity resonators coupled through a common mirror. The analysis of the resonant frequencies of the compound cavity is more tedious than that of the single laser. We will follow a simplified, intuitive kind of analysis to illustrate the principles involved. Our results should contribute to understanding why it works, but are far short of providing design equations.

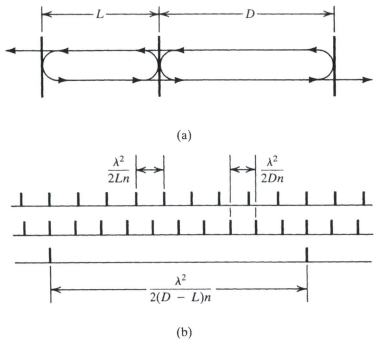

(a)

(b)

Figure 6.2.3 *A compound-cavity resonator. (a) Schematic diagram of the compound-cavity resonator. Two Fabry-Perot interferometers, of lengths L and D, are coupled by light transmission through a common mirror. (b) The resonant frequencies of the L cavity, the D cavity, and the compound cavity. The compound cavity is resonant at wavelengths for which both the L and D cavities are resonant. When L and D are not equal, resonances of the compound cavity can be widely separated making single-mode laser oscillation possible.*

The two resonators of Figure 6.2.3(a), if not coupled, would have resonant frequencies $mc/2Ln$ and $kc/2Dn$, where m and k are integers, L and D are the cavity lengths, and n is the index of refraction. Each couples energy into the other through the common mirror. For resonance, the wave entering either cavity, say cavity D, when returned from the far mirror should reinforce the wave in cavity L, that is, it should have the same phase. The wave propagating in the compound cavity will be stronger if the phase match is exact and will be very weak if the returned phase entering cavity L from cavity D is opposite to that in cavity L. We conclude intuitively that an approximate phase match is needed and that the closer the match, the stronger the oscillation.

The phase match will be closest at any frequencies for which both cavities are at or near a resonance. If L and D are different lengths, as illustrated in Figure 6.2.3(b), there still may be multiple resonances of the compound cavity, but the mode spacing can be made much greater than that of either cavity alone.

The distributed-feedback (DFB) laser uses reflection that is highly frequency selective. At the desired wavelength, the reflectivity of these reflectors is high enough to make α_r small; at other frequencies, the loss is higher than the available gain and the laser does not oscillate.

The frequency-selective reflector is a corrugated grating, as shown in Figure 6.2.4. The wave propagates parallel to the grating. At each discontinuity in the side of the waveguide, some reflection occurs. If the period of the corrugations is equal to one-half of the internal wavelength, then the reflections reinforce. A laser having one side of the active region corrugated in this way and without any end mirrors can oscillate at a frequency defined by the corrugations. Other modes are at frequencies for which the corrugation period is any odd multiple of a half-wavelength. These are not likely to be at frequencies at which the gain is adequate to sustain oscillation.

Distributed reflectors can be placed at the ends of, rather than within, the active region. In this configuration, they serve as reflectors similar in function to the mirrors of the Fabry-Perot resonator, but remain highly frequency selective.

When single-mode operation of the laser has been achieved, the line width of that mode is of concern. The line width is a measure of the degree of coherence of the laser output. The discussion of coherence in Section 6.1.4 pointed out the relationship between line width and degree of coherence. Lasers do not yet have the narrow line widths necessary to make coherent communication techniques practical.

Control of the laser frequency (wavelength) is possible by several means. Equation (6.1.8) shows the parameters that have a direct effect on frequency. They are the length, L, and the index of refraction. In some lasers, control of L by using a piezoelectric transducer to adjust the position of a mirror is feasible and attractive. A more common method for semiconductor lasers is through control of the index of refraction. The index can be influenced by the diode current, by electric fields, and by temperature. An interesting method is the use of a compound cavity, with one cavity providing the gain required for oscillation

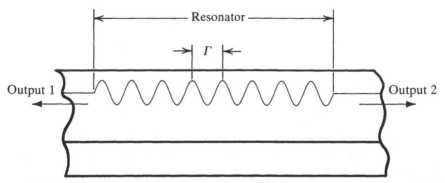

Figure 6.2.4 *A distributed feedback laser. When the corrugation period Γ is one-half the internal wavelength, the corrugated surface will reflect waves traveling parallel to it. Because this reflectivity is highly wavelength selective, the DFB structure makes single-mode laser oscillation possible.*

and the other providing mode-selection and frequency control. A forward-bias current below the lasing threshold current density can offer a relatively easy method for fine-tuning the frequency.

These effects, which are useful for fine-tuning the frequency, are also sources of frequency instability. If the mirror position changes, as by thermal expansion or mechanical vibration, the frequency will change as indicated by Equation (6.2.9). The current, which provides the necessary conduction-band electrons, will affect the index of refraction and thus the frequency. These disturbances in the frequency-controlling parameters can be a serious problem in applications where a stable frequency and narrow line width are necessary.

A final method for frequency control, injection locking, deserves mention because it has received an increasing amount of attention and it may be useful in coherent receivers when the laser technology will support coherent communication techniques. Injection locking is the synchronization of one laser by injecting a light wave from a separate, stabilized laser into the controlled laser's resonator. Applications can include the automatic frequency control of the receiver local oscillator or, in the transmitter, the use of one laser to generate a stable carrier frequency and a second, injection-locked, laser to modulate this carrier. The principles of injection locking are described in [3.11] and the use of injection-locking techniques in semiconductor lasers in [3.12].

6.2.4 *The Double Heterostructure Laser*

Most laser diodes are double-heterostructure (DH) devices. The DH device has made possible lasers with higher efficiencies and lower threshold current densities. Room temperature operation of cw laser diodes would be unlikely without the DH technology.

The conventional *p-n* junction (homojunction) laser has minority-carrier concentration that decreases exponentially from the edge of the depletion region. Recombination and stimulated emission take place within this region of exponentially decreasing excess minority carriers; this region, rather than the depletion region, is the active region in which amplification of the optical wave can take place. The size of this region is ill defined. The optical wave within the laser is in some cases weakly guided by small changes in index of refraction, but is generally not confined to the active region and therefore cannot take full advantage of the amplification available in the active region. The homojunction laser is inefficient. The electron energy level diagram, carrier concentration, index of refraction, and light intensity for a homojunction laser diode are illustrated in Figure 6.2.5.

The double heterostructure laser, illustrated in Figure 6.2.6, takes advantage of two features of the double heterostructure to make possible diode lasers with efficiencies an order of magnitude higher than is achieved in the homostructure laser. In the DH laser, the active region is a narrow-band-gap material sandwiched between two layers of wider-band-gap material(s). The narrow-band-gap material has the higher dielectric constant; the double heterostructure therefore has the structure of a dielectric slab waveguide.

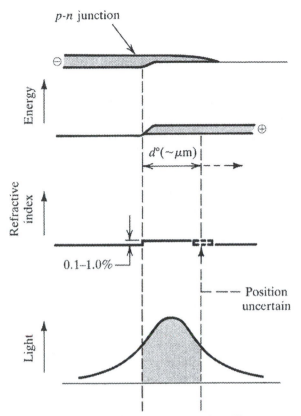

Figure 6.2.5 *Schematic representation illustrating the electron energy-level diagram, index of refraction, and optical field distribution in diffused homostructure laser diodes. The uncertainty in the position and regularity of the right boundary of the active region results from there being no physical structure that determines this position. The energy bands are shown for high forward bias with both the n and p regions heavily doped. Electrons are injected into the p region, but hole injection is negligible. From Panish and Hayashi,* Applied Solid State Science, *vol. 4, Academic Press, 1974 [3.24].*

The carrier-confinement property provided by the narrow-band-gap region results in high concentration of minority carriers in the region between two heterojunctions. The recombination region is thus well defined and fixed in dimension. Because of its wave-guiding property, the DH structure tends to confine the optical wave in the active region, thus providing amplification for a larger fraction of the wave than would be the case with an unguided wave.

Because it has higher efficiency, the DH laser requires smaller input current per unit output power and produces less internal heat. Its threshold current density is smaller. This lower current density, with less heat to be dissipated, makes possible higher average power levels than would be practical with

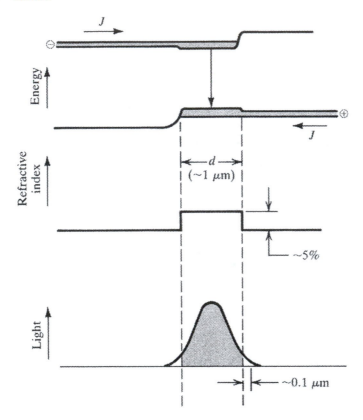

Figure 6.2.6 *Schematic representation illustrating the electron energy-level diagram, index of refraction, and optical field distribution of a double-heterostructure laser diode. The active region is defined by the two heterojunctions. The optical field extends slightly beyond the boundaries of the active region. From Panish and Hayashi,* Applied Solid State Science, *vol. 4, Academic Press, 1974 [3.24].*

lower-efficiency lasers. Continuous operation of the diode laser, a long-sought objective in the early days of optical fiber communications, has been made feasible and readily available by the higher efficiencies of the DH laser.

The stripe geometry can provide a well-defined active region with thickness equal to the thickness of the layer of narrow-band-gap material, width determined by the width of the stripe, and length adequate to provide the necessary gain. A stripe contact laser structure is illustrated in Figure 6.2.7.

A more efficient, and more complex, structure is illustrated in Figure 6.2.8. This is the buried-heterostructure (BH) laser. The width of the active region is defined by wide-band-gap, low-index materials on the sides of the active region. The optical wave is thus guided by a rectangular dielectric waveguide, rather than by the slab of infinite width, thus providing better confinement of the optical wave to the active region and correspondingly better efficiency. The BH

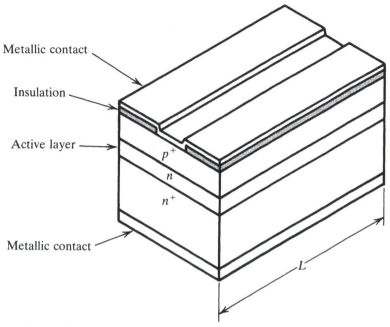

Figure 6.2.7 *The stripe-contact laser. The width of the active region is determined by the path of current flow, which is approximately the width of the gap in the insulating layer.*

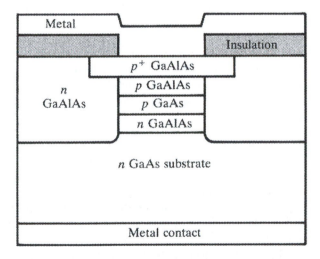

Figure 6.2.8 *A buried-heterostructure laser. The active region, because it is confined on four sides by materials of lower index of refraction, is a rectangular dielectric waveguide.*

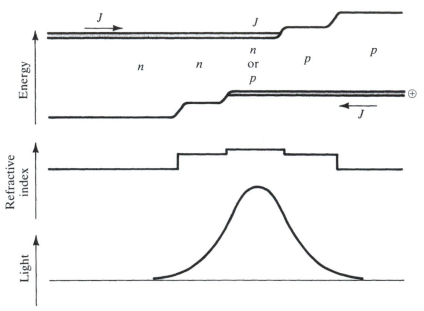

Figure 6.2.9 *The separate confinement heterostructure (SCH) laser, a four-heterojunction structure. The two inner heterojunctions define the active region and the two outer heterojunctions provide optical confinement. From Panish and Hayashi,* Applied Solid State Science, *vol. 4, Academic press, 1974 [3.24].*

laser is still excited by a stripe contact so that the current will be largely confined to the active region.

Some heterostructure lasers have a single heterojunction and some have multiple heterojunctions. The single-heterojunction laser can have attractive properties for high-pulse-power applications. A four-heterojunction laser is illustrated in Figure 6.2.9. The two outside heterojunctions serve as an optical waveguide, making the wave-guiding properties of this structure more effective than those of the DH structure.

6.2.5 *Reliability and Aging*

The discussion of reliability and aging of the LED in Section 5.4.3 is applicable to the semiconductor laser as well. The physical bases for degradation of performance are similar. The manifestations and consequences of aging in laser diodes are different.

A critical parameter in the operation of the laser is its threshold current. Biasing and modulation of the laser are specified in terms of the ratio of the current required and the threshold current. As the laser ages, its threshold current increases. This change must be monitored and compensated so that a drift in

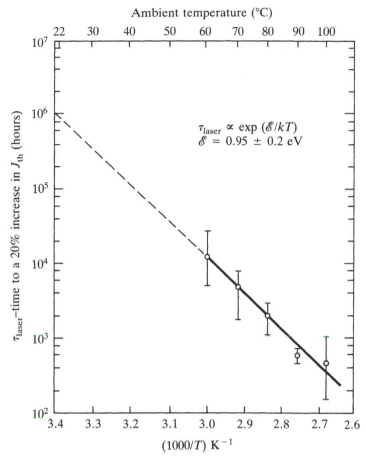

Figure 6.2.10 *Time, at various ambient temperatures, to reach a 20 percent increase in threshold current. The circles are average values determined from devices operated at the indicated temperature; the bars give the range of values. The solid line is a linear "best fit" to the average values; the dashed line is an extrapolation to lower temperatures. From Kressel, Ettenberg, and Ladany,* Applied Physics Letters, *vol. 32, 1978 [3.25].*

J_{th} will not cause the bias point and range of modulating signal current to be moved into regions of nonlinearity or other kinds of degraded performance. Lifetime data for a laser diode are shown in Figure 6.2.10.

In addition to aging, the threshold current is a function of the operating temperature of the laser. As temperature increases, J_{th} increases. Two P versus I graphs for a laser diode showing the effect of temperature are shown in Figure 6.2.11. This change has effects similar to aging, except that they are reversible by reducing the temperature. The designer must be aware that the threshold current is not necessarily stable and must include in the design a tolerance for some drift in J_{th} or some means for monitoring and controlling it.

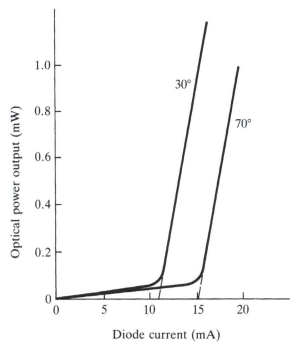

Figure 6.2.11 *Effect of temperature on laser threshold current.*

6.3 Modulation of the Semiconductor Laser

To modulate the semiconductor laser, a current greater than the threshold current is applied and is varied according to the modulating signal. The intensity versus current relationships given in Equation (6.2.8) and Figure 6.2.1 show a region in which this relationship is linear and where conventional intensity modulation is easily achieved. This could be either cw or pulse modulation. However, this linear relationship does not give us any insight concerning the speed of response of the laser output to the modulating current, that is, the rise time for pulse modulation or the modulation bandwidth for cw modulation.

In the following sections, we will use the rate equations to derive the rise time and the modulation bandwidth. Then we will review briefly other light wave modulation techniques that have received some attention in the technical literature and appear to show promise for applications in optical fiber communication systems.

6.3.1 *Pulse Modulation*

The rate equations, (6.2.3) and (6.2.4), can be used to study the dynamic, as well as the steady-state, behavior of the laser. We will seek solutions to these equations when a step function of current is applied to the diode laser.

The simplified rate equations, neglecting the small effect of spontaneous emission, are

$$\frac{dn}{dt} = \frac{J}{qd} - \frac{n}{\tau_{sp}} - Cn\phi$$

and

$$\frac{d\phi}{dt} = Cn\phi - \frac{\phi}{\tau_{ph}}$$

The steady-state solutions to these equations, derived in Section 6.2.2, have established the following relationships:

$$n_{th} = \frac{1}{C\tau_{ph}}$$

$$\frac{J_{th}}{qd} = \frac{n_{th}}{\tau_{sp}}$$

$$\phi_s = \frac{\tau_{ph}}{qd}(J - J_{th})$$

Equation (6.2.8) can be written in several other useful forms by use of (6.2.6), (6.2.7), or other relationships among the steady-state parameters.

First, assume that a step function of J, increasing it from $J_1 < J_{th}$ to $J_2 > J_{th}$, is applied. The initial values for n and ϕ are $n = n_1 < n_{th}$, and $\phi = 0$. The rate equations then reduce to

$$\frac{dn}{dt} = \frac{J}{qd} - \frac{n}{\tau_{sp}} \tag{6.3.1}$$

that is applicable so long as $n < n_{th}$. The solution is

$$n - n_1 = \frac{(J_2 - J_1)\tau_{sp}}{qd}\left[1 - \exp\left(-\frac{t}{\tau_{sp}}\right)\right] \tag{6.3.2}$$

There will be some time delay before n reaches n_{th}. From (6.3.2) we can find this time delay to be

$$t_d = \tau_{sp} \ln\left[\frac{J_2 - J_1}{J_2 - J_{th}}\right] \tag{6.3.3}$$

After n reaches n_{th}, ϕ will start to increase from zero, and n will continue to increase. We must return to Equations (6.2.3) and (6.2.4) to continue the solution.

Let n and ϕ be expressed in terms of their equilibrium values as

$$n = n_{th} + \Delta n \tag{6.3.4}$$

and

$$\phi = \phi_s + \Delta\phi \tag{6.3.5}$$

We expect Δn and $\Delta\phi$, the departures from the equilibrium values of n and ϕ, to decay to zero in a manner to be found by solving the differential equations.

By substituting this n and ϕ into (6.2.3) and (6.2.4), and using (6.2.6) to (6.2.8) to simplify the resulting equations, we can derive the following:

$$\frac{d(\Delta n)}{dt} = -\left[C\phi_s + \frac{1}{\tau_{sp}}\right]\Delta n - Cn_{th}\Delta\phi \tag{6.3.6}$$

$$\frac{d(\Delta\phi)}{dt} = C\phi_s\Delta n \tag{6.3.7}$$

In deriving these equations we have dropped the term $C\Delta n\Delta\phi$, which is assumed to be small.

This pair of equations in Δn and $\Delta\phi$ can be reduced to a single equation in one variable, Δn, by differentiating (6.3.6) and substituting (6.3.7) to eliminate the term involving $\Delta\phi$. Thus

$$\frac{d^2(\Delta n)}{dt^2} + 2\alpha\frac{d(\Delta n)}{dt} + \omega_0{}^2\Delta n = 0 \tag{6.3.8}$$

where

$$2\alpha = C\phi_s + \frac{1}{\tau_{sp}}$$

$$= \frac{\phi_s}{n_{th}\tau_{ph}} + \frac{1}{\tau_{sp}} \tag{6.3.9}$$

and

$$\omega_0{}^2 = C^2 n_{th}\phi_s$$

$$= \frac{\phi_s}{n_{th}\tau_{ph}{}^2} \tag{6.3.10}$$

This differential equation is known to have solutions of the form

$$\Delta n = D\exp(-\alpha t)\sin\omega t \tag{6.3.11}$$

where

$$\omega = [\omega_0{}^2 - \alpha^2]^{\frac{1}{2}} \tag{6.3.12}$$

With this solution for Δn, we can use (6.3.7) to find $\Delta\phi$.

$$\Delta\phi = -\frac{C\phi_s D}{\omega}\exp(-\alpha t)\cos\omega t \tag{6.3.13}$$

where we have assumed that $\exp(-\alpha t)$ is slowly varying with respect to $\cos \omega t$. The coefficient, D, can be determined from initial conditions.

When the initial condition is $n = n_{th}$ and $\phi = 0$ when $t = 0$, the transient solution, applicable only after $n > n_{th}$, is

$$\Delta n = \frac{\omega}{C} \exp(-\alpha t) \sin \omega t \tag{6.3.14}$$

$$\Delta \phi = -\phi_s \exp(-\alpha t) \cos \omega t \tag{6.3.15}$$

The damping constant, α, is usually small compared to ω_0. Then, $\omega \approx \omega_0$ and $\omega/C \approx (n_{th}\phi_s)^{\frac{1}{2}}$. Equation (6.3.14) then becomes

$$\Delta n = (n_{th}\phi_s)^{\frac{1}{2}} \exp(-\alpha t) \sin \omega t \tag{6.3.16}$$

EXERCISE 6.5

The current in a semiconductor diode laser is $J = 0$ for $t < 0$ and $J = 1.5J_{th}$ for $t > 0$. The laser parameters are those given in Exercise 6.3. Calculate α, ω_0, and ω. Sketch graphs of n and ϕ, showing both the time delay and the damped oscillations.

Answers

$3.75 \cdot 10^8 \text{ s}^{-1}$; $5.54 \cdot 10^9 \text{ s}^{-1}$; $5.53 \cdot 10^9 \text{ s}^{-1}$.

Equations (6.3.11) and (6.3.13) describe damped oscillations in both n and ϕ. Graphs of these equations are shown in Figure 6.3.1. The rate equations, coupled by the common term $Cn\phi$, describe a relationship between n and ϕ that has a potential for oscillation. As n increases, the gain and therefore ϕ increase. As ϕ increases, the stronger field intensity depletes the population of electrons and causes n to decrease. The decrease in n leads to a decrease in ϕ, which allows n to increase. This resonant-type relationship, with a time lag between cause and effect, can be expected to oscillate. In the diode laser, the oscillation is damped and ultimately would decay to zero. In some cases, in the laser diode and in other types of laser, the waveshape of the oscillation is more a relaxation oscillation than a sinusoid; the waveshape is different but the phenomena are essentially the same.

Numerical solutions to the more complete rate equations describe transients in n and ϕ that are similar to the graphs in Figure 6.3.1 in most respects. The waveshape of the oscillation may not be sinusoidal, tending toward spikes in ϕ with a corresponding rapid decay in n. These waveforms are sometimes referred to as relaxation oscillations. The time delay, rise time, damping constant, and ringing frequency are similar to those calculated with the simpler rate equations.

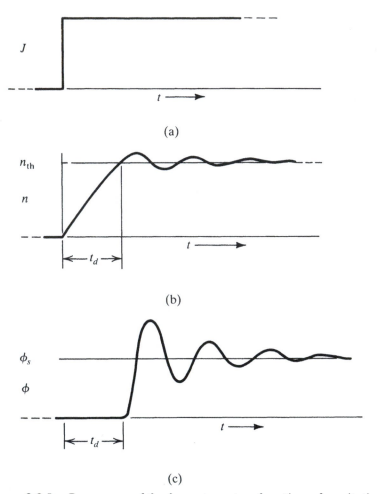

Figure 6.3.1 *Response of the laser to a step function of excitation. The output is characterized by a time delay before any output appears, followed by a damped oscillation.*

6.3.2 *Continuous-Wave Intensity Modulation*

Amplitude and intensity modulation are continuous-wave modulation techniques. The modulating signal and the corresponding variation in light wave intensity are continuous, rather than pulsed, signals. We have seen that the laser has a linear relationship for intensity versus changes in current for currents above threshold.

We have seen in the previous section that the laser responds to changes in excitation at a finite rate. For PCM, the light wave pulses cannot have discontinuous rise and decay, even if the exciting current does. With cw intensity modulation, this finite rate of response has the effect of a limited modulation bandwidth. The bandwidth can be found by using the rate equations with sinusoidal modulation of the photon density. We will assume sinusoidal modulation of the current, J, and calculate the $\phi(t)$ that results. The current is

$$J = J_0[1 + m_J \exp{(j\omega_m t)}] \tag{6.3.17}$$

where m_J is a modulation index.

For reference, let us calculate the modulation that would appear on $\Delta\phi$ with this J when ω_m is 0. From Equation (6.2.8), the modulation on ϕ would be

$$\Delta\phi = \frac{1}{Cn_{th}} \frac{m_J J_0}{qd} \tag{6.3.18}$$

This will be referred to as the low-frequency modulation, $\Delta\phi(0)$.

Equations (6.3.6) and (6.3.7) are still applicable except that one additional term, dJ/dt, must be added to the right-hand side of (6.3.6); in the derivation of (6.3.6) this term did not appear because J was assumed to have its equilibrium value and to be constant. With this change, the equation becomes

$$\frac{d^2(\Delta n)}{dt^2} + 2\alpha \frac{d(\Delta n)}{dt} + \omega_0^2 \Delta n = \frac{j\omega_m m_J J_0}{qd} \exp{(j\omega_m t)} \tag{6.3.19}$$

The particular solution of this differential equation will have the form

$$n = n_{th}[1 + m_n \exp{(j\omega_m t)}] \tag{6.3.20}$$

By substituting this into (6.3.19) we can reduce it to

$$[-\omega_m^2 + j2\alpha\omega_m + \omega_0^2]m_n n_{th} = \frac{j\omega_m m_J J_0}{qd} \tag{6.3.21}$$

The modulation index, m_n, is

$$m_n = \frac{j\omega_m J_0 m_J}{n_{th} qd} \frac{1}{(\omega_0^2 - \omega_m^2) + j2\alpha\omega_m} \tag{6.3.22}$$

The photon density can be found by using this m_n in Equation (6.3.20) and then finding $\Delta\phi$ from Equation (6.3.7).

$$\Delta\phi = \frac{C\phi_s n_{th} m_n}{j\omega_m} \exp{(j\omega_m t)}$$

$$= \frac{C\phi_s J_0 m_J}{\omega_0^2 qd} \frac{1}{\left(1 - \frac{\omega_m^2}{\omega_0^2}\right) + \frac{j2\alpha\omega_m}{\omega_0^2}} \exp{(j\omega_m t)} \tag{6.3.23}$$

Finally, by using (6.2.7) and (6.3.10), this can be reduced to

$$\Delta\phi(\omega_m) = \Delta\phi(0)M(\omega_m)\exp{(j\omega_m t)} \tag{6.3.24}$$

where

$$M(\omega_m) = \frac{1}{1 - \left(\dfrac{\omega_m}{\omega_0}\right)^2 + j\dfrac{2\alpha\omega_m}{\omega_0^2}} \tag{6.3.25}$$

The factor $M(\omega_m)$ gives the relation between the intensity modulation of the output light wave and the modulation of the driving current. If we assume the intensity-modulation process to be linear, as indicated by Equation (6.2.8), then the response to a nonsinusoidal waveform can be found by Fourier transform techniques. $M(\omega_m)$ is a transfer function that is applied to the spectrum of J to find the spectrum of the envelope of the intensity-modulated light wave.

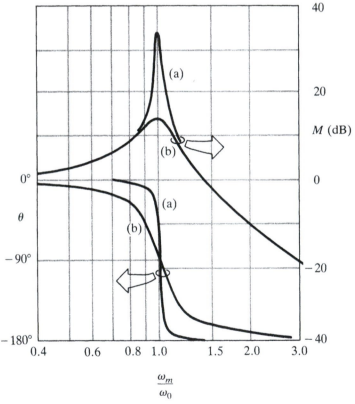

Figure 6.3.2 *Magnitude and phase of the intensity of a modulated laser as a function of the frequency of the modulating signal. The parameters are (a) $\alpha/\omega_0 = 0.01$ and (b) $\alpha/\omega_0 = 0.1$. The bandwidth for modulation is limited to frequencies near or below the natural resonant frequency of the laser.*

Normalized graphs of $M(\omega_m)$ are plotted in Figure 6.3.2 for two values of the parameter α/ω_0. Both show a substantial peak in the magnitude of $M(\omega_m)$ at frequencies near ω_0. This peaking is a result of the same resonant interaction between n and ϕ that causes the self-oscillation in the transient response to a step function of current. It can cause severe distortion to cw modulation waveforms. In many cases, this distortion is unacceptable and must be avoided. It is possible to design lasers with less severe peaking characteristics than those calculated here; current lasers designed for wide-band or high-speed systems applications will have modulation characteristics similar to those shown in Figure 6.3.3. If a severe peak in the modulation transfer function of the laser is unavoidable and unacceptable, the modulation bandwidth must be limited to approximately $\omega_0/2$.

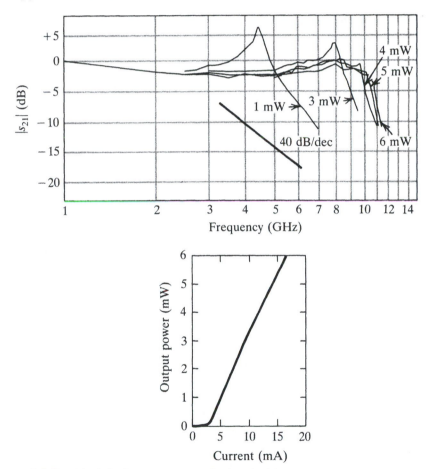

Figure 6.3.3 *Modulation reponse of a buried heterostructure laser with various bias-power levels. The sharply peaked response, illustrated in Figure 6.3.2, is damped and the bandwidth is increased for higher bias power levels. From Lau and Yariv, Chapter 2*, Semiconductors and Semimetals, *vol. 22, Part B., Academic Press, Orlando, 1985 [3.9].*

6.3.3 *External Modulation*

Modulation of the light wave external to the laser is attractive for two reasons:

1. It avoids the degrading effects of direct modulation on laser line width and stability, especially the frequency modulation and mode hopping that are incidental results of direct modulation of the laser.
2. It allows the designer to use modulation schemes that cannot be implemented effectively through direct modulation. For example, phase modulation is relatively simple using external modulation but difficult using direct modulation.

The class of external modulators to be considered here is based on optical integrated circuits that use diffused or implanted dielectric waveguides. Intensity, phase, and frequency modulation, both analog and digital, can be accomplished in this way. There are several other types of light wave modulators that are not discussed here.

Phase modulation appears to be one of the more promising coherent modulation methods. The potential for high receiver sensitivity is the best of any of the standard coherent modulation methods and far better than any incoherent system could offer. A phase modulator can be realized as a dielectric waveguide utilizing the linear electrooptic effect to control the index of refraction of the waveguide. The linear electrooptic effect is one whereby the index of refraction can be changed by the application of an electric field in the dielectric material; the change in index of refraction is proportional to E, the applied electric field intensity. Positive and negative E-fields change the index in opposite directions. A dielectric waveguide phase modulator is illustrated in Figure 6.3.4.

As the index is changed by the modulating signal, the velocity of propagation and the propagation delay change. The propagation time through a waveguide section of length L is nL/c (s). A change of Δn in the index of refraction produces

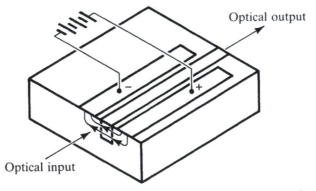

Figure 6.3.4 *An optical phase modulator. Electrodes on the surface of the substrate can produce an electric field in the implanted light guide. This causes a change in the index of refraction and in the velocity of propagation. The propagation time, and therefore the phase delay, is thus controlled by an applied voltage.*

a change in propagation time

$$\Delta t = \frac{\Delta n L}{c} \quad \text{(s)} \tag{6.3.26}$$

and a change in the phase of the output light wave

$$\phi = \omega \cdot \Delta t = \frac{\Delta n \omega L}{c} \tag{6.3.27}$$

The variation in phase is thus proportional to the variation in the index of refraction. Because the Δn is proportional to the applied voltage, the phase modulation, ϕ, is a linear function of applied voltage.

The phase modulator is sometimes characterized by the product of applied voltage and interaction length required to produce a phase shift of π radians in the output light wave. For example, in a phase modulator 5-mm long with $V_\pi L = 20$ V-mm, an applied voltage of 5 volts would produce a phase shift of $5\pi/4$ radians.

EXERCISE 6.6

A LiNbO$_3$ phase modulator has $V_\pi L = 16$ V-mm. What interaction length, L, is required so that an applied voltage of ± 10 V will produce phase modulation of $\pm 2\pi$ radians?

Answer
3.2 mm.

Intensity modulation can be implemented using techniques based on the directional coupler described in Section 4.6. The power division between the two outputs in the conventional directional coupler is fixed by the coupling coefficient and the interaction length; both were considered to be fixed. The coupling coefficient depends on the index of refraction of the waveguide material. Because the index of refraction of an electrooptic material can be controlled to some degree by an applied electric field, the power division can be controlled. If we observe one of the two outputs as we modulate the index of refraction, we see that the intensity varies with the modulating signal. Since the total power output remains constant, the intensities of the two output light waves vary in a complementary manner [3.13, 3.14, 3.15].

LiNbO$_3$ is a very common material for use in optical integrated circuits. Directional couplers can be made by diffusing titanium into the LiNbO$_3$ crystal. The dielectric constant of LiNbO$_3$ can be controlled by an applied electric field, thus varying the coupling and the coupling length of the device.

Figure 6.3.5 shows the structure of a modulator. The unmodulated light wave enters one of the input ports. Modulation is achieved by applying the

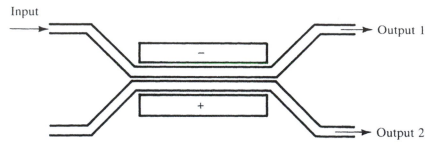

Figure 6.3.5 *A directional-coupler intensity modulator. An applied electric field controls the index of refraction and therefore the coupling coefficient between the two coupled light guides. Because the interaction length is fixed, the energy transferred between the guides varies as the coupling coefficient is varied.*

modulating signal to electrodes, producing an electric field in the dielectric material in and between the two waveguides. The output can be taken from either of the two output ports.

The index-controlled directional coupler can also be used as an optical switch. By switching the index between values that give a 100–0 and 0–100 percent power division, we can turn an output on and off or we can switch the power from one of the output ports to the other. By cascading several such switches, we can build a multi-input, multi-output switch. Figure 6.3.6 shows how a 4 × 4 switch could be made; by the application of the proper control signals to each coupler/modulator, the signal at any input port can be connected to any output port.

A second type of intensity modulator is based on the optical interferometer. The Mach-Zehnder interferometer is the type most commonly used. A conventional Mach-Zehnder interferometer is shown in Figure 6.3.7. The principle of its operation is that the output is the phasor sum of two light waves that travel different paths of potentially different optical path lengths, between a power

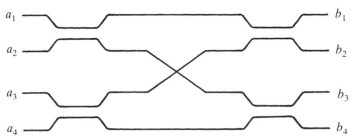

Figure 6.3.6 *A 4 × 4 optical switch, constructed from four directional couplers. Each directional coupler can be controlled so that it can couple all of the power in either input port to either of the two outputs. The four couplers in the 4 × 4 switch can be set so as to connect any one of the inputs to any one of the outputs.*

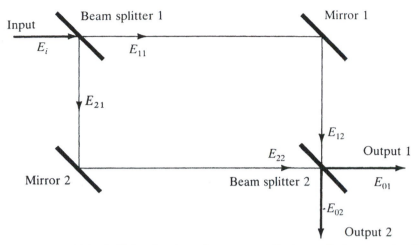

Figure 6.3.7 *A Mach-Zehnder interferometer. An input light wave is divided by beam splitter 1 into two waves that travel different optical paths to the output. The second beam splitter recombines the two waves to produce outputs that are functions of the difference in optical path lengths.*

divider at the input and a combiner at the output. The phasor sum can be relatively large or small, depending on the relative phase of the light waves that have traveled through two different paths. To use the Mach-Zehnder interferometer as an intensity modulator, we place a phase modulator in one arm of the interferometer so as to control the difference in optical path lengths of the two arms.

The beam splitter causes a $\pi/2$ phase difference between the phasors representing the reflected and transmitted waves. In the 50–50 beam splitter, the reflection and transmission coefficients can be represented as

$$r = \frac{-1}{\sqrt{2}} \exp\left(j\frac{\pi}{4}\right) \tag{6.3.28a}$$

and

$$t = \frac{1}{\sqrt{2}} \exp\left(-j\frac{\pi}{4}\right) \tag{6.3.28b}$$

At the first beam splitter, the output electric-field phasors will be $E_{11} = tE_i$ and $E_{21} = rE_i$. These field phasors can be used to confirm that the total power in the two outputs of the beam splitter is equal to the input power.

The phase delays, corresponding to the propagation delay, in the two arms of the interferometer are ϕ_1 and ϕ_2. The phasors at the inputs to the second beam splitter are

$$E_{12} = tE_i \exp(-j\phi_1) \tag{6.3.29a}$$

and

$$E_{22} = rE_i \exp(-j\phi_2) \qquad \text{(6.3.29b)}$$

The two outputs of the second beam splitter must be treated separately. In output 1, the electric-field phasor is

$$\begin{aligned} E_{01} &= rE_{12} + tE_{22} \\ &= rtE_i\left[\exp(-j\phi_1) + \exp(-j\phi_2)\right] \end{aligned} \qquad \text{(6.3.30)}$$

Note that if the two arms have the same optical length, the magnitude of the electric-field phasor is E_i. The power output from output 1 is equal to the power input to the interferometer. When the two arms are not of equal optical length, let the difference in phase delay be $\Delta\phi = (\phi_1 - \phi_2)$ and the average be $\phi_0 = (\phi_1 + \phi_2)/2$. The output electric field is then

$$\begin{aligned} E_{01} &= rtE_i \exp(-j\phi_0)\left[\exp\left(-j\frac{\Delta\phi}{2}\right) + \exp\left(j\frac{\Delta\phi}{2}\right)\right] \\ &= -E_i \cos\left(\frac{\Delta\phi}{2}\right)\exp(-j\phi_0) \end{aligned} \qquad \text{(6.3.31)}$$

The power is proportional to the square of the magnitude of the E field. Then

$$\begin{aligned} P_{01} &= P_{in}\cos^2\left(\frac{\Delta\phi}{2}\right) \\ &= \tfrac{1}{2}P_{in}[1 + \cos(\Delta\phi)] \end{aligned} \qquad \text{(6.3.32)}$$

The power in output 2 of the second beam splitter can be found in a similar manner:

$$\begin{aligned} E_{02} &= tE_{12} + rE_{22} \\ &= E_i\left[t^2 \exp(-j\phi_1) + r^2 \exp(-j\phi_2)\right] \\ &= \tfrac{1}{2}E_i \exp(-j\phi_0)\left\{\exp\left[-j\frac{(\Delta\phi + \pi)}{2}\right] + \exp\left[j\frac{(\Delta\phi + \pi)}{2}\right]\right\} \\ &= -E_i \sin\frac{\Delta\phi}{2}\exp(-j\phi_0) \end{aligned} \qquad \text{(6.3.33)}$$

The power is proportional to the square of the magnitude of E.

$$\begin{aligned} P_{02} &= P_{in}\sin^2\frac{\Delta\phi}{2} \\ &= \tfrac{1}{2}P_{in}[1 - \cos(\Delta\phi)] \end{aligned} \qquad \text{(6.3.34)}$$

Equations (6.3.32) and (6.3.34) describe the division of power between the two outputs as a function of the phase difference in the arms of the interferometer. Note that the total power output is equal to the power input.

Intensity modulation, or switching, is done by placing a phase modulator in one or both arms of the interferometer. The effective phase modulation can be increased by using a push-pull configuration in driving the interferometer. Phase modulators are placed in both arms of the interferometer and are driven with modulating voltages of opposite polarities. The phase is advanced in one arm and retarded in the other. The net phase difference is twice that produced in either arm.

Two ways of building a Mach-Zehnder interferometer in optical integrated circuit form are illustrated in Figure 6.3.8. The first uses two directional couplers, one to divide the input light wave into the two paths, and the other to recombine the waves from the two paths to form the output light wave. The second method, in Figure 6.3.8b, divides and recombines the waves with a Y-branch waveguide structure at each end. In each case, phase modulators in one or both arms of

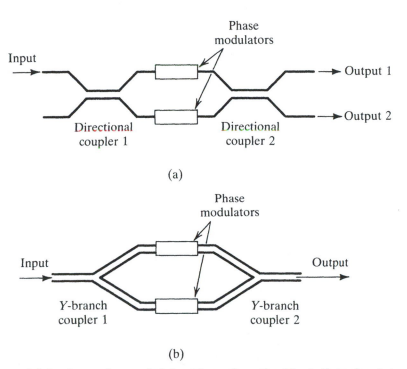

(a)

(b)

Figure 6.3.8   *Intensity modulators based on the Mach-Zehnder interferometer. (a) The interferometer beam splitters are directional couplers. Phase modulators in one or both of the two optical paths can vary the difference in optical path lengths, thus producing intensity modulation of the outputs. (b) The beam splitters are Y-branch waveguide couplers. Phase modulators in the two optical paths can produce phase differences and intensity modulation.*

the interferometer can be used to intensity-modulate the output, as indicated by Equations (6.3.32) and (6.3.34).

Recall that the phase difference, ϕ, is proportional to the applied voltage; the index change is proportional to the voltage and the phase change is proportional to the index change. Because the intensity modulation is represented by $\cos(\phi)$, the transfer function, P/V, is nonlinear. A point of inflection, and therefore the most linear operating point, is at $\phi = \pi/2$.

EXERCISE 6.7

An intensity modulator is made by using a Mach-Zehnder interferometer with a phase modulator in each of its two arms; the phase modulators are driven in a push-pull configuration. Each phase modulator has $V_\pi = 12$ V. The modulating voltage is $V = V_b[1 + mx(t)]$, where V_b determines the operating point and $mx(t)$ represents the modulating signal. (a) Select V_b to set the operating point at the point of inflection in the P/V curve. What other values of bias voltage would place the operating point at a point of inflection? (b) What limits must be placed on the magnitudes of $mx(t)$?

Answers
(a) 3 V; 9 V, 15 V, etc. (b) For $mx(t) \ll 1$, the distortion will be small; for $|mx(t)| > 1$, severe distortion will occur.

This method of phase or intensity modulation is limited in its speed of operation by the transit time of the light wave in the modulator. This can be

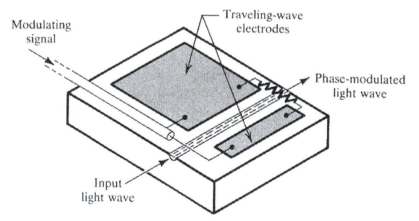

Figure 6.3.9 *A traveling-wave phase modulator. The modulating signal is applied to the electrodes at the input end of the modulator and travels, parallel to the light guide, to the output end. If the velocities of propagation of the light wave and the modulating signal can be made approximately equal, the bandwidth of the phase modulator can be made very large.*

a severe limitation in high-speed systems. The speed, and hence the data rate or modulation bandwidth, can be improved by using a traveling-wave modulator. The electrodes through which the modulating signal is applied can be designed as a short transmission line, fed at one end and terminated with a load resistance at the other end. If the velocity of propagation of the modulating signal on this transmission line can be made equal to the group velocity of the light wave in the waveguide, then the transit time is not a limitation. Wideband, high-speed, traveling-wave modulators are possible and are used when the need for the high data rates they make possible justifies the added cost and complexity that the more complex structure will involve. A traveling-wave modulator is illustrated in Figure 6.3.9.

SUMMARY

The laser consists of an optical cavity resonator and an amplifying mechanism capable of compensating for the losses of the passive resonator. In the semiconductor laser, the optical resonator is normally a Fabry-Perot interferometer with two plane parallel mirrors.

The losses in the passive resonator include absorption, scattering, and transmission through the mirrors. For a laser with internal attenuation of $\alpha_s(m-1)$ and mirror reflectivities R_1 and R_2, the total loss can be expressed

$$\alpha_t = \alpha_s + \frac{1}{2L} \ln \frac{1}{R_1 R_2} \quad (\text{m}^{-1})$$

Amplification in the laser is provided by the stimulated emission of radiation. Electrons at excited energy levels can be stimulated to decay to a lower-energy state by an incident light wave, emitting a photon in the process. The energy of the photon and the frequency of the light wave are related by

$$\mathcal{E}_{ph} = hf = \frac{hc}{\lambda}$$

The amplification of the light wave as it propagates in such an amplifying medium can be expressed

$$P(x) = P(0) \exp{(gx)} \quad (\text{W/m}^2)$$

where

$$g = (n_2 - n_1) \frac{\lambda^2}{8\pi\epsilon_r \tau_{sp}} \quad (\text{m}^{-1})$$

and n_1 and n_2 are the electron densities at energy levels 1 and 2. The photon energy is $\mathcal{E}_{ph} = \mathcal{E}_2 - \mathcal{E}_1$.

One of the requirements for lasing to begin is that $g > \alpha_t$. This requires that $n_2 > n_1$, which is the inverse of their normal relationship; it is called a population inversion. A population inversion requires some kind of nonequilibrium condition to create and sustain it. Means for sustaining a population inversion are called pumping. With an adequate pumping rate, the population inversion will produce enough gain to cause lasing to begin, and the laser field will increase in strength. Saturation ultimately causes the gain to decrease until the gain and total loss are equal; this is the steady-state condition.

The frequency (or wavelength) of oscillation is determined by the resonant frequencies of the optical resonator. The resonant condition is that the length of the resonator is an integral number of internal half-wavelengths.

$$L = \frac{m\lambda}{2n} \quad \text{or} \quad \lambda = \frac{2Ln}{m}$$

The laser will oscillate at all resonant frequencies for which the gain is greater than the total losses. When there are several such modes of oscillation, the spacing between adjacent modes is

$$\Delta f = \frac{c}{2Ln} \quad \text{or} \quad \Delta \lambda = \frac{\lambda^2}{2Ln}$$

In the semiconductor laser, pumping is done by providing a direct current in the forward direction in the junction diode. The photon energy is slightly greater than the band-gap energy, $\mathscr{E}_{ph} \approx \mathscr{E}_g + kT = hc/\lambda$. The wavelength can be expressed

$$\lambda = \frac{1.24}{\mathscr{E}_{ph}}$$

where λ is in μm and $\mathscr{E}_{ph}$ in electron-volts.

The dynamic and steady-state behavior of the laser can be studied using rate equations. These equations express the densities of excess electrons and of photons in terms of the pumping current and parameters of the semiconductor. Solutions to the rate equations can describe the steady-state conditions and the dynamic response of the laser to a disturbance. The steady-state value for photon density is

$$\phi_s = \left(\frac{\tau_{ph}}{qd}\right)(J - J_{th})$$

The power density can be calculated from the photon density.

For PCM transmitters, the laser is modulated with pulses of current. When the current is applied, there is a time delay while the electron density builds up before lasing begins. When the electron concentration reaches its threshold value, lasing begins. The photon density and optical power output increase and will normally exhibit a damped sinusoidal oscillation before reaching the steady-state condition. The resonant frequency and damping constant are

$$\omega_0{}^2 = \frac{\phi_s}{n_{th}\tau_{ph}{}^2}$$

and

$$2\alpha = \frac{\phi_s}{n_{th}\tau_{ph}} + \frac{1}{\tau_{sp}}$$

The frequency of the damped oscillation is

$$\omega = [\omega_0{}^2 - \alpha^2]^{\frac{1}{2}}$$

The damped oscillation of the photon density is

$$\phi = \phi_s[1 - \exp{(-\alpha t)}\cos \omega t]$$

The linear ϕ versus J relationship provides for linear intensity modulation by adding a modulating current to the bias current. The modulation characteristic is sensitive to the frequency of the modulating signal. The low-frequency modulation index must be multiplied by the factor M to find the modulation index at the modulation frequency ω_m.

$$M = \frac{1}{1 - \left(\dfrac{\omega_m}{\omega_0}\right)^2 + j\dfrac{2\alpha\omega_m}{\omega_0{}^2}}$$

This M is sharply peaked at $\omega_m \approx \omega_0$, and is small for $\omega_m \gg \omega_0$. The resonant peak constitutes a form of distortion that can be minimized by restricting the modulation bandwidth to $\omega_m \ll \omega_0$ or by using lasers specifically designed to minimize this resonant-peaking effect.

Direct modulation of the laser has an incidental effect on the frequency and spectral width of the laser output. External modulators can eliminate this un-desired frequency modulation as well as the bandwidth limitations of direct modulation. External intensity modulators are either (1) directional couplers, in which the power division between the two output waveguides is controlled by an applied electric field; or (2) interferometers, in which phase modulators in one or both arms can be used to modulate the intensity of the output. The Mach-Zehnder interferometer, built in the form of an optical integrated circuit with diffused optical waveguides and associated phase modulators, is a common form of interferometer modulator.

The bandwidth of the interferometer is limited by the transit time of the light wave through the interferometer. It can be increased by using a traveling-wave phase modulator in which the velocities of propagation of the modulating signal on a strip-line waveguide and the light wave in the optical waveguide are matched.

PROBLEMS

6.1 The attenuation to a 1.55-μm light wave propagating in an InGaAsP laser material is, without amplification, 575 m^{-1}. The reflection coefficients of the reflectors in the Fabry-Perot interferometer are 0.95 and 0.55. Calculate the gain required in this material in order to make a laser of length $L = 500$ μm.

6.2 A semiconductor laser has $g_{max} = 2000$ m^{-1} and attenuation $\alpha_s = 600$ m^{-1}.
(a) If $R_1 = R_2 = 0.35$, what is the minimum value for the length L?
(b) If $L = 475$ μm, and $R_1 = R_2 = R$, what is the minimum value for R?

6.3 A semiconductor laser has length $L = 350$ μm and index of refraction $n = 3.4$. At what wavelengths near 1.55 μm does this laser satisfy the phase condition for oscillation? What is the spacing, in μm, between these different oscillating modes?

6.4 The light wave inside a laser increases in intensity as it propagates from one mirror toward the other and is decreased in intensity on reflection. For $\alpha_s = 500$ m^{-1}, $R = 0.33$, and $L = 400$ μm, calculate the intensity of the light wave propagating from mirror 1 toward mirror 2 versus the distance from mirror 1. Plot a graph of normalized intensity versus distance, in μm. Show the output intensity on the same graph.

6.5 In a symmetrical laser, with $R_1 = R_2$, there will be a wave similar to that of Problem 6.4 traveling in each direction. The phasor sum of the electric fields of these two traveling waves will produce a standing wave within the laser. For $n = 3.1$ and $\lambda = 1.55$ μm, derive an equation for this standing wave.

6.6 In Equations (6.1.11) through (6.1.17),
(a) What are the units of the A and B coefficients?
(b) Show that these equations are dimensionally consistent.

6.7 A semiconductor laser has length $L = 350$ μm, $R = 0.50$, attenuation $\alpha_s = 300$ m^{-1}, and index of refraction $n = 3.3$. Calculate the photon life-time, τ_{ph}.

6.8 The laser power output is 2 mW when the diode current is 50 mA and is 4.5 mW when the current is 55 mA.
(a) What is the threshold current?
(b) When the current is modulated with a square wave, with $I_{ave} = 50$ mA, $I_{min} = 40$ mA, and $I_{max} = 60$ mA, what will be the max and min output power levels?
(c) If the current is $i(t) = 60[1 + 0.2 \cos \omega t]$ what is $p(t)$?

6.9 An InGaAsP laser diode has dimensions $d = 1$ μm, $w = 10$ μm, and $L = 500$ μm. It operates at a wavelength of $\lambda = 1.3$ μm. The index of refraction is 3.5, the attenuation constant, α_s, is 500 m^{-1}, and the time constant for spontaneous decay, τ_{sp}, is 5 ns. What value of the gain constant, C [in Equation (6.2.3)], is required to sustain an output power of 1 mW when $J/J_{th} = 1.5$? Calculate α_t, τ_{ph}, ϕ_s, J_{th}, n_{th}, and C

6.10 The average photon density in a particular mode at $\lambda = 1.3$ μm is $3 \cdot 10^{18}$ photons/m^3. The dimensions of the laser are $L = 350$ μm, $w = 10$ μm, and $d = 2$ μm. The index of refraction is 3.3 and the mirror reflectivities are $R = 0.9$. What is the power output at each end of the laser?

6.11 A schematic diagram of a laser transmitter is shown in Figure 6P.11. The total laser current is the sum of I_{dc} and i_{ac}. The laser threshold current is $I_{th} = 50$ mA. When $I_{dc} = 75$ mA and $i_{ac} = 0$, the laser power output is $P_0 = 2.5$ mW. What is the ac component of power when $i_{ac} = 10 \cos \omega t$, with $\omega \ll \omega_0$?

Figure 6P.11

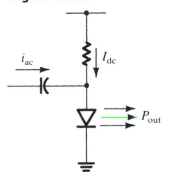

6.12 A semiconductor diode laser has the following parameters:

Wavelength: $\lambda = 1.3$ μm
Dimensions: $L = 300$ μm; $w = 10$ μm; $d = 2$ μm
Index of refraction: $n = 3.3$;
Mirror reflectivity: $R_1 = R_2 = 0.9$
Scattering losses: $\alpha_s = 300$ m^{-1}
Laser gain: $g(\lambda) = 1500 \exp\left[\dfrac{-(\lambda - \lambda_0)^2}{2\sigma^2} \right]$ m^{-1} where $\sigma = 3.5$ nm
Threshold current density: $I_{th} = 50$ mA
Time constant of spontaneous decay: $\tau_{sp} = 5$ ns

At what wavelengths closest to 1.3 μm will the conditions for laser oscillation be satisfied? Calculate one $\lambda < 1.3$ and one $\lambda > 1.3$ μm. Express your answer in nm.

6.13 What is the total attenuation, α_t, for the laser described in Problem 6.12? With the $g(\lambda)$ given, calculate the number of modes that will appear in the laser output.

6.14 Repeat Problem 6.8(c) for $i(t) = 60(1 + 0.01 \cos \omega t)$. The damping constant is $\alpha = 0.1\omega_0$ and the modulating frequency is (a) $\omega = \omega_0/10$; (b) $\omega = \omega_0$; and (c) $\omega = 10\omega_0$.

6.15 For the laser defined in Problem 6.12, calculate the frequency, f, and the damping constant, α, of the transient response in the laser output when the current excitation is a unit step, $i(t) = 85\, u(t)$ mA.

6.16 Show that the frequency and damping constant of the damped oscillations in the laser can be expressed

$$\omega_0 = \left[\frac{1}{\tau_{sp}\tau_{ph}} \left[\frac{J}{J_{th}} - 1 \right] \right]^{\frac{1}{2}}$$

and

$$\alpha = \frac{1}{2\tau_{sp}} \left[\frac{J}{J_{th}} \right]$$

6.17 For a laser with $\tau_{sp} = 4$ ns and $\tau_{ph} = 5$ ps, the input current density is increased from J_1 to $J_2 = 1.75J_{th}$ as a step function. Calculate and plot $\phi(t)$ for $t > 0$ when J_1 is (a) 0, and (b) $0.9J_{th}$. Show the time delay, rise time, and damped oscillation.

6.18 A $LiNbO_3$ dielectric-waveguide phase modulator has $V_\pi = 6$ volts and attenuation of 2 dB. The input light wave has an electric field with peak amplitude $3.5 \cdot 10^5$ V/m. Express the output in the form

$$e(t) = E_0 \cos (\omega t - \phi) \qquad \text{(V/m)}$$

when the modulating voltage is

$$v(t) = 10 \cos \omega_m t \qquad \text{(V)}$$

6.19 An interferometer intensity modulator uses two phase modulators in a push-pull configuration. Each phase modulator has $V_\pi = 8$ volts and each has an applied modulating voltage

$$v_m(t) = 1.0 \cos \omega_m t \qquad \text{(V)}$$

Because the interferometer intensity modulator is nonlinear, there will be harmonics of ω_m in the intensity modulation. Calculate the second and third harmonics of ω_m as percentages of the fundamental component of intensity modulation.

7

Photodetectors

The function of the optical communications receiver that is unique to *optical systems* is photodetection. The photodetector must convert the normally weak optical signal into a correspondingly weak electrical signal. Subsequent stages in the receiver provide amplification and signal processing. The output of the receiver is an electrical signal that meets user-defined specifications concerning signal power, impedance level, bandwidth, and other parameters.

The design of the photodetector subsystem of the receiver includes selecting an appropriate photodetector device, usually a semiconductor photodiode; providing optical components to direct the light wave onto the photodetector device; and designing electrical circuits to couple the output of the photodetector device to the following amplifier or other stages.

The stage following the photodetector is usually an amplifier. The selection of the photodetector device and the design of the first stage of amplification are interdependent decision processes. Because of this interaction between the photodetector and amplifier performance parameters, the first stage of amplification must be regarded as a part of the photodetector subsystem of the optical receiver.

Although they are sometimes used as though they were interchangeable, the words "photodetector" and "photodiode" will have slightly different meanings in our discussions of optical communications receivers. A photodiode is an electronic device, a diode, whose conductivity or *v-i* characteristic is sensitive to the *intensity* of an incident optical wave; the current that flows in the photodiode is a measure of the intensity of the incident wave. The photodetector is that part of the receiver that converts the optical signal into an electrical signal from which the *information* being conveyed by the communication system

205

can be extracted. The photodetector will usually consist of a photodiode and some associated electronic circuits.

This chapter will examine the characteristics of semiconductor photodiodes. The objectives include:

1. Developing an understanding of the physical principles on which the performance of photodiodes is based.
2. Describing the more important characteristics of photodiodes; these include *v-i* characteristics, impedance levels, and response times.

Consideration of the complete photodetector subsystem will be treated in Chapter 8.

Because our objective is to understand the principles of photodiode performance, the physical models used for photodiodes will have simple geometries and the mathematical models will neglect many secondary effects. The student who understands these principles should be able to use manufacturers' published data on commercial photodiodes and to design photodetector subsystems for optical receivers. The introductory treatment offered here is not adequate for designing good photodiode devices. Further advanced study of more complete physical and mathematical models would be required for the more challenging research, development, and design tasks.

7.1 Principles of Semiconductor Photodetectors

As an electromagnetic wave propagates through a semiconductor, the wave may deliver enough energy to a valence electron to free it from the covalent bond that holds it in its place in the crystal structure. When this occurs, an electron-hole pair (EHP) is created and a photon is removed from the wave. The electron that has been excited into the conduction band and the hole in the valence band are now free to move under the influence of an electric field.

If an electric field is present in the region where the EHP has been created, the electron and the hole are accelerated in opposite directions and are swept out of the region. This motion of charges in the semiconductor induces a current that can be detected in the external circuit.

The charges released by the incident photons are called photocarriers. The external current due to the motion of these charges is the photocurrent.

Figure 7.1.1 shows the *v-i* characteristic typical of the semiconductor photodiode. The properties of interest for photodetection are in the third quadrant, corresponding to the diode with reverse bias. There is a small dark current that is due to leakage and to carriers excited into the conduction band by thermal energy. Optical excitation has the effect of producing additional carriers, shifting the entire curve down to larger reverse currents.

In addition to the semiconductor photodiode, there are several other types of photodetector devices, all of which can produce an electrical signal when excited by an optical input. These types include photoconductors, vacuum phototubes, and so forth. For optical fiber communication receivers, the semicon-

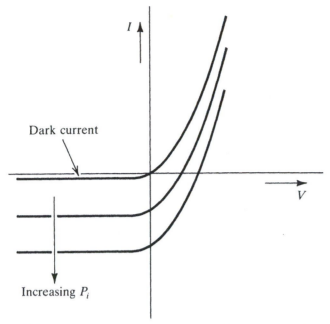

Figure 7.1.1 *Characteristic current-voltage curves for a semiconductor photodiode. Current will flow in the reverse-biased junction diode when energy from an incident light wave is absorbed by the semiconductor material.*

ductor photodiode is the dominant photodetector device. Our attention will be restricted to this class of photodetectors.

The student is assumed to have had some prior familiarity with the physical principles of semiconductor devices. This section will build on that background to study those aspects of the *p-n* junction diode that are most important in understanding the properties of semiconductor photodiodes. For those who need some review of semiconductor devices, references [4.1], [4.2], and [4.3] are suggested.

7.1.1 *Responsivity*

The responsivity $\mathscr{R}_0$ of a photodiode is defined as the output photocurrent produced per unit of incident optical power. Its units are amperes per watt (A/W).

An incident optical power P_i with frequency f is equivalent to P_i/hf photons per second; h is Planck's constant. Let η be defined as the ratio of the average number of electrons excited into the conduction band to the number of incident photons. The average number of electrons per second will then be $\eta P_i/hf$. The photocurrent is

$$I_{ph} = \frac{q\eta P_i}{hf} \quad \text{(A)} \tag{7.1.1}$$

and the responsivity is

$$\mathcal{R}_0 = \frac{I_{ph}}{P_i} = \frac{q\eta}{hf} \quad \text{(A/W)} \qquad (7.1.2a)$$

It is sometimes useful to express $\mathcal{R}_0$ as

$$\mathcal{R}_0 = \frac{\eta q\lambda}{hc} = \frac{\eta\lambda}{1.24} \quad (\lambda \text{ in } \mu\text{m}) \qquad (7.1.2b)$$

EXERCISE 7.1

A silicon photodiode has quantum efficiency $\eta = 0.7$. (a) What is the responsivity of this photodiode for optical excitation having $\lambda = 0.87$ μm? (b) What will be the photocurrent when the incident optical power is 1.5 μW?

Answers
(a) 0.49 A/W. (b) 0.74 μA.

7.1.2 *Quantum Efficiency*

The term η in Equations (7.1.1) and (7.1.2) above is the quantum efficiency of the photodiode. Its value is less than unity and is determined by both the properties of the semiconductor material(s) and the physical structure of the device.

The device in Figure 7.1.2 represents a reverse-biased semiconductor *p-n* junction diode. We assume that the doping levels and cross-sectional area are the same throughout each region and that the junctions are abrupt. In this diode there will be (1) a region around the junction that is completely depleted of free carriers, and (2) an electric field in this depletion region directed from the *n* region toward the *p* region. We will review the characteristics of the depletion region in Section 7.1.4.

Consider an optical wave entering the diode of Figure 7.1.2 from the left through the *n*-type material. W_d is the width of the depletion region and W_1 is the width of the material through which the wave must travel before it reaches the depletion region. The energy that enters the W_2 region has no effect on the photodiode performance; the W_2 term will therefore not enter into the equations.

The optical energy incident on the photodiode can be accounted for with four components:

1. Reflection from the front surface.
2. Transmission through the detector without absorption.
3. Absorption by ionizing collisions in the depletion region.
4. Absorption in other regions and by other means.

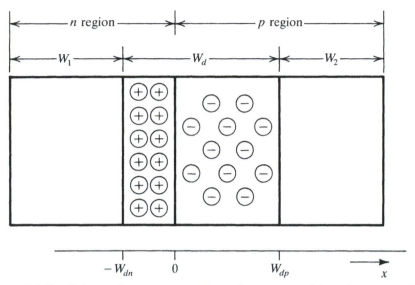

Figure 7.1.2 *Schematic representation of a reverse-biased p-n junction.*

The quantum efficiency η is equal to the ratio of the power absorbed by ionizing collisions in the depletion region to the total incident power.

Reflection at the front surface of the photodiode is represented by a reflectivity R; this factor is defined in Section 3.5. The optical power that passes through the front surface and into the photodetector is $(1 - R)P_i$.

Absorption of energy from the wave as it propagates through the material is represented by an absorption constant α_p. This α_p is essentially the same as the attenuation constant α defined in Section 3.3; if absorption is the only attenuation mechanism present, the two factors are identical. As the wave propagates through the material in the x direction, its power is

$$P(x) = P(0) \exp(-\alpha x)$$
$$= (1 - R)P_i \exp(-\alpha x) \tag{7.1.3}$$

As the wave enters the semiconductor material from the left, it passes first through the n region of thickness W_1 where no electric field exists. Charges produced here will recombine without any significant effect on the photocurrent. However, the wave intensity is attenuated by the factor $\exp(-\alpha W_1)$. The wave then enters the depletion region of width W_d where the useful absorption takes place; the attenuation in this region is represented by $\exp(-\alpha_p W_d)$. The remaining energy leaves the depletion region and makes no further contribution to the current. Combining these factors, we have for the quantum efficiency

$$\eta = (1 - R) \exp(-\alpha W_1) \left[1 - \exp(-\alpha_p W_d) \right] \tag{7.1.4}$$

It is evident from Equation (7.1.4) that for high quantum efficiency we would like R and αW_1 to be small and $\alpha_p W_d$ to be large. Since α and α_p are essentially the same, W_1 should be small and W_d large. For a specific wavelength, R can be made essentially zero by using an antireflection coating on the front surface. A reasonable approximation for quantum efficiency is then

$$\eta = 1 - \exp\left(-\alpha_p W_d\right) \tag{7.1.5}$$

EXERCISE 7.2

The index of refraction of the semiconductor material in a photodiode is $n = 3.5$; the external medium is air. The absorption coefficient of the photodiode material is $\alpha = 10^5 \text{ m}^{-1}$. The width of the depletion region in the photodiode is $W_d = 10 \ \mu\text{m}$ and the thickness of the material through which the light must pass before reaching the depletion region is $W_1 = 10 \ \mu\text{m}$. Find the quantum efficiency of this photodiode.

Answer
$\eta = 0.161$.

7.1.3 *Materials for Photodiodes*

A critical material parameter for the semiconductor photodiode is the band gap—the energy required to move an electron from the valence band to the conduction band. The energy of the photon, hf or hc/λ, must be greater than the band-gap energy $\mathscr{E}_g$. A photodiode is designed to function at a specific wavelength or range of wavelengths. A suitable material must have $\mathscr{E}_g < hc/\lambda$ at all wavelengths in the range.

Band-gap energies and other physical parameters for several important semiconductor materials are given in Appendix C.

The absorption coefficient α will be essentially zero for photon energies less than the band-gap energy. The band-gap energy is a threshold below which the material is transparent—no absorption takes place. Above this threshold the absorption coefficient rises sharply. The wavelength or frequency corresponding to the band-gap energy will sometimes be referred to as the bandedge. Figure 7.1.3 shows curves for the absorption coefficient α_p versus photon energy or λ for several semiconductor materials.

Silicon is the material most commonly used in photodiodes for wavelengths shorter than 1 μm; this covers the wavelengths used in first-generation optical fiber systems. Silicon is not useful for the 1.3- and 1.55-μm wavelengths; its band-gap energy is too large for these longer-wavelength (lower-frequency) spectral regions. From Figure 7.1.3, it appears that germanium should be useful at 1.3 μm and InSb or InAs at 1.55 μm. Other compound semiconductor materials that are finding applications in the longer wavelength spectral regions include InGaAs, GaAlAs, GaAlAsSb, and InGaAsP.

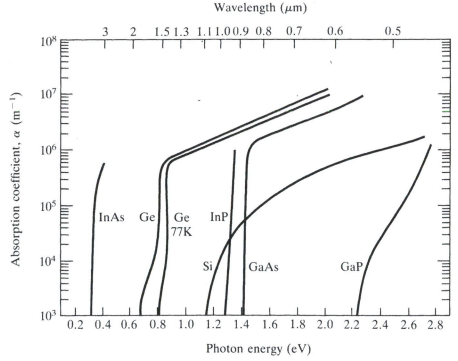

Figure 7.1.3 *The absorption coefficients α for several semiconductor materials. For photon energy less than the band-gap energy, the material is transparent to incident light waves. For photon energy greater than the band-gap energy, energy is absorbed. The absorption coefficient is an exponential attenuation factor that can be used to calculate the energy absorbed. Data are for temperature of 300° K with the exception of one curve for Ge at 77°K. From Stillman and Wolfe, Chapter 5 in* Semiconductors and Semimetals, *vol. 12; Academic Press, 1977 [4.5].*

7.1.4 *The Reverse-Biased p-n Junction*

A semiconductor photodiode is a reverse-biased *p-n* junction diode. Its physical structure can be more complex than the elementary *p-n* junction of Figure 7.1.2, but the internal electric fields that are characteristic of the reverse-biased *p-n* junction are essential for the proper operation of the photodiode. We will examine the elementary *p-n* junction here and two of the more complex structures in Sections 7.2 and 7.3.

A typical *n*-type semiconductor is a crystal of a valence-four element in which valence-five impurity atoms provide some extra electrons beyond those that fit comfortably into the covalent bonds of the natural crystal structure. Most of these extra electrons are thermally excited into the conduction band where, if an electric field is present, they can contribute to current flow. For each free electron in the conduction band, there is a bound positive charge in the nucleus of an impurity atom. Since there are equal numbers of positive and negative charges, the net charge is zero. In the *p*-type semiconductor, the free charges are holes and the bound charges are negative; the net charge is zero. Either type of semiconductor is normally electrically neutral.

In the region around a *p-n* junction, electrons will tend to diffuse from the *n* region into the *p* region. In so doing, they will leave some of the bound positive charges of the *n*-type material uncompensated. Similarly, diffusion of holes from the *p* region toward the *n* region will leave uncompensated negative charges on the *p* side of the junction. There will be some recombination of these electrons and holes in the vicinity of the junction, reducing the total number of free carriers in the device. Thus, there will be bound positive charges on the *n* side that are not compensated by electrons and bound negative charges on the *p* side not compensated by holes. Since the recombination of free electrons and holes will eliminate equal numbers of each, the numbers of uncompensated positive and negative bound charges will be equal. The resulting charge distribution is illustrated in Figure 7.1.4(a). In those regions where a net uncompensated bound charge exists, this charge density will be equal to the doping level, N_D or N_A; N_D is the density of donor atoms in the *n*-type material and N_A is the density of acceptor atoms in the *p*-type material.

There will exist an internal electric field in the junction region due to the uncompensated bound charges. The direction of this field will be from the positive charges to the negative charges, that is, from the *n* region toward the *p* region. This field opposes the tendency of electrons and holes to diffuse. An equilibrium will be reached with a region including the *p-n* junction that is depleted of free carriers and with the net current flow across the junction equal to zero. This depletion region is critical to the operation of photodiodes.

The potential drop across the depletion region that results from the equilibrium E field is the diffusion potential V_d. The polarity of this diffusion potential is in the direction of a reverse-bias voltage across the junction; V_d is therefore negative. The magnitude of V_d is equal to the difference between the quasi-Fermi levels of the *p*- and *n*-type materials divided by q, or it can be calculated from[1]

$$V_d = -\frac{kT}{q} \ln \frac{N_A N_D}{n_i^2} \quad \text{(V)} \tag{5.1.10}$$

When an external voltage is applied to the diode with its positive polarity on the *n* side of the junction, the width of the depletion region will increase. The total voltage across the depletion region is the applied reverse-bias voltage plus the diffusion potential. A very small reverse saturation current due to leakage and thermally generated carriers will flow. In the photodiode this reverse saturation current is the "dark current," the residual current that flows when no light is incident on the input facet of the photodiode.

Among the characteristics of the depletion region that will be of interest in studying photodiodes are (1) the strength of the internal electric field, (2) the

---

[1] This equation was presented and discussed briefly in Section 5.1.3

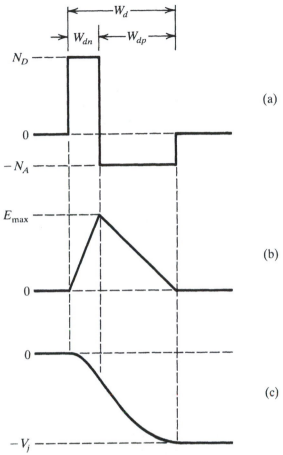

Figure 7.1.4 *Charge distribution, electric field intensity, and potential difference for the p-n junction diode of Figure 7.1.2.*

width of the depletion region, and (3) the magnitude of the applied voltage. These three are interrelated. For given device dimensions and doping levels we can develop the relationships between E and V by using Poisson's equation. The schematic representation of Figure 7.1.2 describes the structure to be analyzed.

Poisson's equation, for variation only in the x direction, is

$$\nabla^2 V = \frac{d^2 V}{dx^2} = -\frac{\rho}{\epsilon} \tag{7.1.6}$$

Since

$$E = E_x = -\frac{dV}{dx} \tag{7.1.7}$$

then

$$\frac{dE}{dx} = -\frac{d^2V}{dx^2} = \frac{\rho}{\epsilon} \qquad (7.1.8)$$

In the depletion region, ρ is the bound charge density; it is equal to the density of the impurity atoms. Thus, the charge density ρ in Poisson's equation is equal to the doping level in the p or n region under consideration. In the n-type material

$$\frac{dE}{dx} = \frac{\rho}{\epsilon} = \frac{qN_D}{\epsilon} \qquad (7.1.9)$$

$$E_n(x) = \int_{-W_{dn}}^{x} \left(\frac{qN_D}{\epsilon}\right) dx$$

$$= \left(\frac{qN_d}{\epsilon}\right)(W_{dn} + x)$$

$$= E_{max}\left(1 + \frac{x}{W_{dn}}\right) \qquad -W_{dn} < x < 0 \qquad (7.1.10)$$

where

$$E_{max} = \frac{qN_DW_{dn}}{\epsilon} \qquad (7.1.11)$$

This equation shows linear variation of E from $E = 0$ at the edge of the depletion region to its maximum value at the junction. A corresponding equation can be developed for the field in the p region.

$$E_p(x) = E_{max} - \frac{qN_A}{\epsilon}x = E_{max}\left(1 - \frac{x}{W_{dp}}\right) \qquad 0 < x < W_{dp} \qquad (7.1.12)$$

These E fields are shown in Figure 7.1.4b.

To find the voltage across the junction, substitute E_n and E_p from Equations (7.1.10) and (7.1.12) into Equation (7.1.7) and integrate. The result is

$$V_j = -\frac{q}{2\epsilon}(N_DW_{dn}^2 + N_AW_{dp}^2) \quad \text{(V)} \qquad (7.1.13)$$

This result could be found directly from Figure 7.1.4 by interpreting the integral of $E(x)$ as the area under the $E(x)$ curve.

The V_j of Equation (7.1.13) is equal to the voltage across the depletion region. Recall that this includes both the applied voltage V and the diffusion potential V_d.

$$V_j = V + V_d \qquad\qquad (7.1.14)$$

If our objective is to establish a specific E_{max} in the depletion region, we can find V_j, then subtract V_d to find the applied voltage V.

EXAMPLE 7.1

A silicon *p-n* junction diode has $N_D = 10^{22}$ (m^{-3}) and $N_A = 10^{21}$ (m^{-3}). For $E_{max} = 10^7$ (V/m), calculate the width of the depletion region and the voltage across it.

Solution

In the *n* region,

$$\frac{dE}{dx} = \frac{qN_D}{\epsilon} = \frac{1.6 \cdot 10^{-19} \cdot 10^{22}}{11.8 \cdot 8.854 \cdot 10^{-12}} = 1.53 \cdot 10^{13} \ \ (\text{V/m}^2)$$

The width required to reach $E = 10^7$ with this slope is

$$W_{dn} = \frac{10^7}{1.53 \cdot 10^{13}} = 0.65 \ \ (\mu\text{m})$$

Similarly, in the *p* region,

$$\frac{dE}{dx} = -1.53 \cdot 10^{12} \quad \text{and} \quad W_{dp} = 6.53 \ \ (\mu\text{m})$$

The total width of the depletion region is

$$W_d = W_{dn} + W_{dp} = 7.18 \ \ (\mu\text{m})$$

If we use Equation (7.1.13), the voltage across the depletion region is

$$V_j = -\frac{q}{2\epsilon}[N_D W_{dn}^2 + N_A W_{dp}^2] = -35.9 \ \ (\text{V})$$

This V_j across the junction is the sum of the applied voltage and the diffusion potential. For this diode, the diffusion potential is -0.6 V. The applied voltage is then

$$V = -35.9 + 0.6 = -35.3 \ \ (\text{V})$$

7.2 The PIN Photodiode

The PIN diode is a *p-n* junction structure with a very lightly doped "intrinsic" region placed between the normal *p-* and *n*-type regions. Typical doping levels and electric field intensities are illustrated in Figure 7.2.1.

By specifying appropriate values for the doping levels and widths of the *p*, *i*, and *n* regions, the intrinsic region can be designed to have essentially complete depletion and high electric field intensity. Both of these properties can be advantageous in photodiodes.

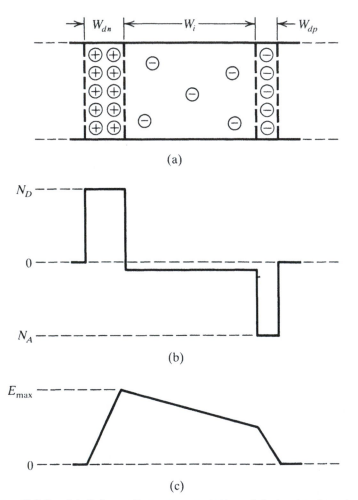

Figure 7.2.1 (a) *Schematic representation,* (b) *doping levels, and* (c) *electric field intensity of reverse-biased PIN photodiode. The relatively low doping level in the center, intrinsic, region provides for a high electric field intensity through most of the depletion region.*

7.2.1   *The Depletion Region of the PIN Diode*

In the PIN diode, the *p*- and *n*-type regions usually have relatively high doping levels and are separated by a region with a much lower doping level. This central region is referred to as "intrinsic," implying no doping at all; in practice, this doping level is a few orders of magnitude smaller than that of the end *p* and *n* regions, but it is not necessarily zero.

For the PIN diode of Figure 7.2.1, the relationships established in Section 7.1.1 for dE/dx and $V(x)$ still hold. To use them, we must apply them to the slightly more complex structure of the PIN device; there are now three regions, rather than just two, in which dE/dx is not zero and across which a voltage drop will be found. As before, we will assume that the doping is uniform throughout each region and that the change in doping at the edge of each region is abrupt.

Figure 7.2.2(c) represents the *E* fields in an n^+-*p*-p^+ PIN diode. This notation indicates that the intrinsic region has some *p*-type impurity atoms in the crystal and that the doping levels of the end n^+ and p^+ regions are considerably higher than that of the *i* region. Expressions for the value of *E* at each end of the *i* region are indicated on Figure 7.2.2(c).

If we use methods from Section 7.1.4, the voltage across the depletion region can be shown to be

$$V_{\text{PIN}} = -\frac{q}{2\epsilon}\left[N_D W_{dn}(W_{dn} + W_i) + N_A W_{dp}(W_{dp} + W_i)\right] \tag{7.2.1}$$

If both W_{dn} and W_{dp} are small compared to W_i, then

$$V_{\text{PIN}} = -\frac{qW_i}{2\epsilon}\left[N_D W_{dn} + N_A W_{dp}\right] \tag{7.2.2}$$

This V_{PIN} corresponds to the V_j of Section 7.1.1 and is the sum of the applied voltage and the diffusion potential. The diffusion potential for the PIN diode can be found from the doping levels N_D and N_A of the n^+ and p^+ regions using Equation (7.1.5) or from the quasi-Fermi levels of these end regions; if the intrinsic region is completely depleted, it has no effect on the diffusion potential.

The normal mode of operation of the PIN photodiode is with the intrinsic region completely depleted, as illustrated in Figures 7.2.1 and 7.2.2. When we know $|E|_{\max}$, W_{dn}, or W_{dp}, we can calculate the applied voltage necessary to establish such an *E*-field pattern. When the applied voltage is substantially less than this, the depletion region may not reach through the *i* region. In such cases the device would behave more like an n^+-*p* diode than a PIN diode. The advantages of the PIN diode for photodetection would be lost. The applied voltage for which the depletion region just reaches the boundary between the *i* and *p* regions is called the "reach-through" voltage. Normal operation would use an applied voltage substantially higher than the reach-through voltage.

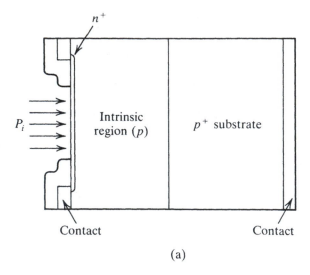

(a)

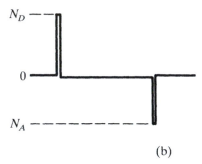

(b)

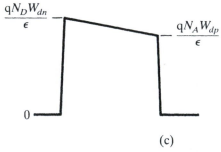

(c)

Figure 7.2.2 (a) *Schematic diagram,* (b) *doping levels, and* (c) *electric field intensity of PIN photodiode. The thin n$^+$ region results in low attenuation of the wave between the input facet and the depletion region, thus improving quantum efficiency and responsivity.*

EXAMPLE 7.2

The doping levels in a silicon n^+-p-p^+ PIN diode are 10^{22}, 10^{19}, and 10^{21} m^{-3} respectively. The width of the intrinsic region is 30 μm. What applied voltage is required to make the maximum voltage in the depletion region of this diode 10^6 V/m? What is the field intensity at the other end of the intrinsic region?

Solution

In the n^+ region, $dE/dx = qN_D/\epsilon = 1.53 \cdot 10^{13}$ V/m^2. Similarly, in the p and p^+ regions $dE/dx = -1.53 \cdot 10^{10}$ and $1.53 \cdot 10^{12}$, respectively.

The width of the depleted region in the n^+ material is

$$W_{dn} = \frac{10^6}{1.53 \cdot 10^{13}} = 0.065 \ \mu\text{m}$$

The ΔE across the intrinsic region is $W_i \cdot dE/dx = 0.46$ MV/m. The field intensity at the other end of the intrinsic region is 1 MV/m, the maximum, less the change across the intrinsic region. This is $1.0 - 0.46 = 0.54$ MV/m.

The width of the p^+ region is 0.352 μm. The voltage across the junction can now be found by using Equation (7.2.1). Thus, $V_{\text{PIN}} = -23.2$ V.

7.2.2 Equivalent Circuit

The equivalent circuit for a *p-n* junction diode or a PIN photodiode consists of a resistance and a capacitance in parallel. The values we should use for R and C are the dynamic values, $R = dV/dI$ and $C = dQ/dV$, that represent the response of the circuit to *signals* superimposed on the static voltage and current of the operating point.

The equivalent resistance of the reverse-biased *p-n* junction can be found from the slope of the characteristic *v-i* curve. This curve is practically flat, giving a very high equivalent resistance. The actual equivalent resistance of a photodiode is more likely to be determined by leakage resistances than by the diode characteristic. In either case it is very high and is usually neglected.

The equivalent capacitance can be found from the charge that moves into or out of the diode in response to a small change in terminal voltage.[2] We can find a functional relationship between Q and V and then differentiate to find dQ/dV.

---

[2] For other discussions of the capacitance of *p-n* junction devices, see [4.1, p. 144–147] or [4.2, p. 173–176].

When the voltage applied across the *p-n* junction diode changes, the width of the depletion region changes. This change in width is equivalent to adding (or removing) a discrete number of conduction electrons and holes to (or from) the *n* and *p* regions, thus changing the number of uncompensated bound charges in the depletion region. The quantity of charge added (or removed) can be calculated from the change in uncompensated bound charges; the change is equal to the doping level (in C/m^3) multiplied by the change in volume of the depletion region.

The total uncompensated bound charge in the *n* region is

$$Q_n = qN_DW_{dn}A \quad (C)$$ (7.2.3)

The total charge in the *p* regions will be the sum of the charges in the *i* and p^+ regions and will be equal to Q_n.

$$Q_p = Q_{p^+} + Q_i = Q_n = Q$$ (7.2.4)

We will assume that $W_i \gg W_{dn}, W_{dp}$ and that Q_i is very small; then $Q_p = Q_n = Q$ and

$$N_DW_{dn} = N_AW_{dp} = \frac{Q}{qA}$$ (7.2.5)

We substitute this into Equation (7.2.2) to get a relationship between Q and V.

$$V = \frac{qW_i}{2\epsilon}\frac{2Q}{qA} = \frac{W_i}{\epsilon A}Q$$

Then

$$C = \frac{dQ}{dV} = \frac{\epsilon A}{W_i}$$ (7.2.6)

The negative sign that appears in Equation (7.2.2), which corresponds to defining reverse bias as a negative applied voltage, has been dropped. It can be shown by examining the polarities of the charges and voltage that the equivalent capacitance is positive.

It should not be surprising to find that this is the equation for the capacitance of a parallel plate capacitor. Our assumption of heavy doping in the n^+ and p^+ regions means that the depletion region does not extend far into these regions. All of the change in charge takes place in these very thin regions that are separated by the much greater distance W_i. Hence, this PIN diode is physically similar to a parallel plate capacitor with plate area A and spacing W_i.

We may derive more exact equations for C by using Equation (7.2.1) rather than (7.2.2) to get the relationship between Q and V. We find that V is equal to a quadratic equation in Q. If we now assume that the n^+ and p^+ doping levels are equal and are much greater than that of the i region, we can find that

$$C = \frac{\dfrac{\epsilon A}{W_i}}{\left[1 + \dfrac{4\epsilon}{NqW_i^2}V\right]^{\frac{1}{2}}} \qquad (7.2.7)$$

Thus, for a sufficiently large V, C varies as $V^{-\frac{1}{2}}$.

A still more general result has the form

$$C = \frac{C_0}{(1 + kV)^m} \qquad (7.2.8)$$

where C_0 and k are functions of dimensions and doping levels, and m is a number in the range $\frac{1}{3}$ to 3. This equation is not restricted to uniform doping and abrupt junctions. For a linearly graded junction, $m = \frac{1}{3}$; for large V, C varies as $V^{\frac{1}{3}}$.

7.2.3 Quantum Efficiency

Quantum efficiency was defined in Section 7.1.2 as the ratio of the optical power absorbed by ionizing collisions in the depletion region to the total incident power. To maximize the quantum efficiency and hence the responsivity, we want to maximize the photon absorption in the depletion region and minimize it elsewhere. In the PIN photodiode, there are two features that can be used for this purpose.

The depletion region includes the entire intrinsic region, with relatively small doping, and relatively small sections of the p^+ and n^+ regions; this tends to be true regardless of the width W_i. By making W_i large, we can ensure that $\exp(-\alpha_p W_i)$ is small, thus minimizing the energy that leaves the depletion region without producing photocarriers. The absorption coefficient α_p should also be large; here we may be limited to the choice of one of only a few materials that will absorb effectively at the wavelength of interest.

The quantum efficiency can also be increased by using heterostructures similar to those described in Section 5.3. If the n^+ material through which the optical wave enters the device can be selected to have a band gap $\mathscr{E}_g > hf$, it will not absorb energy from the wave as it passes through this section of the photodiode. Absorption is thus confined to the region where it is effective in producing photocarriers.

EXERCISE 7.3

A PIN photodiode, with $R = 0.1$ and $\alpha = 10^5 \text{ m}^{-1}$, has $W_{dn} = 3 \ \mu\text{m}$, $W_i = 50 \ \mu\text{m}$, and $W_{dp} = 2 \ \mu\text{m}$, and $W_1 = 2 \ \mu\text{m}$. Find the quantum efficiency.

Answer
$\eta = 0.734$.

7.2.4 *Speed of Response*

The rate at which the photodiode can respond to changes in the intensity of the optical input signal is a measure of the maximum information rate or bandwidth of the photodetector. We will examine the speed of response of the photodiode in the time domain by finding its impulse response. We can then use this result to calculate the frequency response of the photodetector.

The speed of response of the photodiode is limited by the photocarrier transit times in the absorbing region of the device. The carriers are accelerated by the electric fields in the depletion region and will move out of this region with a finite velocity. The transit time is the time required for a photocarrier to move from the point at which it was created to the edge of the region.

The velocities with which the photocarriers move through the depletion region depend on properties of the semiconductor material and on the strength of the electric field. The velocities for electrons and holes, v_e and v_h respectively, are not necessarily equal. As E increases, the velocities will increase until a saturation velocity is reached; above this threshold collisions with the host lattice will prevent further increase in average velocity. Curves for average velocity versus E are given in Figure 7.2.3.

The photocurrent that flows in the external circuit results from the motion of photocarriers in the depletion region. Each individual charge carrier (electron

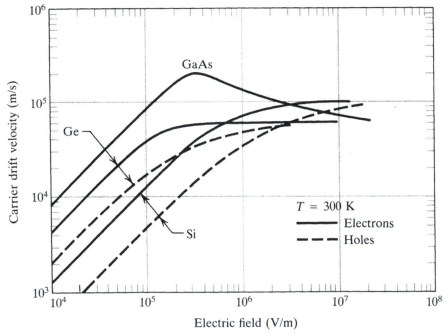

Figure 7.2.3 *Carrier drift velocity versus electric field intensity. From Sze, Physics of Semiconductor Devices, second edition; copyright © 1981, John Wiley & Sons, Inc [3.1].*

or hole) in the depletion region has some small effect on the distribution of free carriers in the electrically neutral n and p regions of the device. As this carrier moves through the depletion region, it will cause some small redistribution of charges in the neutral regions and in the external circuit. Each EHP created by photoionization in the depletion region will ultimately cause one electron to leave the n terminal of the diode and one electron to enter the p terminal. Thus, each EHP will contribute one electron to the external photocurrent.

The impulse response of the photodetector is the photocurrent $i_{ph}(t)$ that flows in the external circuit due to an impulse of optical input. The impulse response is determined by the number of photocarriers produced by the impulse of excitation, the locations at which these photocarriers are produced, and the velocities with which they move out of the depletion region.

Consider the photodiode of Figure 7.2.2. Light enters from the n^+ region at the left. In the depletion region, electrons will move to the left with velocity v_e and holes will move to the right with velocity v_h. The transit times for electrons and holes that must move through the entire distance W_d are $\tau_e = W_d/v_e$ and $\tau_h = W_d/v_h$, respectively. The duration of the impulse response is the larger of these two transit times.

As an electron that was generated at the right-hand side of the depletion region moves from right to left with velocity v_e, it will generate a current pulse of duration $\tau_e = W_d/v_e$ and area $\int i \, dt = q$. The current pulse produced by this electron has average magnitude q/τ_e. Current pulses due to electrons produced elsewhere in the depletion region will have the same average magnitude but shorter durations. If at some time t there are N_e free electrons in the depletion region, then the total photocurrent due to these electrons will be

$$i_e(t) = N_e \frac{q}{\tau_e} = \frac{N_e q v_e}{W_d} \quad \text{(A)} \tag{7.2.9a}$$

Similarly, for holes,

$$i_h(t) = \frac{N_h q v_h}{W_d} \quad \text{(A)} \tag{7.2.9b}$$

Both $i_e(t)$ and $i_h(t)$ will be decreasing functions of time since N_e and N_h will decrease as electrons and holes leave the depletion region. The total photocurrent is the sum of $i_e(t)$ and $i_h(t)$.

$$i_{ph}(t) = \frac{q}{W_d} [N_e(t)v_e + N_h(t)v_h] \quad \text{(A)} \tag{7.2.10}$$

For a first example of an impulse response, assume that the density of EHP generation is uniform throughout the depletion region. The initial values of N_e and N_h are equal.

$$N_e(0) = N_h(0) = N_0 \tag{7.2.11}$$

and

$$i_{ph}(0) = \frac{qN_0}{W_d}(v_e + v_h) \quad \text{(A)} \tag{7.2.12}$$

As electrons are swept to the left with velocity v_e, $N_e(t)$ decreases.

$$N_e(t) = N_0\left(1 - \frac{t}{\tau_e}\right) \quad (0 < t < \tau_e) \tag{7.2.13}$$

$$i_e(t) = \frac{qN_0v_e}{W_d}\left(1 - \frac{t}{\tau_e}\right) \quad (0 < t < \tau_e) \tag{7.2.14a}$$

Similarly,

$$i_h(t) = \frac{qN_0v_h}{W_d}\left(1 - \frac{t}{\tau_h}\right) \quad (0 < t < \tau_h) \tag{7.2.14b}$$

In this example, based on uniform initial distribution of photocarriers, each type carrier produces a triangular current pulse. The $i_e(t)$ pulse and the $i_h(t)$ pulse

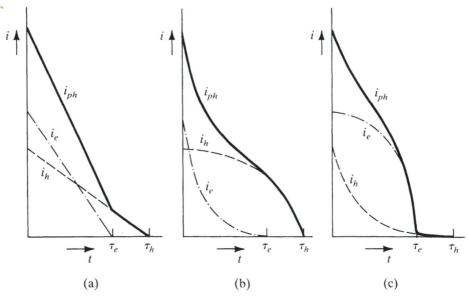

Figure 7.2.4 *Photodiode impulse response. (a) Light is absorbed and EHP produced uniformly throughout the depletion region. (b) Light enters the n-side of the junction and is attenuated exponentially. (c) Light enters the p-side of the junction and is attenuated exponentially. In all three cases, the electron velocity is greater than the hole velocity.*

have different initial amplitudes and different durations. The total impulse response is the sum of these two pulses. This impulse response is shown in Figure 7.2.4(a).

Note that the integral $\int i(t)\, dt = qN_0$. This could have been predicted from the fact that each EHP causes one electron to move through the external circuit.

For a second, more realistic example, assume an exponential absorption characteristic such as was described in Section 7.1.2. The initial distribution of electrons and holes will follow the same exponential equation that describes the optical intensity.

$$N(x) = N(0) \exp(-\alpha x) \quad (\text{m}^{-3}) \tag{7.2.15}$$

The total number of EHP created in the depletion region is

$$N_e = N_h = N_0 = A \int_0^{W_d} N(x)\, dx$$

$$N_0 = N(0)[1 - \exp(-\alpha W_d)] \tag{7.2.16}$$

The calculations for electron and hole currents will be similar for those with uniform initial distributions of photocarriers, except that the $N_e(t)$ and $N_h(t)$ expressions will be different. The initial photocurrent is

$$i_{ph}(0) = \frac{qN_0}{W_d}(v_e + v_h)$$

$$= \frac{qI_0}{W_d hf}[1 - \exp(-\alpha W_d)](v_e + v_h) \quad (A) \tag{7.2.17}$$

where I_0 is the energy of the impulse incident on the depletion region. The duration of the impulse response will again be the larger of τ_e and τ_h. The area under the $i_{ph}(t)$ curve will be

$$\int i_{ph}(t)\, dt = qN_0 = \frac{qI_0}{hf}[1 - \exp(-\alpha W_d)] \quad (C) \tag{7.2.18}$$

To find $N_e(t)$, note that the size of the region containing electrons is decreasing as the entire population of free electrons moves to the left with velocity v_e. This is equivalent to having the left-hand boundary of the region move to the right with velocity v_e. Thus

$$N_e(t) = A \int_x^{W_d} N(x)\, dx \qquad x = v_e t$$

or

$$N_e(t) = N(0)v_e A \int_t^{\tau_e} \exp(-\alpha v_e t)\, dt \qquad 0 < t < \tau_e \tag{7.2.19}$$

Evaluating this integral and substituting the resulting $N_e(t)$ into Equation (7.2.29a), we get for the current due to electrons

$$i_e(t) = \frac{qI_0v_e}{W_dhf} \exp\left(-\alpha W_d\right) \left\{ \exp\left[\alpha W_d\left(1 - \frac{t}{\tau_e}\right)\right] - 1 \right\} \quad (A) \qquad (7.2.20a)$$

An equivalent calculation for the current due to holes yields

$$i_h(t) = \frac{qI_0v_h}{W_dhf} \left\{ 1 - \exp\left[-\alpha W_d\left(1 - \frac{t}{\tau_h}\right)\right] \right\} \quad (A) \qquad (7.2.20b)$$

The impulse response of this photodiode is

$$h_{ph}(t) = \frac{i_e(t) + i_h(t)}{I_0} \qquad (7.2.21)$$

This current, with its electron and hole components, is illustrated in Figure 7.2.4(b).

The shape of this current pulse can be interpreted as follows. The higher density of both electrons and holes is on the left side, that is, the side on which the optical wave entered the active region. Because electrons move to the left and holes to the right, the electrons exit the depletion region starting on the high-density end of the electron population; holes exit starting on the low-density end. Thus, in this case, hole current dominates the total. If the optical impulse had entered on the right rather than the left, the impulse response would have the shape shown in Figure 7.2.4(c).

The impulse response of the photodetector must include the effects of the circuit into which the photocurrent flows as well as the shape of the photocurrent pulse. This circuit includes the equivalent capacitance of the photodiode and external circuit elements. In an elementary photodetector, the load impedance consists of an R and a C in parallel. The photocurrents given by Equations (7.2.20) flow through this Z_{RC} to produce an output voltage. The impulse response of this photodetector can be found using Laplace transform methods.

The Laplace transform of $i_e(t)$, from Equation (7.2.20a), is

$$I_e(s) = \frac{A}{s + \beta} \left\{ 1 - \exp\left[-(s + \beta)\tau\right] \right\}$$

$$- \frac{A \exp\left(-\beta\tau\right)}{s} \left[1 - \exp\left(-s\tau\right)\right] \qquad (7.2.22)$$

where $\beta = \alpha W_d/\tau_e$ and $A = qI_0/\tau_ehf$. The impedance of the RC load is

$$Z(s) = \frac{1}{C(s + \gamma)} \quad \text{where} \quad \gamma = \frac{1}{RC} \qquad (7.2.23)$$

The component of the photodetector impulse response due to electron current is then

$$V_e(s) = \frac{A}{C} \left\{ \left[\frac{1}{(s + \beta)(s + \gamma)} \right] \{1 - \exp[-(s + \beta)\tau]\} \right. $$
$$\left. - \frac{\exp(-\beta\tau_e)}{s(s + \gamma)} [1 - \exp(-s\tau_e)] \right\} \tag{7.2.24}$$

The inverse Laplace transform gives, for $0 < t < \tau_e$

$$v_e(t) = \frac{A}{C} \left\{ -\frac{\exp(-\beta\tau_e)}{\gamma} + \frac{1}{\gamma - \beta} \exp(-\beta t) \right. $$
$$\left. - \left[\frac{1}{\gamma - \beta} - \frac{\exp(-\beta\tau_e)}{\gamma} \right] \exp(-\gamma t) \right\} \tag{7.2.25a}$$

Similarly, the output due to hole current is

$$v_h(t) = \frac{B}{C} \left\{ \frac{1}{\gamma} - \frac{\exp(-\delta\tau_h)}{(\gamma + \delta)} \exp(\delta t) \right. $$
$$\left. + \left[\frac{\exp(-\delta\tau_h)}{(\gamma + \delta)} - \frac{1}{\gamma} \right] \exp(-\gamma t) \right\} \tag{7.2.25b}$$

where $\delta = \alpha W_d/\tau_h$ and $B = qI_0/\tau_h hf$. The impulse response of the photodetector is the sum of these two voltages.

The reader is here cautioned to note that two similar symbols, v_e and $v_e(t)$, represent the drift velocity of electrons and the photodetector output voltage due to the electron current component of photocurrent, respectively. Equivalent symbols, v_h and $v_h(t)$, represent hole velocity and output voltage due to the hole component of photocurrent.

For $t > \tau$, there will be no further current flowing into the load. The output voltage will be the residual voltage on the capacitor, which will decay exponentially with time constant RC. Thus, for $t > \tau$

$$v(t) = [v_e(\tau) + v_h(\tau)] \exp(-\gamma t) \tag{7.2.26}$$

EXAMPLE 7.3

A silicon PIN photodiode has $W_d = 30\ \mu m$, $\alpha_p = 7 \cdot 10^4\ m^{-1}$, and $v_e = v_h = 10^5\ m/s$. It is excited by an impulse of light of unit energy (one joule). (a) Calculate and plot the impulse response of the photodiode. (b) This photodiode has a load impedance consisting of an R and a C in parallel. $R = 100$ ohms; $C = 2$ pF. The photodiode and its load impedance constitute a photodetector. Calculate and plot the impulse response of this photodetector for $\lambda = 0.85\ \mu m$.

Solution ✱

(a) The impulse response of the photodiode is the photocurrent due to the impulse of optical excitation. This will be the sum of the electron and hole currents, given by Equations (7.2.20) and (7.2.21). The transit times for electrons and holes will be equal, since the velocities are equal. These transit times are

$$\tau_e = \tau_h = \frac{W_d}{v_e} = \frac{30 \cdot 10^{-6}}{10^5} = 3 \cdot 10^{-10} \text{ s} = 0.3 \text{ ns}$$

From Equations (7.2.20a) and (7.2.20b) we have, for $0 < t < \tau$,

$$i_e(t) = 2.79 \cdot 10^8 \left[\exp\{7(\tau - t)\} - 1\right]$$

$$i_h(t) = 2.28 \cdot 10^9 \left[1 - \exp\{-7(\tau - t)\}\right]$$

where t and τ are in ns.

From (7.2.21), the impulse response is

$$i_{ph}(t) = i_e(t) + i_h(t)$$

This magnitude, 4 GA, seems very large. However, the excitation, one joule, is also very large. A more realistic excitation, for example 1 μW $\times$ 1 ns $= 10^{-15}$ joule, will give currents of the order of 1 μA, a more believable current.
(b) We can find the impulse response of the photodetector circuit, which consists of the photodiode and an RC load impedance, from Equations (7.2.25a), (7.2.25b), and (7.2.26). In addition to the parameters used in part (a), we will need the following.

$$\gamma = \frac{1}{RC} = 5 \cdot 10^9$$

and

$$\beta = \delta = \alpha v = 7 \cdot 10^9$$

Then, for $0 < t < \tau$

$$v(t) = 200 - 570 \exp(-7t) - 12 \exp(7t) + 382 \exp(-5t) \text{ GV}$$

where t is in ns. For $t > \tau$ we have, from (7.2.26),

$$v(t) = 121 \exp\left[-5(t - \tau)\right] \text{ GV}$$

These solutions are plotted in Figure 7.2.5.

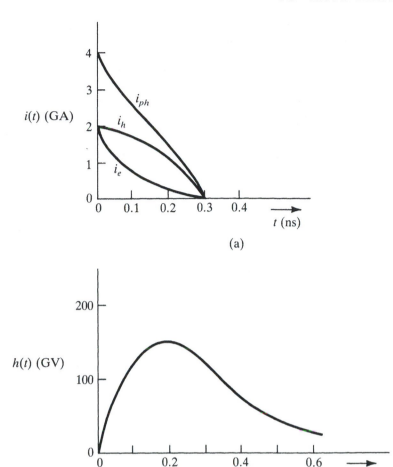

(a)

(b)

Figure 7.2.5 *Photodetector impulse response; see Example 7.3. (a) Photo-currents. (b) Photodetector output voltage. This voltage is developed by the photocurrent of (a) flowing through the parallel R-C photodector load impedance.*

When the excitation has sinusoidal modulation or when the spectrum of the signal is of interest, the bandwidth of the photodiode is an important parameter. Bandwidth can be considered by letting the modulation be a sine wave and finding the effects of its frequency on the magnitude of the response. Yariv [2.4, pp. 369–374] has analyzed bandwidth in this way.

The photocurrent at any time t is proportional to the number of carriers moving through the depletion region and to their velocities. Because of the time required for these carriers to move through and out of the depletion region, the current at time t is due to the sum, or integral, of charges produced during some time interval preceding t. When the period $T = 1/f$ of the modulation is of the same order of magnitude as the transit time, this sum corresponds to an integral over a significant fraction of the modulation period T. The result is that this integral will have magnitude less than it would have when the modulating frequency is very small.

Yariv assumed that a layer of carriers is generated at one edge of the depletion region and moves at the drift velocity to the other edge. This produces a square-wave of current. The magnitude factor derived by Yariv has the form $\sin(\omega_m \tau_d/2)/(\omega_m \tau_d/2)$. If bandwidth is defined as the frequency at which the magnitude is reduced by 3 dB, then

$$B_{3\,\text{dB}} = \frac{0.45}{\tau_d}\,\text{Hz} \tag{7.2.27}$$

This would be the 3-dB bandwidth if all carriers were in transit for the same time τ_d. If this τ_d is W_d/v_e, then it is a maximum, not an average, transit time. Carriers generated within, rather than at the edge of, the depletion region will be in transit for shorter time intervals; these carriers would lead to a greater bandwidth than that given by Equation (7.2.27). Since the total current is due to charge carriers produced throughout the depletion region, we can conclude that the bandwidth will be larger than that given by Equation (7.2.27).

An alternative approach to estimating bandwidth is to find, or estimate, the impulse response, find its Fourier transform, and calculate the half-power bandwidth of the Fourier transform. As an example, assume the impulse response of Figure 7.2.4(a) with the carrier velocities, and therefore the transit times, equal. The impulse is then a triangular pulse of duration τ.

$$h(t) = 1 - \frac{t}{\tau} \qquad 0 < t < \tau \tag{7.2.28}$$

The Fourier transform of this $h(t)$ is

$$H(\omega) = \frac{j}{\omega}\left(\left|\frac{\sin\frac{\omega\tau}{2}}{\frac{\omega\tau}{2}}\right|\exp\left(-j\frac{\omega\tau}{2}\right) - 1\right) \tag{7.2.29}$$

The 3-dB bandwidth of this $H(\omega)$ is

$$B_{3\,\text{dB}} = \frac{0.55}{\tau}\,\text{Hz} \tag{7.2.30}$$

This is slightly larger than the bandwith found from Yariv's result.

7.2.5   *Noise*

The generation of photocurrent is a sequence of discrete events that includes the creation of EHP and the motion of these charges under the influence of the local electric field. Each EHP will result in a pulse of current. The total current is the sum of many pulses. The total current is not a smooth continuous flow, but has variations about an average value. This variation is the "shot noise" described in Section 2.3.

The mean-squared value of the shot noise associated with the photocurrent is

$$\langle i_{sh}{}^2 \rangle = 2qIB \qquad (2.3.3)$$

where I is $I_d + I_{ph}$, the sum of the dark current and the photocurrent. This noise is unavoidable. If the noise current is large compared to the signal current, then the signal current may be masked by the noise and therefore becomes useless. If the noise current is relatively small, it may have a negligible effect. It will be important in system design to determine the signal-to-noise ratio and to ensure that this ratio is high enough that the system will perform adequately.

As an example, assume that the light wave is intensity modulated with modulation index m.

$$P_i = P_0[1 + mf(t)] \qquad (W) \qquad (7.2.31)$$

The photocurrent is

$$i_{ph} = \mathscr{R}_0 P_0[1 + mf(t)] \qquad (A) \qquad (7.2.32)$$

The average photocurrent is $I_{ph} = \mathscr{R}_0 P_0$, and the signal component of the photocurrent is

$$i_s(t) = I_{ph}mf(t) \qquad (A) \qquad (7.2.33)$$

The signal-to-noise ratio is defined as the ratio of signal power to noise power or the ratio of the squared currents. Then

$$\begin{aligned}
\frac{S}{N} &= \frac{\langle i_s{}^2(t) \rangle}{\langle i_n{}^2(t) \rangle} \\
&= \frac{I_{ph}{}^2 m^2 \langle f^2(t) \rangle}{2q(I_d + I_{ph})B}
\end{aligned} \qquad (7.2.34)$$

If $I_{ph} \gg I_d$, then

$$\frac{S}{N} = \frac{I_{ph}}{2qB} m^2 \langle f^2(t) \rangle \qquad (7.2.35)$$

This S/N applies to the photocurrent. There are other noise sources in the receiver that must be considered. These will be treated more completely in the following chapter.

EXAMPLE 7.4

The peak power of a PCM pulse at the input of a photodiode is 2.5 μW. The responsivity of the diode is 0.4 A/W. The load impedance consists of a 1-kohm resistor in parallel with a 5-pF capacitance; the temperature of the resistance is 300°K. What is the signal-to-noise ratio in dB of the voltage developed across this RC load impedance?

Solution

At the peak of the signal pulse, the signal current in the photodiode is 0.4 A/W $\times$ 2.5 μW = 1.0 μA.

The noise current will be the sum of shot noise and thermal noise currents; these can be calculated by using Equations (2.3.1b) and (2.3.3). These equations require that we know the bandwidth, which is not given; the "noise-equivalent bandwidth" can be calculated in terms of the R and C of the load impedance.

The output of the photodetector is the voltage developed across the RC load. The total mean-squared noise voltage can be calculated by integrating the noise spectral density over the appropriate frequency interval. If we interpret the shot noise and thermal noise formulas using df in place of B and multiply the noise current by the load impedance, which is a function of frequency, we obtain the mean-squared noise voltage in the frequency interval df. We can interpret $2qI$ and $4kTG$ as spectral densities, with units A^2/Hz. Since the shot noise and thermal noise spectral densities are not functions of frequency, the only variable part of the integral will be the load impedance. The noise equivalent bandwidth is then defined as the integral of the magnitude of the impedance squared divided by the square of the dc load impedance (i.e., the resistance). This calculation will show that the noise-equivalent bandwidth of the RC parallel load is $B = 1/4RC$. For the 1 kohm/5 pF load, $B = 50$ MHz.

With this value for B, and using Equations (2.3.1b) and (2.3.3), the noise currents are found to be

$$\langle i_{sh}{}^2 \rangle = 2qIB = 1.6 \cdot 10^{-17} \ \text{A}^2$$

and

$$\langle i_{th}{}^2 \rangle = 4kTGB = 8.28 \cdot 10^{-16} \ \text{A}^2$$

The total noise current is

$$\langle i_n{}^2 \rangle = \langle i_{sh}{}^2 \rangle + \langle i_{th}{}^2 \rangle = 8.44 \cdot 10^{-16} \ \text{A}^2$$

The signal current is $I_s{}^2 = 10^{-12}$ A^2.
The signal-to-noise ratio is

$$\frac{S}{N} = 10 \log \frac{I_s{}^2}{\langle i_n{}^2 \rangle} = 10 \log 1.185 \cdot 10^3 = 30.74 \text{ dB}$$

7.3 The Avalanche Photodiode

The photocurrents produced in the photodiode of an optical communications receiver are usually very small. Amplification of these weak electrical signals is necessary before useful signal levels are established. In some cases the photodiode is followed by conventional electronic amplifier stages. Another option is the use of an avalanche phenomenon to provide current amplification within the photodiode. The device that provides this current amplification is the avalanche photodiode (APD). The acronym APD is widely used in referring to this class of devices.

Useful references for supplementary reading on avalanche photodiodes include [4.4] and [4.5].

7.3.1 *Avalanche Current Multiplication*

Free carriers in a photodiode acquire kinetic energy from the internal electric field. When these carriers collide with the crystal lattice, they will lose some energy to the crystal. If the kinetic energy of a carrier is greater than the band gap energy of the valence electrons, the collision can free a bound electron. The free electron and the hole thus created can themselves acquire enough kinetic energy to cause further impact ionization. The result is an avalanche, with the numbers of free carriers growing exponentially as the process continues.

One effect of this cumulative impact ionization is that the total number of free carriers in the depletion region of the photodiode can be greater than the number produced by photoionization. The total current is correspondingly greater than the primary photocurrent. The ratio of total current to primary photocurrent is the current amplification of the device, represented by the symbol M.

$$M = \frac{I_{\text{APD}}}{I_{ph}} \tag{7.3.1}$$

The responsivity of the avalanche photodiode is

$$\mathscr{R}_{\text{APD}} = M \mathscr{R}_0 \tag{7.3.2}$$

Figure 7.2.3 indicates that as the electric field strength is increased a limiting value for drift velocity is reached. This might suggest that no higher velocities, and hence no higher kinetic energies, can be achieved; such is not the case. The

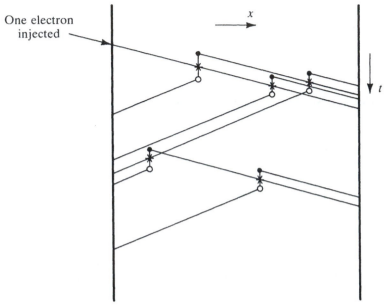

Figure 7.3.1 *Schematic diagram of avalanche current multiplication process. For this example, α ≈ 2β. One injected electron results in five impact ionization events. Since one primary electron produces six total electrons, M = 6.*

drift velocities and saturation velocities in Figure 7.2.3 are *average* velocities. Some carriers will have higher velocities and some lower when collisions occur. The maximum velocities will continue to increase with increasing E even though the average velocity does not.

Figure 7.3.1 shows a schematic diagram of the avalanche multiplication process. For the case shown, the probability of an electron causing impact ionization is approximately twice that for holes. Note that the duration of the avalanche-induced current is substantially longer than that due to transit times of electrons or holes across the avalanche region. In Figure 7.3.1, each electron injected into the avalanche region produces five additional EHP; the multiplication factor M equals 6.

The structure of an avalanche photodiode is similar to that of the PIN photodiode with an avalanche region added on one end of the intrinsic region. The avalanche region must have a high internal electric field.

Doping levels and electric field patterns suitable for an avalanche photodiode are shown in Figure 7.3.2. The intrinsic absorption region is similar to that of the PIN photodiode. The avalanche region has a higher electric field and is placed on one end of the absorbing region so that the photocarriers injected into the avalanche region are of one type only, either electrons or holes, but not both. The reason for preferring this means for initiating the avalanche is given in Section 7.3.3.

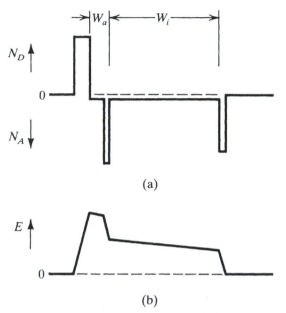

(a)

(b)

Figure 7.3.2 (a) *Doping levels and* (b) *electric field intensity patterns for avalanche photodiode. A high electric field intensity at one end of the depletion region causes current multiplication by impact ionization.*

As is evident in Figure 7.3.1, the total carrier density and hence the total current increase exponentially with x. The coefficient that describes the rate of exponential growth is the ionization coefficient α.[3] If only electrons produce impact ionization, then

$$I_{\mathrm{APD}} = I_{\mathrm{ph}} \exp (\alpha W_a) \quad \text{(A)} \tag{7.3.3}$$

and

$$M = \exp (\alpha W_a) \tag{7.3.4}$$

For the general case, let α and β represent the ionization coefficients for electrons and holes, respectively. Both α and β will be functions of the material, the electric field strength, and the temperature. In the equations below, $k = \beta/\alpha$. The current multiplication factors are [4.4, p. 301][4.5, p. 383]:

$$M_e = \frac{(1 - k) \exp \left[\alpha W_a (1 - k)\right]}{1 - k \exp \left[\alpha W_a (1 - k)\right]} \tag{7.3.5a}$$

[3] To conform to general practice, the same symbol, α, is used for the impact ionization coefficient as is used for the photo-ionization and attenuation coefficients. All three coefficients represent exponential growth (or decay) of the magnitude of a propagating quantity. The reader can usually determine from context which of these coefficients α represents.

for injected electrons and

$$M_h = \frac{\left(1 - \frac{1}{k}\right) \exp\left[\beta W_a\left(1 - \frac{1}{k}\right)\right]}{1 - \frac{1}{k} \exp\left[\beta W_a\left(1 - \frac{1}{k}\right)\right]} \tag{7.3.5b}$$

for injected holes.

It can be shown (see Problem 7.19) that

$$\text{when} \quad k \ll 1, \text{then} \quad M_e \approx \frac{\exp\left[\alpha W_a\right]}{1 - k \exp\left[\alpha W_a\right]} \approx \exp\left(\alpha W_a\right) \tag{7.3.6a}$$

and

$$\text{when} \quad k = 1, \quad M_e \approx \frac{1}{1 - \alpha W_a} \tag{7.3.6b}$$

One may suspect that if both α and β are high enough, the avalanche could sustain itself without any further injection of carriers into the avalanche region. An initial injected electron accelerated to the right produces holes and electrons.

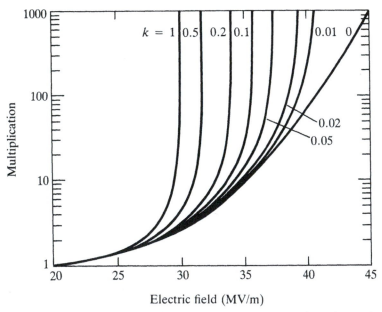

Figure 7.3.3 *Calculated multiplication factor as a function of electric field intensity. These curves illustrate the charactistics of avalanche breakdown and the sensitivity of M to variations in voltage for conditions near break-down; the numerical scales are based on specific values for the ionization coefficient and device dimensions. From Webb, McIntyre, and Conradi, RCA Review, vol. 35, pp. 234–278, June 1974 [4.16].*

These holes, when accelerated to the left, produce more electrons, which produce more holes, and so forth, ad infinitum. This condition can be thought of as infinite gain; the conditions under which it can occur can be found by setting the denominators of Equations (7.3.5) and (7.3.6) equal to zero.

The avalanche breakdown, or infinite gain, phenomenon can be represented by Equation (7.3.7), which expresses M as a function of the applied voltage V [4.6].

$$M = \frac{1}{1 - \left(\dfrac{V}{V_B}\right)^n} \tag{7.3.7}$$

where V_B is the breakdown voltage, and n is an adjustable parameter selected to match experimental data. As V is increased, the gain factor M increases until breakdown occurs as V approaches V_B.

Figure 7.3.3 illustrates the M versus E characteristic and the avalanche breakdown condition. It is evident that the avalanche photodiode should be operated well below the breakdown voltage not only to avoid breakdown but also to avoid the large fluctuations in gain that would result from small fluctuations in voltage if the operating point were in the region of high dM/dV.

EXERCISE 7.4

The thickness of the avalanche region of an APD photodiode is $W_a = 10 \ \mu m$. The avalanche is initiated by electrons. The ratio of ionization coefficients is $k = 0.1$. For what α, the ionization coefficient for electrons, is the avalanche multiplication factor $M_e = 100$?

Answer

$\alpha = 2.46 \cdot 10^5 \ m^{-1}$.

7.3.2 Bandwidth

Estimation of the response time of the avalanche photodiode is complicated by the fact that the avalanche process can extend the duration of the impulse response well beyond that attributable to the one-pass transit times of electrons and holes. For the case of electron injection, the total duration is the sum of three times: the time required for the most distant primary electron to reach the avalanche region, the duration of the avalanche process, and the time required for all holes produced during avalanche to move back across the avalanche and intrinsic regions.

Studies of the gain and bandwidth of the avalanche process indicate that when $M > 1/k$, the duration of the avalanche process can be represented by [4.7]

$$\tau_a = M\tau_1 \tag{7.3.8}$$

where $\tau_1 = Nk\tau_c$ is an effective transit time of the avalanche region, $\tau_c = W_a/v_c$ is the transit time of the initiating carrier through the avalanche region, and N is a numerical parameter derived empirically from numerical solutions. N ranges from $\frac{1}{3}$ to 2 as k ranges from 1 to 10^{-3}; it will be treated as a constant in the following discussions.

The total duration of the impulse response is then

$$\tau = \frac{W_i}{v_e} + M\,\frac{NkW_a}{v_e} + \frac{W_a + W_i}{v_h} \tag{7.3.9}$$

This is the maximum duration, measured at the base of the current pulse. The shape of this pulse is highly asymmetric.

The avalanche multiplication factor M has been expressed as a function of frequency [4.7]:

$$M(\omega) = \frac{M_0}{1 + jM_0\tau_1\omega} \tag{7.3.10}$$

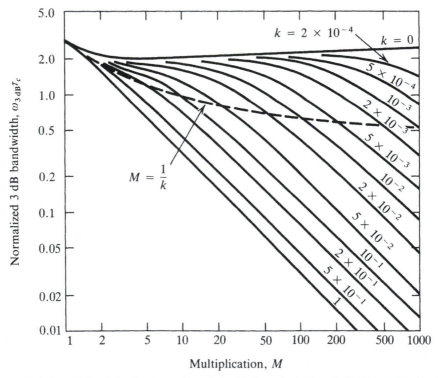

Figure 7.3.4 *Calculated normalized 3-dB bandwidth of APD multiplication factor as a function of k and low-frequency multiplication factor M. τ_c is the transit time through the avalanche region. From Emmons,* Journal of Applied Physics, *vol. 38, p. 3705, 1967 [4.7].*

where M_0 is the current multiplication as defined in Equations (7.3.1) and (7.3.5). From this expression we can determine the 3-dB bandwidth of the avalanche process to be

$$B_{3\,\text{dB}} = \frac{1}{2\pi M_0 \tau_1} \quad (\text{Hz}) \tag{7.3.11}$$

The bandwidth is inversely proportional to the low-frequency gain M_0. The gain-bandwidth product, $M_0 B_{3\,\text{dB}} = \frac{1}{2\pi\tau_1}$.

Curves for $B_{3\,\text{dB}}$ versus M for several values of k are given in Figure 7.3.4; these curves were calculated from a more detailed analysis of the processes than is appropriate here. The approximation $M > 1/k$, used in Equations (7.3.9) and (7.3.10), is represented by the dashed curve in this figure; the condition $M > 1/k$ corresponds to the region below the dashed curve. It is evident from Figure 7.3.4 that for $M > 1/k$ the inverse relationship between $B_{3\,\text{dB}}$ and M is a good approximation; above the $M = 1/k$ curve, the normalized bandwidth $B_{3\,\text{dB}}\tau_c$ tends to be constant, independent of M. In the limit as k decreases, $2\pi B_{3\,\text{dB}} \approx \frac{2}{\tau_c}$.

7.3.3 *Noise*

The avalanche current-multiplication process is a random process in that the M factor used to represent it is an *average*, but not an absolutely fixed, multiplication factor. The number of secondary electrons and holes that results from any individual injected electron or hole may differ from this average value M. The multiplication process has its own fluctuations that are superimposed on any fluctuations (noise) inherent in the primary photocurrent. The noise associated with I_{APD} will consist of the sum of the amplified shot noise of the primary photocurrent and the excess noise produced by the multiplication process.

The excess noise is usually represented by an excess noise factor F. This F is defined as the ratio of the total noise associated with I_{APD} to the noise that would exist in I_{APD} if the multiplication process produced no excess noise at all. F is the total noise divided by the multiplied shot noise. The mean-squared total noise current is F times the multiplied shot noise

$$\langle i \rangle_n^2 = 2q I_{ph} M^2 F B \tag{7.3.12}$$

where $I_{ph} = \mathscr{R}_0 P_i$ is the primary photocurrent and B is the bandwidth.

Figure 7.3.5 shows curves of F versus M for various values of k. These curves are based on the following equations developed by McIntyre [4.8]:

$$F_e = M_e \left[1 - (1 - k)\frac{(M_e - 1)^2}{M_e^{\,2}} \right] \tag{7.3.13a}$$

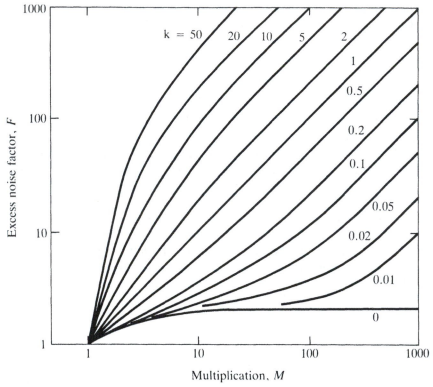

Figure 7.3.5 *Excess noise factors calculated from Equations (7.3.13a) and (7.3.13b). From Stillman and Wolfe, Chapter 5 in Semiconductors and Semimetals, vol. 12; Academic press, 1977 [4.5].*

when the avalanche is initiated by electrons, and

$$F_h = M_h \left[1 - \left(1 - \frac{1}{k} \right) \frac{(M_h - 1)^2}{M_h^2} \right] \tag{7.3.13b}$$

when it is initiated by holes. It is evident from Figure 7.3.5 that for a high-gain avalanche photodiode, small values for k give the smallest values for F.

EXERCISE 7.5

What is the excess noise factor, F, for the APD of Exercise 7.4?

Answer

$F = 11.8$.

EXAMPLE 7.5

The optical power input to an APD detector is 1 μW and the noise-equivalent bandwidth is 1 GHz. The APD parameters are $\mathscr{R}_0 = 0.55$, $M = 25$, and $k = 0.05$. What is the rms noise current in the output of this photodiode?

Solution

$$I_{ph} = 0.55 \cdot 10^{-6} = 5.5 \cdot 10^{-7} \text{ (A)}$$

$$F_e = 25 \left[1 - 0.95 \left(\frac{24}{25} \right)^2 \right] = 3.1$$

and

$$\langle i_n^2 \rangle = 2 \cdot 1.6 \cdot 10^{-19} \cdot 5.5 \cdot 10^{-7} \cdot 25^2 \cdot 3.1 \cdot 10^9$$

$$= 3.4 \cdot 10^{-13} \text{ (A}^2)$$

$$i_{rms} = [3.4 \cdot 10^{-13}]^{\frac{1}{2}} = 0.58 \text{ (}\mu\text{A)}$$

SUMMARY

A light wave propagating in a semiconductor can be considered to be a beam of photons, each with energy hf. If the photon energy is greater than the band-gap energy of the semiconductor material, the photons can free electrons from the crystal bonds, creating conduction electrons and holes in the semiconductor. The charge carriers produced by the absorption of photons are called photocarriers. In a reverse-biased junction diode, the electric field in the depletion region will cause the carriers in the depletion region to move out of the depletion region, producing a current that can be measured in the external circuit. This current is a photocurrent. Current produced in this way is the basis for photodetection in semiconductors.

The efficiency of a semiconductor *p-n* junction in producing photocarriers is its quantum efficiency, η. η can be interpreted as the probability that an incident photon will produce an electron-hole pair (EHP), or the average electrons per photon when the number of photons is large. The quantum efficiency is less than unity. The quantum efficiency is maximized by having as much of the incident optical energy as possible absorbed in the depletion region.

The responsivity of the photodiode, $\mathscr{R}_0$, is the average photocurrent divided by the incident optical power; the units are amperes per watt.

$$\mathscr{R}_0 = \frac{q\eta}{hf} = \frac{\eta\lambda}{1.24} \quad \text{(A/W)}$$

where λ is in μm.

The PIN photodiode is a *p-n* junction diode with an intrinsic region between the *p* and *n* regions. With a suitable applied reverse-bias voltage, the depletion region includes all of the intrinsic region and small parts of the *p* and *n* regions. This structure has two advantages over the elementary *p-n* junction photodiode—the quantum efficiency is higher and the electric field strength in the depletion region is high. The high electric field strength reduces transit times and thus increases bandwidth.

The avalanche photodiode (APD) is a photodiode with an intrinsic region for high quantum efficiency and an avalanche region on one end of the intrinsic region. In the avalanche region, high electric fields can cause ionization by collision, providing a mechanism for current multiplication. With a current multiplication factor M, the output photocurrent is

$$I_{APD} = M\mathcal{R}_0 P_i$$

The current multiplication has the effect of increasing the responsivity of the photodiode by the factor M.

Noise in photodiodes is conventional shot noise

$$\langle i_{sh}^2 \rangle = 2qIB$$

where I is the sum of the photocurrent and the dark current. In the APD, additional noise is produced by the avalanche current multiplication process. This is represented by an excess noise factor, F. The total noise in the APD current is F times the total shot noise.

$$\langle i_n^2 \rangle = 2q\mathcal{R}_0 P_i B M^2 F$$

PROBLEMS

7.1 Plot the responsivity, $\mathcal{R}_0$, versus wavelength, λ, for photodiodes having quantum efficiencies of 0.4, 0.7, and 1.0. The range of wavelengths should be $0.6 < \lambda < 1.6 \ \mu\text{m}$.

7.2 A germanium photodiode is intended for use at $\lambda = 1.3 \ \mu\text{m}$. What additional parameters must be specified, and what values should these parameters have, if the responsivity is to be 0.75?

7.3 (a) Show that the E field in the p region of Figure 7.1.4 is given by Equation (7.1.12).
(b) Show that the total charge in the p and n parts of the depletion region are equal, that is, $N_D W_{dn} = N_A W_{dp}$.

7.4 A free electron is accelerated by an electric field of $5 \cdot 10^6$ V/m. What distance must this electron travel before reaching a velocity of 10^5 m/s?

7.5 A silicon *p-n* junction diode has $N_D = 5 \cdot 10^{21}$ and $N_A = 7 \cdot 10^{20}$. The external applied voltage is -25 V.
(a) What is the width of the depletion region?
(b) What is the maximum electric field intensity in the depletion region?

7.6 The width of the intrinsic region in a silicon n^+-p-p^+ PIN photodiode is $W_i = 25$ μm. The electric field intensity will be $E = 10^6$ V/m at one end of the intrinsic region 2.5 · 10^6 V/m at the other end.
   (a) What doping level is required in the intrinsic region?
   (b) If the doping levels in the n^+ and p^+ regions are equal and are ten times that of the intrinsic region, what is the total width of the depletion region?
   (c) What is the reach-through voltage of this photodiode?

7.7 The width of the depletion region in a germanium PIN photodiode is $W_i = 10$ μm. Assuming no reflection from the end facets of the photodiode and no absorption before reaching the depletion region, what is its quantum efficiency for $\lambda = 1.3$ μm?

7.8 A silicon PIN photodiode has $\mathscr{R}_0 = 0.25$ for $\lambda = 0.8$ μm. Use the absorption data from Figure 7.1.3 to plot $\mathscr{R}_0$ versus λ for this photodiode for $0.6 < \lambda < 1.6$ μm. The dimensions of the absorbing regions are $W_1 = 5$ μm and $W_d = 20$ μm.

7.9 The incident impulse of optical excitation has energy I_0 (joules) as it enters the depletion region. The attenuation and absorption coefficients α and α_p are equal. Show that

$$N_0 = \frac{I_0}{hf} [1 - \exp(-\alpha W_d)]$$

and

$$N(x) = \frac{\alpha I_0}{Ahf} \exp(-\alpha x) \quad (\text{m}^{-3})$$

7.10 A silicon PIN photodiode with $W_i = 10$ μm has the following doping levels: in the n^+ region, $N_D = 10^{21}$ m^{-3}; in the p region, $N_A = 10^{19}$ m^{-3}; and in the p^+ region, $N_{A+} = 10^{20}$ m^{-3}. The E field in the center of the intrinsic region is 7.5 · 10^5 V/m. Calculate (a) E_{max}, (b) the width of the depletion region, and (c) the applied voltage.

7.11 The doping levels in the n^+ and p^+ regions of Problem 7.10 are increased to $N_D = N_{A+} = 10^{22}$ m^{-3}. The doping level and width of the depletion region are unchanged. (a) Repeat the calculations of 7.10. (b) Estimate the bandwidth of the photodiode.

7.12 A silicon PIN photodiode is to be designed to have, at room temperature and with $\lambda = 0.85$ μm, $\eta > 0.6$ and $B > 500$ MHz. The applied voltage cannot exceed 6 volts. Specify suitable dimensions, doping levels, and other parameters as required. The doping levels cannot exceed 10^{22} m^{-3}.

7.13 A germanium PIN photodiode for use at $\lambda = 1.3$ μm has $W_d = 5$ μm and $E = 10^5$ V/m. Plot its impulse response $i_{ph}(t)$.

7.14 The light wave incident on a photodiode is intensity-modulated with a sinusoidal modulating signal:

$$P_i = P_0[1 + 0.7 \sin(\omega_m t)], \quad \text{with} \quad P_0 = 1 \ \mu\text{W}$$

The modulating frequency is $f_m = 1$ MHz and the noise-equivalent band-width is 1.7 MHz. The responsivity of the photodiode is $\mathcal{R}_0 = 0.8$. The dark current, I_d, is negligible.

(a) What is the noise spectral density, in A^2/Hz, in the output of this photodiode?

(b) What is the signal-to-noise ratio?

7.15 The silicon in an APD has ionization coefficients $\alpha = 10^6$ m^{-1} for electrons and $\beta = 10^5$ m^{-1} for holes. The width of the avalanche region is 2 μm. Find M_e and M_h for APD devices made using these parameters.

7.16 An APD device has doping and electric field intensity patterns similar to those of Fig. 7.3.2. Which current multiplication factor, M_e or M_h, should be used for this APD?

7.17 A silicon APD has an ionization coefficient $\alpha = 10^6$ m^{-1}, $W_a = 2.5$ μm, $k = 0.05$, and $v_e = 10^5$ m/s. The avalanche is initiated by electrons. Estimate the 3-dB bandwidth of the avalanche process.

7.18 The photodiode of Problem 7.14 is replaced with an APD having $M = 25$ and $F = 5$; other parameters remain the same as in Problem 7.14. What is the S/N in the output of the APD?

7.19 Show that when $k \ll 1$,

$$M_e \approx \frac{\exp(\alpha W_a)}{1 - k \exp(\alpha W_a)} \approx \exp(\alpha W_a)$$

and that when $k = 1$,

$$M_e \approx \frac{1}{1 - \alpha W_a}$$

7.20 Show that (a) for $M \gg 1$, $F_e \approx kM$, and (b) for a very small k, $F_e \approx 2$.

CHAPTER

Optical Receivers

The primary performance parameters that determine the usefulness of the optical receiver in a communication system are the wavelengths it will receive, the sensitivity, and the bandwidth. Other considerations, such as dynamic range, linearity, and reliability, may be important but will not be considered in this chapter. Physical characteristics, such as size, weight, power consumption, and constraints on the operating environment, are also important but are outside the scope of this book.

The sensitivity of the receiver is the minimum input optical power that is adequate to produce a useful output signal from the receiver. The receiver noise will tend to mask small signals and thus increase the minimum input power required.

The performance requirements of the receiver subsystem can be divided into four principal functions:

1. Photodetection—conversion of an optical input signal into an electrical signal from which the information being communicated can be extracted.
2. Amplification—increasing the amplitude of the electrical signal from the photodetector to the levels required for effective utilization of these signals.
3. Filtering—limiting the bandwidth of the receiver to that required for the signal spectrum and designing the frequency response $H(j\omega)$ of the receiver to optimize its performance.
4. Signal processing—processing the amplified and filtered signal to provide the output signal characteristics required by the ultimate user. The signal processing provides the interface between the receiver and the user's system.

Several stages of amplification are normally needed to bring the signal amplitude to the level required for further signal processing. The amplification function can be separated into two parts, a preamplifier section and the main amplifier section; each section can have several stages.

The preamplifier brings the signal and noise amplitudes to a level where the noise produced in subsequent stages has a negligible effect on the overall signal-to-noise ratio. In this section the receiver's signal-to-noise ratio, and therefore its sensitivity, are determined. Noise sources in the photodetector and preamplifier are reviewed in Section 8.1.1. The receiver signal-to-noise ratio is studied in Section 8.1.3.

The main amplifier provides the additional amplification required. It consists of conventional electronic amplifier stages and will not be treated in detail here. The filtering function, discussed in Section 8.1.2, is usually associated with the amplifier circuits and can be placed in either or both sections.

8.1 Noise in the Receiver

The primary factor in determining the receiver sensitivity is the noise produced in the receiver. Each active device and each resistance in the receiver will produce some noise. To determine the total noise, the signal-to-noise ratio, and the receiver sensitivity, we will examine a reasonably complete equivalent circuit, identify the noise sources in this circuit, and add the effects of all such sources.

A simplified circuit diagram of the first stage of an optical receiver is shown in Figure 8.1.1(a); the simplifications are the omission of biasing and other components that have no direct effect on the signals. An equivalent circuit suitable for analyzing signals and noise is shown in Figure 8.1.1(b).

8.1.1 *Receiver Noise Sources*

The photodetector consists of either an APD or a PIN photodiode with a parallel R-C load impedance. Both the photodiode and the resistance are sources of noise.

The PIN photodiode produces shot noise given by[1]

$$\langle i_{sh}{}^2 \rangle = 2q(I_{ph} + I_d)\,\Delta f \qquad (\text{A}^2) \tag{8.1.1}$$

where I_{ph} is the average photocurrent and I_d is the dark current. If the photodiode is an APD, the shot noise is

$$\langle i_{sh}{}^2 \rangle = 2q(I_{ph} + I_d)\,M^2 F \Delta f \qquad (\text{A}^2) \tag{8.1.2}$$

---

[1] This equation can be interpreted as a spectral density $2qI$ multiplied by a bandwidth Δf. In some cases the Δf is interpreted as a noise-equivalent bandwidth B. In other cases it becomes df and the resulting expression is intergrated over frequency to find the total noise.

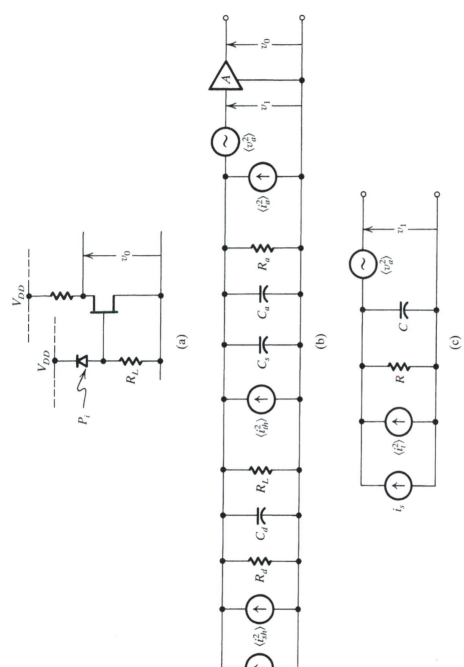

Figure 8.1.1 *Photodetector and preamplifier stage. (a) Simplified circuit diagram. The photodiode may be either a PIN or an APD. The transistor may be a BJT or a JFET. (b) Equivalent circuit for signal and noise analysis. R_d and C_d are the equivalent resistance and capacitance of the photodiode. R_a and C_a represent the input impedance of the amplifier. C_s is miscellaneous parasitic capacitance. Since all of these R's and C's are in parallel, they can be represented by a single R and single C, as shown in (c), a simplified equivalent circuit for the photodetector.*

247

We will use Equation (8.1.2) for either photodiode by setting $M = 1$ and $F = 1$ for the PIN photodiode.

The thermal noise is represented by a current source i_{th} in parallel with the resistance.

$$\langle i_{th}{}^2 \rangle = 4kTG\Delta f \quad (A^2) \tag{8.1.3}$$

The first stage of amplification can take any of several forms, each with its own characteristic noise sources. First we will analyze a normal transistor amplifier without feedback. This amplifier is often called a high-input-impedance, or simply high-impedance, amplifier. A second popular amplifier circuit, the transimpedance amplifier, will be treated in Section 8.1.4.

The active device, usually a transistor, is represented by an equivalent circuit that has two noise sources [4.9][4.12]. The two noise sources can be placed on the input side of the transistor equivalent circuit, as in Figure 8.1.1(b), or one can be at the input and the other at the output side.[2]

The active device used in the preamplifier can be either a bipolar junction transistor (BJT) or a field effect transistor (FET). The junction field effect transistor (JFET) is often preferred over other FET devices because it tends to be less noisy.

For a first-order, low-frequency approximation, the transistor noise sources can be represented to be shot noise with spectral density $2qI$. One such noise current source is placed across the input terminals and the other across the output terminals. For the BJT, the spectral densities of the noise current sources are $2qI_B$ between the base and emitter terminals, and $2qI_C$ between the collector and emitter. To refer the collector noise source to the input side of the transistor, we divide by $g_m{}^2$. The two noise sources for the BJT amplifier are

$$\langle i_a{}^2 \rangle = 2qI_B\Delta f \quad (A^2) \quad \text{and} \quad \langle v_a{}^2 \rangle = \frac{2qI_C}{g_m{}^2}\Delta f \quad (V^2) \tag{8.1.4a}$$

For the JFET amplifier, the corresponding equations are

$$\langle i_a{}^2 \rangle = 2qI_G\Delta f \quad \text{and} \quad \langle v_a{}^2 \rangle = \frac{2qI_D}{g_m{}^2}\Delta f \tag{8.1.4b}$$

At high frequencies there can be additional terms in these equations. For a more thorough discussion of noise in transistors the reader should consult references [4.9] to [4.12].

It is convenient to introduce a symbol, $\{x^2\}$, for the spectral density of x^2. Then $\{i_{sh}{}^2\} = 2qI$ and $\langle i_{sh}{}^2 \rangle = \{i_{sh}{}^2\}\Delta f$. The units are (A^2/Hz) for $\{i^2\}$ and (V^2/Hz) for $\{v^2\}$.

---

[2] These two noise sources are treated here as independent, but in general they are not. In a more accurate analysis the correlation between them would be recognized when the total noise is calculated.

With the equivalent circuit of Figure 8.1.1(c) and Equations (8.1.2) and (8.1.3), we can write the following equations for the total noise.

$$\{i_t^2\} = 2qIM^2F + 4kTG + \{i_a^2\} \quad \text{(A}^2\text{/Hz)} \tag{8.1.5}$$

$$\{v_n^2\} = \{i_t^2\}|Z_L|^2 + \{v_a^2\} \quad \text{(V}^2\text{/Hz)} \tag{8.1.6}$$

where

$$Z_L = \frac{1}{(G + j\omega C)} = \frac{R}{(1 + j\omega RC)} \quad \text{(ohms)} \tag{8.1.7}$$

By substituting Equations (8.1.5) and (8.1.7) into (8.1.6), we can get an expression for the total noise referred to the input of the amplifier.

$$\{v_n^2\} = [2qIM^2F + 4kTG + \{i_a^2\}]\frac{R^2}{1 + \omega^2R^2C^2} + \{v_a^2\} \quad \text{(V}^2\text{/Hz)} \tag{8.1.8}$$

Equation (8.1.8) represents the spectral density[3] of the noise voltage at the input to the amplifier. If the voltage transfer function of the amplifier is $H(j\omega)$, the mean-squared output voltage will be

$$\langle v_n^2 \rangle = \int_0^\infty \{v_n^2\}|H(j\omega)|^2 df \quad \text{(V}^2\text{)} \tag{8.1.9}$$

EXERCISE 8.1

A PIN photodiode has $\mathcal{R}_0 = 0.7$ A/W; its dark current is 10 nA. The load resistance is $R = 800$ ohms and the total capacitance in parallel with R is 10 pF. What is the rms noise voltage developed across the RC load when the input optical power is 0.1 μW?

Answer
20.4 μV.

8.1.2   *Equalization*

The load resistance R in Figure 8.1.1(c) affects the signal-to-noise ratio in three ways:

1.   The signal voltage at the input to the amplifier is proportional to R.
2.   The spectral density of the thermal noise current is inversely proportional to R.

---

[3] A mean-squared voltage (or current) spectral density is proportional, but not equal, to the power spectral density. When we use such terms to find the signal-to-noise ratio, the constant of proportionality will cancel and we will get a correct result. Reference to a $\{v^2\}$ or $\{i^2\}$ as a power spectral density is conventional and convenient but not strictly accurate.

3. The bandwidth of the photodetector load impedance is inversely proportional to R.

From items 1 and 2 above, we can conclude that increasing R will increase the signal-to-noise ratio. The price we must pay for increasing R is that the bandwidth of the photodetector load impedance is reduced. The bandwidth requirements for the receiver thus limit the extent to which we can increase S/N by increasing R.

The bandwidth limitation due to the RC load impedance is represented by a 20-dB-per-decade roll-off of the magnitude of the photodetector load impedance. This roll-off can be compensated to some degree by placing a rising frequency response characteristic, with the same break point, elsewhere in the receiver. To the extent that we can provide such compensation, R can be increased, thus increasing the S/N of the photodetector. Such compensation is referred to as equalization. It is commonly used in optical receivers.

The desired equalizer has the following transfer function:

$$H_{eq}(j\omega) = 1 + j\omega RC \qquad \omega < 2\pi B \qquad \textbf{(8.1.10)}$$

so that

$$H_{eq}(j\omega) Z_L(j\omega) = R \qquad \textbf{(8.1.11)}$$

If this condition can be satisfied for all frequencies within the signal bandwidth B, then the frequency roll-off produced by the photodetector load impedance is completely compensated.

The frequency response of the amplifier should be adequate to pass all essential parts of the signal spectrum. It will also pass any noise that occupies the same spectrum as that required for the signal. If the amplifier pass band is greater than that required for the signal spectrum, this excess bandwidth will result in an increase in the noise level with no increase in signal level; the signal-to-noise ratio is thus decreased.

The amplifier frequency response should include (1) equalization to allow higher values for the photodetector load resistance, and (2) a cut-off characteristic designed to optimize the signal-to-noise ratio.

8.1.3 *Signal-to-Noise Ratio*

The signal-to-noise ratio of the photodetector and preamplifier will be calculated using the equivalent circuit of Figure 8.1.1(c). This procedure is based on the assumption that the noise of the second and subsequent amplifier stages will be small compared to the amplified noise from sources shown in this equivalent circuit. If this assumption is not valid, the calculation of S/N must include the gain of the first stage and noise sources of the second stage.

If we use the noise spectral density from Equation (8.1.8), the equalization characteristic of Equation (8.1.10), and a sharp cut-off filter with bandwidth B, the mean-squared noise voltage referred to the input of the preamplifier is[4]

$$\langle v_n^2 \rangle = \int_0^B \{v_n^2\} |H_{eq}(j\omega)|^2 \, df \quad (V^2)$$

$$= \left[\left(2qIM^2F + \frac{4kT}{R} + \{i_a^2\} \right) R^2 \right.$$

$$\left. + \{v_a^2\} \left(1 + \frac{(2\pi RCB)^2}{3} \right) \right] B \qquad \qquad \textbf{(8.1.12)}$$

The signal-to-noise ratio can be calculated as it was in Section 7.2.5. There we found that for an intensity-modulated optical input signal, the signal current produced by the photodiode is $I_{ph}mf(t)$. The signal voltage referred to the input of the preamplifier is

$$v_s(t) = MI_{ph}Rmf(t) \quad \text{(V)} \qquad \qquad \textbf{(8.1.13)}$$

where M is included to represent APD current amplification; if a PIN is used, $M = 1$. The mean-squared signal voltage is the square of the right-hand side of Equation (8.1.13). The signal-to-noise ratio is then

$$\frac{S}{N} = \frac{\langle v_s^2 \rangle}{\langle v_n^2 \rangle}$$

$$= \frac{I_{ph}^2 m^2 \langle f^2(t) \rangle}{\left[2qI_{ph}F + \dfrac{4kT}{M^2R} + \dfrac{\{i_a^2\}}{M^2} + \dfrac{\{v_a^2\}}{M^2} \left(\dfrac{1}{R^2} + \dfrac{(2\pi BC)^2}{3} \right) \right] B} \qquad \textbf{(8.1.14)}$$

This equation is in a form that can give us insight into the effects of the parameters that the designer may be able to specify or control.

Note that as R is made larger, S/N increases. Equation (8.1.14) places no limit on how large R can be. A practical limit is that as R becomes larger, effective equalization becomes more and more difficult. The difficulty of equalization or the reduction of photodetector bandwidth will limit the size of R.

Increasing M will reduce all noise terms except that representing the shot noise in the primary photocurrent. Making M very large has the effect of amplifying the primary photocurrent, both signal and noise, so that all other noise terms are negligible in comparison. Since, as M increases, F also increases, there will be an optimum value for M that will produce a maximum signal-to-noise ratio. This is illustrated in graphs of F and S/N versus M shown in Figure 8.1.2.

We also see that S/N is increased by making T, C, $\{i_a^2\}$, $\{v_a^2\}$, or B smaller. The designer may have some control over each of these parameters. The extent to which we attempt to make one or more of these parameters smaller is

---

[4] The following development and the resulting Equation (8.1.14) are similar to those found in references [4.11] and [5.1].

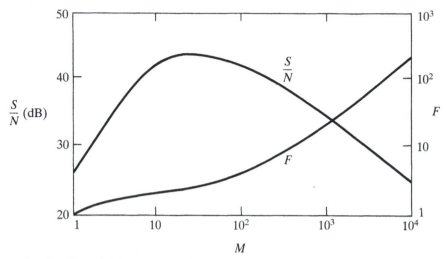

Figure 8.1.2 *F and S/N versus M for mf(t) = 1.0 μA. S/N has its maximum value for M = 24.5.*

determined largely by the value of better sensitivity and the extent to which we are willing to increase the cost and complexity of the receiver to achieve it.

EXERCISE 8.2

An optical receiver uses an APD photodiode that has $M = 50$, $F = 5$, and $\mathcal{R}_0 = 0.7$; the dark current is 20 nA. The photodetector load impedance consists of $R = 800$ ohms and $C = 10$ pF in parallel. The amplifier has noise sources, referred to the input, of $\{i_a^2\} = 0$ and $\{v_a^2\} = 10^{-17}$ (V²/Hz). The amplifier transfer function includes equalization, and its bandwidth is $B = 50$ MHz. What is the total rms noise voltage referred to the input of the amplifier when the input optical power is 0.1 μW?

Answer
117.3 μV.

8.1.4 *The Transimpedance Amplifier*

Another amplifier circuit commonly used in optical receivers is the transimpedance amplifier. It uses negative feedback and is similar in principle to the familiar operational amplifier. A schematic circuit diagram and an equivalent circuit are shown in Figure 8.1.3.

The noise sources defined in Equations (8.1.1) to (8.1.4) are all present in the transimpedance amplifier. The noise current $\{i_t^2\}$ in Figure 8.1.3 includes

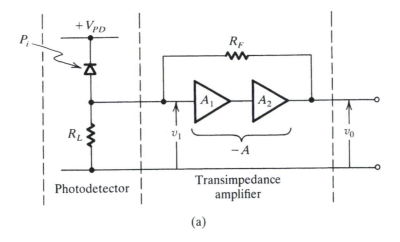

(a)

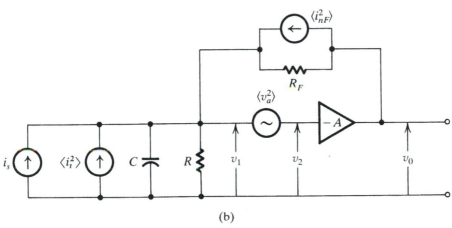

(b)

Figure 8.1.3   *Transimpedance amplifier. (a) Schematic circuit diagram. A_1 is a low-noise, high-input-impedance preamplifier. (b) Equivalent circuit.*

the photodiode, load resistance, and amplifier $\{i_a{}^2\}$ noises. There is also a noise source $\{i_{nF}{}^2\}$ associated with the feedback resistance R_F.

Analysis of the equivalent circuit for signal gain is the same as the analysis of the operational amplifier circuit. We will assume that the input current of the amplifier A is very small (its input impedance is very high) and that its gain A is high.

The signal current i_s and the feedback current i_F both flow through the RC load impedance to produce the amplifier input voltage v_1. Since $i_F = (v_o - v_1)/R_F$, $v_o = -Av_1$, and $i_s = i_{ph}M$, we can find

$$v_o(t) = \frac{-i_{ph}(t)MR_F}{1 + \dfrac{1}{A} + \dfrac{R_F}{AR} + j\,\dfrac{\omega R_F C}{A}} \qquad \text{(V)} \qquad \text{(8.1.15)}$$

When $A \gg (1 + R_F/R)$ and $f \ll A/2\pi R_F C$, we have the low-frequency response given by

$$v_o(t) = -MR_F i_{ph}(t) \quad \text{(V)} \tag{8.1.16}$$

The mean-squared output noise voltage can be found by calculating the spectral density, $\{v^2\}$, for each noise source, adding these spectral densities, and integrating over the amplifier bandwidth B.

The noise sources $\{i_t^2\}$ and $\{i_{nF}^2\}$ can be considered to be in parallel; in the equations describing the transimpedance amplifier performance, we will assume that $\{i_t^2\}$ includes the $4kT/R_F$ term representing $\{i_{nF}^2\}$. Since these current sources are in parallel with the signal, the same gain function derived for the signal will apply. The component of output noise due to the sum of all noise currents is

$$\{v_{n1}^2\} = \frac{R_F^2\{i_t^2\}}{\left[1 + \dfrac{1}{A} + \dfrac{R_F}{AR}\right]^2 + \left[\dfrac{\omega R_F C}{A}\right]^2} \quad \text{(V}^2\text{/Hz)} \tag{8.1.17}$$

To find the output noise due to $\{v_a^2\}$, we can write three equations relating the voltages v_a, v_1, v_2, and v_o.

$$v_1 = v_o \frac{Z_{RC}}{R_F + Z_{RC}} = v_o \frac{1}{1 + \dfrac{R_F}{R} + j\omega R_F C}$$

$$v_2 = v_1 + v_a$$

$$v_o = -Av_2$$

By eliminating v_1 and v_2 and using the voltage spectral densities, we can show that

$$\{v_{n2}^2\} = \{v_a^2\} \frac{\left[1 + \dfrac{R_F}{R}\right]^2 + [\omega R_F C]^2}{\left[1 + \dfrac{1}{A} + \dfrac{R_F}{AR}\right]^2 + \left[\dfrac{\omega R_F C}{A}\right]^2} \quad \text{(V}^2\text{/Hz)} \tag{8.1.18}$$

The total output noise mean-squared voltage spectral density is

$$\{v_n^2\} = \{v_{n1}^2\} + \{v_{n2}^2\}$$

$$= \frac{R_F^2\{i_t^2\} + \left[\left(1 + \dfrac{R_F}{R}\right)^2 + (\omega R_F C)^2\right]\{v_a^2\}}{\left[1 + \dfrac{1}{A} + \dfrac{R_F}{AR}\right]^2 + \left[\dfrac{\omega R_F C}{A}\right]^2} \tag{8.1.19}$$

For large A, the denominator reduces to unity.

Finally, we integrate the numerator of Equation (8.1.19) (the denominator is assumed to be unity) over the frequency range from 0 to B to get $\langle v_n^2 \rangle$ and find the ratio of mean-squared signal and noise voltages to be

$$\frac{S}{N} = \frac{I_{ph}^2 m^2 \langle f^2(t) \rangle}{\left\{ 2qI_{ph}F + \frac{4kT}{M^2}\left(\frac{1}{R} + \frac{1}{R_F}\right) + \frac{\{i_a^2\}}{M^2} + \frac{\{v_a^2\}}{M^2}\left[\left(\frac{1}{R} + \frac{1}{R_F}\right)^2 + \frac{(2\pi BC)^2}{3}\right] \right\} B}$$

(8.1.20)

This is the same equation for S/N as (8.1.14), derived for the conventional amplifier, except that the R of (8.1.14) is replaced by R and R_F in parallel. In the transimpedance amplifier, we should expect to make R large so that it has little effect on the gain and noise properties of the amplifier; R is in effect replaced by R_F in the S/N equation.

Equation (8.1.20) gives the S/N for the transimpedance amplifier without equalization. The fact that equalization is not needed is one of the advantages of the transimpedance amplifier for optical receivers.

The conditions that must be satisfied in using the simplified Equation (8.1.20) are

$$A \gg 1 + \frac{R_F}{R}$$

(8.1.21a)

and

$$B \le \frac{A}{2\pi R_F C}$$

(8.1.21b)

The first condition should be easy to satisfy so long as $A \gg 1$. In the second condition, the 3-dB bandwidth of the $R_F C$ circuit multiplied by A should be greater than the width of the signal spectrum. The design problem may reduce to designing the A section to have a constant high gain over the bandwidth B.

8.2 Receiver Sensitivity

Receiver sensitivity is defined as the minimum received signal power that is adequate to enable the receiver output signal to satisfy the needs of the user. It is usually expressed in dBm. Better sensitivity is equivalent to lower minimum signal power; smaller is better. When expressed in dBm, sensitivity is usually a negative number, that is, the sensitivity is less than 1 milliwatt.

Sensitivity is not an absolute quantity that can be computed solely in terms of the receiver components and circuit configuration. It is based on subjective performance criteria, whatever the user believes is necessary. A receiver may have adequate sensitivity for one user and not for another.

Sensitivity specifications can be reduced, for the purposes of system design, to a signal-to-noise ratio specification. The user may define a minimum acceptable probability of error for a digital system, an intelligibility criterion for a speech system, and so forth. These performance measures can be reduced to a minimum receiver signal-to-noise ratio specification. It is adequate in studying communication system design to limit our attention to signal-to-noise ratio and to the received power level necessary to achieve a specified S/N.

8.2.1 *Analog Receivers*

In an analog system, the intensity, phase, or frequency of the optical carrier is varied in proportion to a modulating signal. The current state of the art in optical communications is adequate for intensity modulation, but will not support high performance phase or frequency modulation. Our analysis will be limited to intensity modulation.

Equation (8.1.14) can represent the S/N of an analog receiver. If the minimum acceptable S/N is specified, the minimum I_{ph} can be found. Then, $P_i = I_{ph}/\mathcal{R}_0$. P_i is the sensitivity of this receiver for intensity-modulated optical signals. If the received optical signal power has P_i greater than this, the receiver output signal will be adequate to satisfy the user's needs. If the P_i is less than this, we can examine Equation (8.1.14) to see if there are parameters under the control of the designer that can be used to improve the sensitivity.

One caution is in order: Equation (8.1.14) includes only those noise sources associated with the receiver. Some noise from other sources will be received with the signal; other sources include the optical transmitter, the optical transmission channel, and background light that enters the receiver. Although the receiver will often be the dominant noise source, the system designer should not ignore other noise sources unless and until they have been shown to be small.

8.2.2 *Digital Receivers*

In a digital system the variable optical carrier parameter, here assumed to be the intensity, can have one of a limited number of discrete values. The binary digital system is by far the most common. It can be characterized by setting $f(t) = 1$ for one binary state and $f(t) = 0$ for the other. Thus, "optical signal present" can represent the binary *1* and "no optical signal present" the binary *0*.

Because noise is a random variable, it is characterized by an average value and a range of most probable values; we cannot say with certainty that the value of the noise at a specific time will lie within any specified range. The range of possible values of signal plus noise, representing the binary *1*, will overlap the range for noise alone, representing the binary *0*. Decision errors cannot be reduced to zero. The relationship between signal-to-noise ratio (S/N) and probability of error was discussed in Chapter 2 and illustrated in Figures 2.3.4 and 9.1.2.

The performance specification for a digital system is usually given as a minimum acceptable probability of error or error rate. This can be related to the S/N at the point in the receiver at which the *1/0* decisions are made. The sensitivity of the receiver is the minimum P_i for which the probability of error will be acceptable to the user.

EXERCISE 8.3

A PCM receiver must have S/N of at least 18 dB in order to function with an adequate probability of error. The receiver uses a PIN photodiode with $\mathcal{R}_0 = 0.7$, $R = 800$ ohms. The rms noise in the receiver is 7 μV, due primarily to thermal noise; the shot noise and amplifier noise are negligible. What is the sensitivity of this receiver?

Answer
-40 dBm.

EXAMPLE 8.1

For an example of the use of the signal and noise relationship developed above, consider a receiver that uses a JFET operating in the saturation region with g_m of 4000 μS and has a signal bandwidth of 25 MHz and a total photodetector load capacitance of 15 pF. We will compare the PIN and APD photodetectors, both with $\mathcal{R}_0 = 0.5$ A/W, and the APD with $M = 100$ and $k = 0.02$. The input optical power level is 1 μW.

Solution

If the 3-dB cutoff frequency of the photodetector load impedance were required to be 25 MHz, the load resistance must be $R = 1/2\pi BC = 425$ ohms. If we choose to make $R = 1000$ ohms, we can compensate for the lower photodetector bandwidth (10.6 MHz) by using an equalizer in a later amplifier interstage network. The equalizer must have a breakpoint at 10.6 MHz and a frequency response that rises at 20 dB per decade between 10.6 and 25 MHz. At 25 MHz its response will be 7.4 dB above the low-frequency response.

In Equation (8.1.14), let $I_{ph} = 0.5P_i$ and $mf(t) = 1$. For $\{i_a^2\}$ and $\{v_a^2\}$, use Equations (8.1.4b) with $\{v_a^2\} = 8kT/3g_m$ and $I_G = 0[4.9][4.11]$.

$$\frac{S}{N} = \frac{P_i^2}{\left\{ qFP_i + \dfrac{4kT}{M^2R} + \dfrac{2.8kT}{M^2g_m}\left[\dfrac{1}{R^2} + \dfrac{(2\pi BC)^2}{3}\right] \right\} 4B} \tag{8.2.1}$$

For this example, this equation can be reduced to

$$\frac{S}{N} = \frac{P_i^2}{1.6 \cdot 10^{-11}\, FP_i + \dfrac{1.66 \cdot 10^{-15}}{M^2} + \dfrac{8.26 \cdot 10^{-16}}{M^2}} \tag{8.2.2a}$$

The last two terms in the denominator, representing resistance noise and transistor noise, are of the same order of magnitude. Their sum is $2.49 \cdot 10^{-15}/M^2$.

$$\frac{S}{N} = \frac{P_i^2}{1.6 \cdot 10^{-11}\, FP_i + \dfrac{2.49 \cdot 10^{-15}}{M^2}} \qquad \text{(8.2.2b)}$$

For the PIN receiver, with $M = 1$, $F = 1$, and $P_i = 1\ \mu W$, the $S/N = 399$ (26 dB). For the APD, with $M = 100$ and $k = 0.02$, we first calculate $F = 4$ using Equation (7.3.13); then from Equation (8.2.2b) above we find $S/N = 1.556 \cdot 10^4$ (41.9 dB). The S/N at the output of the APD photodetector is 15.9 dB higher than that of the PIN photodetector.

In the analysis above, a fixed value for M was specified. It may be possible to realize a higher S/N or lower P_i with some other value for M. Examining Equation (8.2.2b), we can see that as M is increased, one of the noise terms decreases and the other increases. Since either decreasing M to very low values or increasing it to very high values is expected to cause the total noise to increase, there must be an intermediate value for which the noise has a minimum value. The M for which the denominator is minimum is the optimum value. M_{opt} could be calculated either by differentiating the denominator of Equation (8.2.1) with respect to M, or by plotting a graph of S/N versus M. For the numerical example above, this graph is shown in Figure 8.1.2. The maximum S/N is achieved with $M = 24.5$. With this

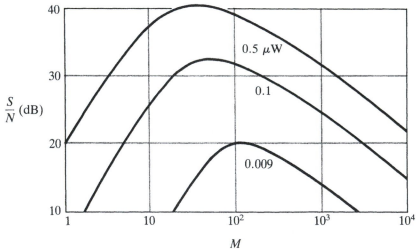

Figure 8.2.1 *Sensitivity of APD photodetector. Graphs of S/N versus M show that 0.009 μW is the smallest P_i for which a 20-dB S/N can be achieved. The $M = 1$ axis represents the PIN photodetector; $P_i = 0.5\ \mu$W gives $S/N = 20$ dB.*

M, we calculate $F = 2.4$ and $S/N = 43.7$ dB. By reducing M from 100 to 24.5 we have increased S/N by 1.8 dB.

We now wish to compare the sensitivities available from the PIN and APD photodetectors; M remains fixed at $M = 100$. The sensitivity of the receiver is defined as the minimum P_i for which a specified S/N can be realized. Let us specify 20 dB as the minimum acceptable S/N. For the PIN receiver, we then have a quadratic equation that can be solved for P_i; this P_i is the minimum acceptable input optical power and is therefore the sensitivity. For $S/N = 100$ (20 dB), $P_i = 0.5\ \mu$W (-33 dBm). In a similar manner, the sensitivity for the APD receiver with $M = 100$ can be found to be 9.1 nW (-50.4 dBm).

We will now remove the constraint that M be fixed. We wish to find the minimum P_i for the APD receiver to realize an S/N of 20 dB. We can use either analytical or numerical methods. The analytical approach will solve the equations above for M_{opt} in terms of P_i and other parameters, then use this M_{opt} in the equation for F, and finally substitute these equations for M and F into the S/N equation. We then have S/N expressed as a function of P_i; to solve this for P_i would probably require numerical or graphical methods. An alternative procedure is to use graphical or numerical methods from the start. The graph of Figure 8.2.1 illustrates this method. The sensitivity of the APD receiver is 9.0 nW (-50.5 dBm), and the associated APD parameters are $M = 120$ and $F = 4.4$. The sensitivity of the APD receiver is 17.5 dB better than that of the PIN receiver.

8.2.3 APD Versus PIN Photodetectors

The choice between the APD and PIN photodiodes will often be made on the basis of the cost versus the value of the higher sensitivity normally available from the APD photodetector.

Equation (8.1.14) indicates that as M is increased, the photodiode noise term increases with increasing F, while the amplifier and resistance noise terms decrease as $1/M^2$. There is an optimum value for M that gives the lowest total noise and the highest S/N for each P_i. Since F approaches kM for large M, one could conclude that the APD S/N would become poorer than that of the equivalent PIN photodetector as M becomes very large. In most cases, however, with other factors being the same, an APD photodetector will have higher sensitivity than a PIN photodetector.

Some types of APD have higher dark currents than the corresponding PIN device; in digital receivers this higher noise in the "no optical signal present" condition will increase the probability of error. The APD requires a substantially higher bias voltage. Since M is sensitive to temperature and bias voltage, automatic gain control may be required. In general, the APD is more difficult to use. It is the appropriate choice when the advantages it offers are needed, but the PIN is usually preferable if its characteristics are adequate for the system under consideration.

8.3 Heterodyne Detection

Coherent communication techniques take advantage of the fact that the sinusoidal carrier signal has a known frequency and that the frequency and phase of this carrier are constant and stable. For example, phase modulation requires a phase reference at the receiver to recover information from the phase variations of the received carrier. Coherent communication systems of other types can use a stable local reference signal to improve the sensitivity or other performance measures of the receiver.

Heterodyne detection is an important class of coherent communication techniques. We will examine the principles of heterodyne detection, the application of heterodyne techniques in optical receivers, and some of the advantages that can result from its use. Other aspects of coherent communication systems are discussed in Chapter 9.

8.3.1 Principles of Heterodyne Detection

In designing a radio communication receiver, the first consideration is the selection of a desired signal from all of the signals in the electromagnetic spectrum that are received by the antenna. This is a filtering problem; it can be done in the antenna coupling circuits and the first stage(s) of amplification following the antenna. A second problem is to provide filters that optimize the signal-to-noise ratio or other frequency-sensitive characteristics of the signal. This class of filter is often difficult to design for radio frequencies, both because the desired frequency response characteristic may be difficult to realize at high frequencies and because the radio-frequency filters are usually variable—they must be able to be tuned so that different radio frequency signals can be selected.

The heterodyne detector, as used in radio and radar receivers, is shown schematically in Figure 8.3.1. The r-f amplifier includes the variable-frequency filters that select the desired r-f signal; in some receivers this section can be omitted. The local oscillator provides the locally generated sinusoidal signal. The mixer generates the product of the local and received signals. To illustrate the nature of the signals generated in the mixer, assume an amplitude-modulated

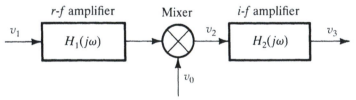

Figure 8.3.1 *Schematic diagram of receiver front end using heterodyne detector. The combination of a local oscillator and mixer to produce a product of the r-f and local oscillator signals results in a heterodyne detector. A receiver with a heterodyne detector and a fixed-frequency i-f amplifier is a superheterodyne receiver.*

received signal, represented as

$$v_1(t) = V_1[1 + mf(t)] \cos(\omega_1 t) \tag{8.3.1}$$

and a local oscillator signal

$$v_0(t) = V_0 \cos(\omega_0 t) \tag{8.3.2}$$

The mixer will produce a product term

$$
\begin{aligned}
v_2(t) &= v_1(t) \cdot v_0(t) \\
&= V_1 V_0 [1 + mf(t)] \cos(\omega_1 t) \cos(\omega_0 t) \\
&= \tfrac{1}{2} V_1 V_0 [1 + mf(t)] \{\cos[(\omega_1 + \omega_0)t] + \cos[(\omega_1 - \omega_0)t]\}
\end{aligned} \tag{8.3.3}
$$

This last form of the product expression shows two amplitude modulated terms, one at $(\omega_1 + \omega_0)$ and one at $(\omega_1 - \omega_0)$. The filters that follow the mixer stage will select one of these two terms; we will select for this example the difference-frequency term. The filters in the intermediate-frequency amplifier will pass this term and discriminate against all others. The output of the *i-f* amplifier is then

$$v_3(t) = V_3[1 + mf(t)] \cos[(\omega_1 - \omega_0)t] \tag{8.3.4}$$

This is identical in form to the original $v_1(t)$ except that the carrier frequency has been translated from ω_1 to $(\omega_1 - \omega_0)$. Note that if the local oscillator frequency can be varied, then its frequency can be adjusted to place the *i-f* frequency within the *fixed* passband of the *i-f* amplifier. This feature, that of a fixed intermediate frequency at which the critical filtering can be done, is one of the advantages of the heterodyne receiver.[5]

The frequency translation property of the heterodyne detector can be put into a more general and more compact form by using the Fourier transform. If $x(t)$ is a modulated carrier, with Fourier transform $X(\omega)$, and the local-oscillator signal is $\cos(\omega_0 t)$, then the output of the heterodyne detector will be

$$y(t) = x(t) \cdot \cos(\omega_0 t) \tag{8.3.5a}$$

$$Y(\omega) = \tfrac{1}{2}\{X(\omega + \omega_0) + X(\omega - \omega_0)\} \tag{8.3.5b}$$

Either the sum frequency or the difference frequency will be selected with a bandpass filter. The spectrum is like $X(\omega)$, the carrier-frequency spectrum, in all respects except that its frequency has been translated to the intermediate-frequency band. Thus, any modulation, noise, interference, or other components of the carrier frequency spectrum will appear in the intermediate-frequency band.

---

[5] This is more properly called a superheterodyne receiver. The combination of a local oscillator and mixer to produce sum and difference frequency terms is a *heterodyne detector*. A receiver using a heterodyne detector followed by a fixed frequency (i.e., intermediate frequency) amplifier is a *superheterodyne receiver*.

8.3.2 *Optical Heterodyne Detection*

The schematic diagram of an optical heterodyne detector is shown in Figure 8.3.2. The received optical signal and a local laser signal are combined (added) in a beam splitter or directional coupler. The beam splitter reflects part of the incident power and transmits the rest through the beam splitter. The output of the beam splitter is the optical input to the photodetector. The photocurrent is

$$i_{ph} = \mathscr{R}_0 P_i = \mathscr{R}_e \langle e_i^2 \rangle \tag{8.3.6}$$

where P_i is the total optical input power and $e_i(t)$ is the instantaneous electric field intensity; the constant $\mathscr{R}_e$ includes the responsivity and the constants necessary in the conversion from watts to $(V/m)^2$. Recall that the sum of two electromagnetic waves must be represented in terms of the sum of their field intensities. We will assume that the two input electric field vectors have the same polarization.

The input optical field is the sum of the modulated carrier, e_1, and the local laser, e_0; the photocurrent is

$$i_{ph} = \mathscr{R}_e \langle e_i^2 \rangle = \mathscr{R}_e \langle (e_1 + e_0)^2 \rangle \tag{8.3.7a}$$

$$= \mathscr{R}_e \langle (e_1^2 + 2e_1 e_0 + e_0^2) \rangle \tag{8.3.7b}$$

where

$$e_1 = E_1 [1 + mf(t)] \cos \omega_1 t$$

and

$$e_0 = E_0 \cos \omega_0 t$$

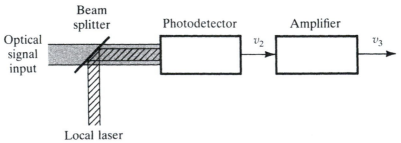

Figure 8.3.2 *Optical heterodyne detector. The beam splitter is necessary to combine the signal and local laser beams so that they are both incident on the photodetector from the same direction. The amplifier may be either an i-f amplifier or a baseband amplifier. When the local laser frequency is equal to the received carrier frequency, then v_2 is at baseband; this case is called homodyne detection.*

Note that e_1 has its amplitude, rather than intensity, modulated. We will consider intensity modulation briefly at the end of this section.

We can derive an equation for the photocurrent in the photodiode by substituting these expressions for e_1 and e_0 into Equation (8.3.7b), expanding the product of cosine terms, and neglecting the high frequency terms.

$$i_{ph} = \mathscr{R}_e \{ \tfrac{1}{2} E_1{}^2 [1 + mf(t)]^2 + \tfrac{1}{2} E_0{}^2$$
$$+ E_0 E_1 [1 + mf(t)] \cos (\omega_1 - \omega_0) t \} \qquad (8.3.8)$$

In the heterodyne detector, we expect to have $E_0 \gg E_1$. Then,

$$i_{ph} = \mathscr{R}_e \{ \tfrac{1}{2} E_0{}^2 + E_0 E_1 [1 + mf(t)] \cos (\omega_1 - \omega_0) t \} \qquad (8.3.9)$$

This current consists of a dc term and an intermediate-frequency term. The *i-f* term would be amplified and then recovered in a demodulator; an envelope detector would produce an output voltage of the form

$$v = A \mathscr{R}_e E_0 E_1 mf(t) \qquad (8.3.10)$$

Referred to the photodiode circuit, the equivalent photocurrent is

$$i = \mathscr{R}_e E_0 E_1 mf(t) = 2 \mathscr{R}_0 [P_0 P_1]^{\frac{1}{2}} mf(t) \qquad (8.3.11)$$

This equation indicates that the signal amplitude in the receiver is proportional to the amplitude of the local laser input as well as to that of the received light wave. It is possible to improve the sensitivity of the receiver by using heterodyne detection with a strong local laser signal. However, there are limits to the improvement available in this way. We will find in Section 8.3.4 that as the laser power is increased beyond a certain level, receiver noise tends to increase at the same rate as the detected signal level increases, giving no further improvement in signal-to-noise ratio.

The homodyne detector is a heterodyne detector in which the local oscillator frequency is equal to the optical carrier frequency; the intermediate frequency is therefore zero. Equation (8.3.9) indicates that the photocurrent consists of the sum of a signal current and a dc component.

The homodyne detector recovers the amplitude, not the intensity, modulation of the carrier. If the input light wave were intensity modulated, the output signal would be

$$v_3 = V_3 [1 + mf(t)]^{\frac{1}{2}} \qquad (8.3.12)$$

This output does not have the waveform, $[1 + mf(t)]$, of the modulating signal; the homodyne detector produces distortion when the light wave is intensity modulated. When the degree of modulation is small, that is, $|mf(t)| \ll 1$, the distortion is small and $[1 + mf(t)]^{\frac{1}{2}} \approx 1 + m/2 f(t)$. With this approximation, Equation (8.3.11) remains valid with m replaced by $m/2$.

8.3.3 *Balanced Heterodyne Detector*

A balanced optical heterodyne detector is shown in Figure 8.3.3. To study the performance of the balanced heterodyne detector, it is necessary to determine the relative phase angle between the two photodetector output voltages v_{2a} and v_{2b}. There is a $\pi/2$ phase difference between the transmitted and coupled waves of the directional coupler. This can be represented by writing the transmitted wave as a sine wave and the coupled wave as a cosine wave. The input power is divided equally between the two outputs. Then, from Figure 8.3.3, the input to photodetector (*a*) is $E_1 \sin \omega_1 t + E_0 \cos \omega_0 t$, and the input to photodetector (*b*) is $E_1 \cos \omega_1 t + E_0 \sin \omega_0 t$. The product terms in the mixer and the photodetector output voltages are

$$i_a = 2\mathscr{R}_e E_1 E_0 \sin \omega_1 t \cdot \cos \omega_0 t \tag{8.3.13}$$

$$v_{2a} = E_1 E_0 \sin (\omega_1 - \omega_0)t \tag{8.3.14}$$

for photodetector (*a*), and

$$i_b = 2\mathscr{R}_e E_1 E_0 \cos \omega_1 t \cdot \sin \omega_0 t \tag{8.3.15}$$

$$v_{2b} = E_1 E_0 \sin (\omega_0 - \omega_1)t = -v_{2a} \tag{8.3.16}$$

for photodetector (*b*). The two output voltages are then similar in form but opposite in sign. The combiner (Σ) function in Figure 8.3.3 is shown with one positive input and one negative input, indicating that the circuit will form the difference between the two inputs. The output of the balanced optical heterodyne detector can be represented as

$$v_3 = v_{2a} - v_{2b} = 2v_{2a} \tag{8.3.17}$$

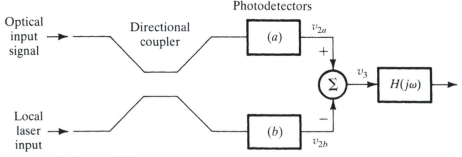

Figure 8.3.3 *A balanced optical heterodyne detector. Photodetectors a and b have identical detection characteristics. These photodetectors, the local laser, the beam splitter, and the Σ circuit constitute a balanced optical heterodyne detector.*

This is twice the voltage, or four times the power, available from either of the single photodetectors. The output signal is 6 dB higher than that of either single photodetector.

8.3.4 *Noise in Heterodyne Detectors*

The noise produced in the photodetector is given by Equation (8.1.12). The total noise consists of the sum of photodiode shot noise, thermal noise in the resistance(s), and noise in the preamplifier. In the heterodyne detector, an advantage can be gained because of the effects of a large local laser signal power. We have seen in Equation (8.3.11) that the detected signal amplitude increases as the laser amplitude is increased. We can also see in Equation (8.3.9) that the photodiode dc current increases with the laser power; the photodiode shot noise is proportional to the dc current and therefore to the total optical power. So long as other noise sources in the receiver are significant relative to the shot noise, increasing laser power will increase the signal strength without a corresponding increase in total noise; thus, the signal-to-noise ratio can be improved. When the shot noise becomes large enough to be the dominant noise, then further increase in laser power yields no further improvement in S/N.

Under the assumption that all noise sources other than shot noise are negligible, Equation (8.1.14) gives, for the total noise,

$$\langle i_n^2 \rangle = 2qIB \tag{8.3.18}$$

We have from Equation (8.3.9), $I = \mathcal{R}_e E_0^2/2 = \mathcal{R}_0 P_0$. If the signal power is taken to be the square of Equation (8.3.11), the signal-to-noise ratio is

$$\frac{S}{N} = \frac{\mathcal{R}_0^2 P_0 P_1 m^2 \langle f^2(t) \rangle}{q \mathcal{R}_0 P_0 B} = K P_1 m^2 \langle f^2(t) \rangle \tag{8.3.19}$$

where $K = \mathcal{R}_0/qB$. Note that P_0, the laser power, cancels; for large laser power, the signal and noise both increase with increasing laser power. Increasing this power beyond the point at which other noise sources become negligible is not profitable.

In the balanced heterodyne detector, the noise produced in the two photodetectors will be added in the Σ circuit. The total noise power will be twice, or 3 dB higher than, that of a single photodetector. Since the signal level is 6 dB higher and the noise level is only 3 dB higher, the signal-to-noise ratio of the balanced heterodyne detector can be 3 dB higher than that of either single heterodyne detector.

Among the practical problems that must be recognized in heterodyne detection systems is the requirement for *stable* frequencies in both the transmitter laser and the local receiver laser. Both of these frequencies must be essentially constant so that the difference frequency will be essentially constant. Although the quality of the semiconductor laser as a source of constant frequency is being improved through research, instability of the laser frequency remains a problem.

For heterodyne detection of intensity-modulated light waves, laser instabilities must be accommodated by making the intermediate frequency bandwidth large enough to allow the intermediate frequency to vary without moving some of the modulation spectrum outside the pass band. Some other coherent communication systems, such as phase modulation, cannot become practical for wide application until more stable lasers are available.

EXAMPLE 8.2

Calculate the S/N in an optical heterodyne detector as a function of local laser power using the photodiodes and signal parameters defined in Example 8.1.

Solution

The S/N can be calculated using Equation (8.2.2b) if we substitute for the P_i terms in that equation the equivalent terms applicable to the heterodyne detector. The term in the numerator of Equation (8.2.2b) comes from the signal current; in the heterodyne circuit the equivalent terms are

$$I_s = \mathscr{R}_e E_0 E_1 = \mathscr{R}_0 [P_0 P_1]^{\frac{1}{2}}$$

$$I_s^2 = \mathscr{R}_0^2 P_0 P_1$$

P_i^2 in the numerator is therefore replaced by $P_0 P_1$. The P_i in the denominator comes from a $2qI$ shot noise term. For the heterodyne detector the equivalent current is due to the total optical power input to the photodiode, $P_0 + P_1$. Then P_i is replaced by $P_0 + P_1$.

With these substitutions, the S/N equation becomes

$$\frac{S}{N} = \frac{P_0 P_1}{1.6 \cdot 10^{-11} F(P_0 + P_1) + \dfrac{2.49 \cdot 10^{-15}}{M^2}} \qquad (8.3.20)$$

The signal power input is 1 μW. Since half of the input power appears at each output of the directional coupler, the signal power incident on the photodiode is 0.5 μW; $P_1 = 0.5$ μW.

Consider first a PIN photodiode. Then, $F = M = 1$. The S/N equation then reduces to a function of P_0 alone.

$$\frac{S}{N} = \frac{P_0}{3.2 P_0 \cdot 10^{-5} + 5.0 \cdot 10^{-3}}$$

where P_0 is in μW. When P_0 becomes large, $S/N = 1/3.2 \cdot 10^{-5}$ or 44.9 dB. This is $44.9 - 26.0 = 18.9$ dB higher than the S/N found in Example 8.1 for the same PIN photodiode.

Next, consider the APD photodetector in the heterodyne detector circuit. From Example 8.1, we have $F = 4$ and $M = 100$. When we substitute these into Equation (8.3.20) above the APD heterodyne detector, S/N becomes

$$\frac{S}{N} = \frac{P_0}{(12.8\,P_0 + 6.45)\,10^{-5}}$$

When P_0 becomes large, $S/N = 1/12.8 \cdot 10^{-5}$ or 38.9 dB. This is 3 dB less than that for the APD detector in Example 8.1. This 3 dB can be accounted for by the reduction in input power due to the power division in the beam splitter. We can expect that with the same power input to the photodiode, the APD and heterodyne detectors will have approximately the same S/N because in both cases the advantage in S/N is due to neglecting all noise sources except the photodiode shot noise.

It is perhaps surprising to find that in this example calculation the PIN has a 6 dB higher S/N than the APD. The explanation of this result is left as an exercise for the student.

In both the PIN and APD heterodyne detectors, the S/N could be increased by an additional 3 dB by using a balanced heterodyne detector.

SUMMARY

The primary functions of the optical receiver are photodetection, amplification, filtering, and signal processing. The photodetector converts the received optical signal into an electrical signal. Some amplification is usually required to bring the electrical signal up to a useful level. Both photodetection and amplification introduce noise; filters are required to minimize the detrimental effects of noise and to increase the signal-to-noise ratio. In some receivers, additional signal processing is used to convert the received signal into the form required by the ultimate user of the signal.

All resistances and active devices are sources of noise. In the optical receiver, these sources include the photodetector, load resistance, and the first stages of amplification. After some amplification, the signal levels are high enough that noise sources at that stage have a negligible effect on the signal-to-noise ratio.

The noise is reduced, and the signal-to-noise ratio increased, by increasing the load resistance. This has the undesired effect of lowering the bandwidth of the photodetector load impedance. The bandwidth of the photodetector load can be compensated, to a limited extent, by equalization. When the photodetector load consists of an R and C in parallel, and the equalizer transfer function is $H(j\omega) = 1 + j\omega RC$, the overall transfer function is R, independent of frequency. Because complete equalization at all frequencies is not practical, and because the bandwidth of the signal is limited, the frequency response of the equalizer amplifier will not extend much beyond the highest frequency in the signal spectrum.

The signal-to-noise ratio of the optical receiver, with equalization, is

$$\frac{S}{N} = \frac{I_{ph}^2 m^2 \langle f^2(t) \rangle}{\left\{ 2qI_{ph}F + \frac{4kT}{M^2R} + \frac{\{i_a^2\}}{M^2} + \frac{\{v_a^2\}}{M^2}\left[\frac{1}{R^2} + \frac{(2\pi BC)^2}{3}\right]\right\} B}$$

The transimpedance amplifier, without equalization, has the same equation for S/N if R is interpreted to be the load resistance and the feedback resistance in parallel. This equation includes the APD parameters M and F; it can represent a PIN receiver by letting $M = 1$ and $F = 1$.

Receiver sensitivity is defined as the minimum received signal power that is adequate to enable the receiver to provide a required signal-to-noise ratio. For a given transmitter power, the lower the sensitivity the longer the distance over which the system can provide an adequate signal-to-noise ratio. An APD photodetector can usually offer a lower, that is, better, sensitivity than can a PIN photodetector.

Heterodyne detection uses a locally generated sinusoidal signal to translate the received signal spectrum from the optical carrier frequency to a lower intermediate frequency at which amplification, filtering, and detection can be done with electronic circuits. It has a signal-to-noise ratio advantage over direct photodetection because the use of a strong local oscillator signal serves to make all receiver noise sources, other than photodetector shot noise, negligible. Balanced heterodyne detection can provide S/N that is 3 dB higher than a single-ended heterodyne detector.

PROBLEMS

8.1 The photocurrent in a PIN photodiode is 100 μA. The load resistance through which this current flows is 100 ohms; the temperature of the resistance is 300°K. The voltage developed across the load resistance is applied to the input of an FET preamplifier. The FET parameters are $g_m = 8000$ μS, $I_G = 50$ nA, and $I_D = 5$ mA. The effective bandwidth is $B = 200$ MHz. What is the total rms noise voltage referred to the input of the preamplifier?

8.2 Repeat Problem 8.1 using an APD with the same parameters as the PIN and with $k = 0.01$ and $M = 50$. Note that the photocurrent, $I_{ph} = \mathcal{R}_0 P_i = 100$ μA, is the current induced by the optical input before avalanche multiplication.

8.3 If the received optical signal is intensity modulated with $m = 0.5$ and $\langle f^2(t) \rangle = 0.3$, compare the S/N available from the PIN and APD receivers of Problems 8.1 and 8.2. Show that the difference (in dB) between these two S/N figures is independent of the signal parameters m and $\langle f^2(t) \rangle$.

8.4 Show that the noise-equivalent bandwidth for a parallel RC load impedance is $B = 1/4RC$ Hz.

8.5 A PIN photodetector has $I_{ph} = 300$ μA, $R = 2200$ ohms, and $C = 5$ pF. The temperature of the resistance is 350°K.
(a) What is the 3-dB bandwidth of this photodetector?

(b) What is its noise-equivalent bandwidth?

(c) What is the rms noise voltage developed across the load impedance?

8.6 The photodetector of Problem 8.5 is used with an FET preamplifier having $I_G = 50$ nA, $I_D = 5$ mA, and $g_m = 5000$ μS. The amplifier does not include equalization; its transfer function is that of an ideal low-pass filter with a sharp cut-off at $B = 20$ MHz. How much does this preamplifier change the S/N calculated in problem 8.5?

8.7 The load impedance of a PIN photodetector has $R = 1000$ ohms and $C = 5$ pF. An equalizer is needed to extend the 3-dB bandwidth of the photodetector to 100 MHz. Sketch the magnitude of the equalizer transfer function (in dB) versus log ω.

8.8 A photodetector load impedance consists of a 250-ohm resistance in parallel with total shunt capacitance of 7 pF. The detected signal spectrum extends from 0 to 200 MHz.

(a) Calculate the 3-dB bandwidth of the load impedance.

(b) Plot the magnitude of the load impedance versus log ω.

(c) If the receiver frequency response is to include equalization, a 3-dB cutoff frequency of 200 MHz, and frequency response above 200 MHz corresponding to a 3-pole low-pass filter, plot the magnitude of the amplifier frequency response versus log ω.

8.9 The photodiode in Problem 8.8 is used with a transimpedance amplifier. The effective bandwidth is 200 MHz. Write specifications (gain, feedback resistance, frequency response, etc.) for the transimpedance amplifier.

8.10 The photodetector in Problem 8.8 is made to have a bandwidth of 200 MHz by reducing the value of R without changing C. What value for R will this require? What effect will this change in R have on the S/N?

8.11 The beam splitter in an optical heterodyne detector has reflectivity and transmissivity R and T respectively. The signal power and the local oscillator power at the inputs to the beam splitter are fixed. Show that the output signal strength, as given by Equation (8.3.10), is maximum when $R = T$.

8.12 The optical power into a balanced optical heterodyne detector is 0.1 μW and the local laser power is 1.0 mW. Let $mf(t) = 0.8 \cos(\omega_m t)$. The noise-equivalent bandwidth is 200 MHz. The beam splitter reflects 60 percent and transmits 40 percent of the input power. The photodetectors use PIN photodiodes with responsivity of 0.55 A/W and dark current of 4 nA. The load resistance has $R = 1000$ ohms; its temperature is 350°K. Calculate the S/N for each individual photodetector and that of the balanced detector.

8.13 A photodetector uses an APD with $\mathscr{R}_0 = 0.55$, $k = 0.1$, and $M = 80$. The received optical signal has a carrier power of 50 nW and is intensity modulated with $m = 0.3$ and $f(t) = 2 \cos \omega_m t$. The photodetector load impedance consists of an $R_L = 6000$ ohms and $C = 2$ pF in parallel. The preamplifier is a transimpedance amplifier with $R_F = 1200$ ohms, $B = 300$ MHz, $\{i_a^2\} = 0$, and $\{v_a^2\} = 2.2 \cdot 10^{-18}$ V^2/Hz. $f_m \ll B$. The temperature of the resistors is 300°K. What is the S/N at the output of the preamplifier?

CHAPTER 9

9

Optical Fiber Communication Systems

In Chapters 4 through 8 we studied the component subsystems that make up the optical fiber communication system. We gave particular attention to those characteristics that can limit the information capacity and transmission path length of such systems. These characteristics include the power/attenuation, the bandwidth and related time-domain response times, and noise. We will now examine the same characteristics from the point of view of the system as a whole.

We will begin with a point of view that is strongly biased toward digital systems. This is because such systems are very important—almost dominant in current commercial systems—and because they are in many ways simpler to analyze and understand. However, much of what we learn about digital systems can be useful in analyzing or designing analog systems as well. In Section 9.2, some of the special considerations for analog systems will be recognized. Coherent optical-fiber communication systems are discussed in Section 9.3, and two classes of optical fiber systems, undersea systems as an example of long-haul systems and local-area networks, are described in Section 9.4.

9.1 Long-Distance Digital Systems

By definition, a long-distance[1] communication system is one in which the distance between terminals is an important consideration. In many such systems it is not feasible to transmit signals from the point of origin directly to the ultimate destination. The total distance that must be covered is usually broken into several shorter paths in tandem so that transmission of adequate quality can be achieved. Each of the shorter paths is connected to the next through a repeater that can provide amplification and perhaps other processing to maintain adequate signal quality. In digital systems, the repeater will usually include a receiver to detect and regenerate the binary digital signal and a transmitter, modulated by this signal, to drive the following fiber link.

The maximum length of an optical fiber path is limited either by the signal attenuation or by the bandwidth. It was shown in Chapter 4 that both the attenuation and the bandwidth of the optical fiber are expressed in terms of the fiber length. Typical units are dB/km and MHz-km; attenuation increases with length and bandwidth decreases with length.

In designing a multilink system, the total system performance specifications must be reduced to a specification for each link. For example, if the overall system can have no more than one binary error in one billion bits, then each individual link must be better than this; how much better depends on how many links there will be and perhaps on other details. On the basis of performance specifications such as this, the designer can first determine bandwidth and signal-to-noise ratio requirements for each link and then receiver sensitivity and transmitter power requirements.

9.1.1 *Probability Of Error In Binary Decisions*

A consideration that is unique to, and characteristic of, digital system design is the binary-decision process and the probability of error in these binary decisions.[2] The signal transmitted can be in either of two states. In the receiver, the signal plus noise is converted into an electrical signal, probably a voltage, before the binary decision is made. This voltage is applied to a threshold circuit. If the voltage falls on one side of the threshold voltage, a binary *1* decision is registered; if it falls on the other side, a binary *0* is registered. Because signal levels are not necessarily stable, and because the inevitable noise can cause the signal-plus-noise voltage to fall on the wrong side of the threshold, some binary decisions will be in error. In this section, we will examine the decision process for binary PCM and determine means for calculating the relationship between signal-to-noise ratio and probability of error.

Before calculating the effects of noise, we will define and calculate the quantum limit for binary decisions. The quantum limit is a result of the quantum

---

[1] The designation "long-haul" system is commonly used by communication system engineers.
[2] Our consideration will be limited to the binary digital system. It is adequate for the understanding of digital systems in general, and it is the dominant digital system in current use.

nature of very weak optical signals. In the limit, the received signal may consist of only a few photons. Furthermore, even though the average number of photons per second, or photons per pulse, is known, the exact number in any specific pulse is uncertain. The uncertainty is represented by a statistical distribution function; in this case, the Poisson distribution is the appropriate function. The Poisson distribution,

$$P(n) = \frac{N^n \exp{(-N)}}{n!} \qquad \textbf{(9.1.1)}$$

gives the probability of receiving exactly n photons during a pulse interval T when the average number of photons received during this interval is N.

Consider a system that transmits a pulse of power P to represent a binary *1* and transmits nothing to represent a *0*. If the pulse interval is T and the total attenuation is represented by an attenuation constant α, then the average received energy representing a binary *1* is $\mathscr{E} = PT \exp{(-\alpha L)}$ and the average number of photons per pulse is $N = \mathscr{E}/(hf)$. Since no energy is transmitted for the binary *0*, then none is received and $N = 0$. However, even for the binary *1*, there may be no photons. The probability of receiving nothing when a *1* is transmitted is $P(0) = \exp{(-N)}$. Thus, for binary PCM with no noise considered, we expect some decision errors.

A common PCM performance specification is that the average probability of error must not exceed 10^{-9}. We assume that *1* and *0* are equally probable, that there is no noise, and that one or more received photons will be interpreted as a *1*. When a *0* is transmitted, no photons will be received and the binary decision will be made with probability of error equal to zero. If the average probability of error cannot exceed 10^{-9}, the probability of error in detecting the *1* cannot exceed $2 \cdot 10^{-9}$. The *1* will be received in error only when the number of photons received is zero, that is, $n = 0$. With $n = 0$ and $P(0) = 2 \cdot 10^{-9}$, the Poisson formula, Equation (9.1.1), gives $N = 20$. The quantum limit for average probability of error of 10^{-9} is 20 photons per binary *1* or an average of 10 photons per bit.

Although the quantum limit is of interest as a basis for evaluating the performance of specific systems, it is not a realistic measure of the sensitivity of such systems. Two assumptions are unrealistic: (1) that when no signal is transmitted, the output of the detector will be zero, and (2) that when a signal is transmitted, the only output is due to signal photons from the transmitter. In both of these cases, noise will produce electrical signals in the receiver that have no relationship to the received signal. The effect of noise on the probability of binary-decision errors is an important aspect of system and receiver design.

For purposes of studying the probability of error in binary decisions, as well as for many other purposes, noise is characterized with a Gaussian probability density function:

$$f(x) = \frac{1}{[2\pi\sigma^2]^{\frac{1}{2}}} \exp{\left[\frac{-(x-m)^2}{2\sigma^2}\right]} \qquad \textbf{(9.1.2)}$$

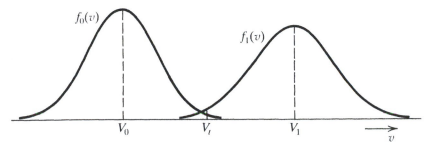

Figure 9.1.1 *Gaussian distribution function for binary PCM signals. A signal voltage V_1 with rms noise voltage σ_1 is received for a binary 1, and signal voltage V_0 with rms noise voltage σ_0 for a binary 0. $\sigma_0 = 0.20V_1$ and $\sigma_1 = 0.24V_1$. V_t is the threshold voltage.*

Here m is the mean value of x and σ is the rms deviation from the mean, that is, the standard deviation. To find the probability that x lies between two specified values, say x_1 and x_2, we must integrate $f(x)\,dx$ between the limits x_1 and x_2.

In using the Gaussian density function to study the probability of error in binary decisions, σ represents the noise, m the signal, and x the sum of signal and noise. In the case of PCM, a binary *1* would be represented by a signal voltage of mean value V and a binary *0* by zero volts. In each case, noise will cause an rms deviation from the signal voltage of σ volts. The Gaussian distributions representing the two possible PCM signal levels are shown in Figure 9.1.1.

The decision circuits in the receiver must measure the received signal during each pulse interval and determine whether a binary *1* or *0* was most likely transmitted during that interval. This can be done by establishing a threshold voltage between the two signal levels and comparing the received voltage with this threshold voltage. It is clear from Figure 9.1.1 that no matter where the threshold is placed, some of the area under the curve is on the wrong side of the threshold. Thus, some of the binary decisions will be in error. However, we can calculate the probability of error and we can design the system so as to make this probability as small as it needs to be, but never quite zero.[3]

To study methods for calculating the probability of error, consider again Figure 9.1.1; let $V_0 = 0$. The cases for the two possible binary signals are considered separately. The threshold is placed at V_t, between V_1 and zero. The probability that when the signal voltage is zero the noise voltage alone will exceed the threshold is equal to the integral of a Gaussian density function from V_t to infinity. The probability that when the signal voltage is V_1 the sum of signal plus noise will not exceed the threshold is the integral of another Gaussian density function from negative infinity to the threshold voltage, V_t. In general, the two Gaussian density functions are not identical. For the case illustrated in

---

[3] When the probability of error is made very small, the difficulty and cost of making it still smaller can be substantial. We therefore learn to be realistic about how small it needs to be.

Figure 9.1.1, the noise associated with the binary *0* is less than that for the binary *1*.

The error function is an integral of the Gaussian probability density function from which the probability of error can be found. It is a well-known, tabulated function, defined as

$$erf(x) = \int_{-x}^{x} \frac{1}{\sqrt{\pi}} \exp\left[-y^2\right] dy \tag{9.1.3a}$$

This integral is over the center portion of the Gaussian density function. The probability of error is the integral over one of the infinite tails of the distribution. The complementary error function is defined as

$$erfc(x) = 1 - erf(x). \tag{9.1.3b}$$

It is the integral over both infinite tails. The probability of error is then one-half of the complementary error function. A graph of $erfc(x)$ is shown in Figure 9.1.2.

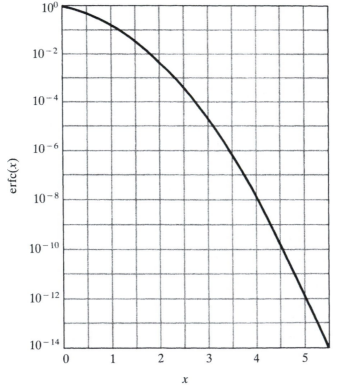

Figure 9.1.2 *The complementary error function, defined in Equation (9.1.3).*

The probability of error when a binary *1* is transmitted is

$$P(e|1) = \int_{-\infty}^{V_t} \frac{1}{\sqrt{2\pi}\sigma_1} \exp\left[\frac{-(v - V_1)^2}{2\sigma_1^2}\right] dv \qquad (9.1.4)$$

From the symmetry of the Gaussian density function, this can be reduced to

$$P(e|1) = \tfrac{1}{2}erfc\left[\frac{(V_1 - V_t)}{\sqrt{2}\sigma_1}\right] \qquad (9.1.5a)$$

Similarly,

$$P(e|0) = \tfrac{1}{2}erfc\left[\frac{V_t}{\sqrt{2}\sigma_0}\right] \qquad (9.1.5b)$$

The average probability of error in binary decisions is a weighted average of $P(e|0)$ and $P(e|1)$.

$$P(e) = P(1)P(e|1) + P(0)P(e|0) \qquad (9.1.6)$$

When the binary *1* and *0* are equally probable, the average probability of error is the arithmetic average of the individual probabilities of error.

EXERCISE 9.1

A binary PCM system has equal probabilities of *1* and *0*. The decision threshold is placed at $V_1/2$. The average probability of error in binary decisions in the receiver is to be 10^{-9}, and $\sigma_1 = \sigma_0 = \sigma$. Show that $V_1 = 12\sigma$ and $V_t = 6\sigma$.

The usual way for representing these probabilities in binary PCM systems is to plot the average probability of error versus S/N. Figure 9.1.4 is such a graph. This is simply an *erfc* graph with the scales labeled in the units of the communication system designer.

It is common practice to use the terms error rate, or bit error rate, as synonyms for probability of error. A probability of error of 10^{-9} means that an average of one bit in 10^9 will be in error. Strictly speaking, the bit error rate (BER) should be the average number of errors per unit time. However, we will conform to common practice and interpret BER to be equal to the probability of error. When we wish to refer to bit errors per second, it will be necessary to make that clear by including the units, errors per second, with the number.

9.1.2 Attenuation and Power Requirements

We have seen in earlier chapters that there is a minimum received signal power required to sustain an adequate signal-to-noise ratio. Since total fiber attenuation is proportional to fiber length, there is a maximum path length beyond which the received signal strength will be inadequate to sustain proper operation of the communication system.

If, in system design, we begin with specifications that include the transmitted power and the receiver sensitivity, we can calculate the maximum attenuation. This attenuation includes losses in coupling the light wave into and out of the fiber at the ends of the link, losses at each splice along the length of the fiber, and the distributed attenuation of the fiber itself. In addition to these average losses, a dB margin is also usually added to protect against aging of components, losses due to an occasional misalignment or minor damage to the fiber, or other excess losses.

EXERCISE 9.2

The transmitter output power in an optical fiber communication system is -0.5 dBm and the receiver sensitivity is -41.5 dBm. The average fiber attenuation, including the losses due to splices, is 1.6 dB/km. The loss in coupling the laser output into the fiber is 3 dB and the coupling loss at the receiver is 2 dB. If the design is to include a 3-dB margin, what is the maximum distance between the transmitter and the receiver?

Answer
20.6 km

EXERCISE 9.3

The distance between the transmitter and the receiver in an optical fiber link is 35 km. The transmission path is made by splicing similar fibers, each of which has maximum length of 3 km. The attenuation of this fiber is 0.6 dB/km. The losses due to splices average 0.1 dB per splice. The coupling losses are 2 dB at each end, and the design margin is 3 dB. The receiver sensitivity is -30 dBm. What transmitter power output is required to meet these specifications?

Answer
-0.9 dBm

The receiver sensitivity, a critical parameter in link design, is determined by the acceptable probability of error (BER) (or another measure of quality of performance) and by the receiver design. The receiver sensitivity required, for a particular BER and λ, is illustrated in Figure 9.1.3. This figure shows the quantum limit of 10 photons per bit, as discussed in Section 9.1.1; this is easily converted to watts when the bit rate and wavelength are specified. The second graph is for an arbitrarily selected 1000 photons per bit. The points added to the figure represent actual receivers that have been described in the technical

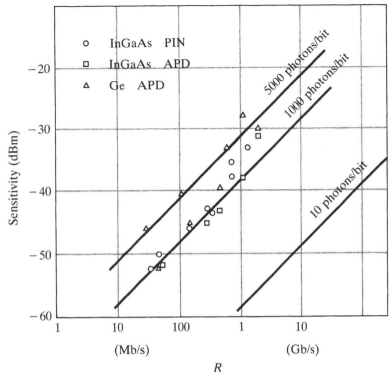

Figure 9.1.3 *Receiver sensitivity versus data rate for λ = 1.3 μm. The points shown are sensitivities of experimental receivers, taken from Table IV, reference [4.17].*

literature [4.17]. It is evident that a practical limit on receiver sensitivity appears to be approximately 1000 photons per bit; only a few APD receivers have sensitivity better than this. When heterodyne detection becomes practical, sensitivities of a few hundred photons per bit will probably be achieved.

For the purpose of studying multilink systems, two types of repeater will be considered, the digital regenerator and the linear amplifier. In both types, the receiver must include a photodetector and an optical transmitter. The signal from the receiver is processed in the repeater and then used to modulate the transmitter. In the case of the digital regenerator, the processing consists of making binary *1-0* decisions on the detected signal and using the regenerated digital signal to modulate a PCM transmitter. In the case of the linear amplifier, the signal from the receiver is amplified and used to intensity-modulate the transmitter; here the PCM signal is handled just as an AM signal would be, without taking any advantage of the digital nature of the signal.

Consider first the linear amplifier repeater. It is assumed that the gain in the repeater is exactly equal to the losses in the previous length of fiber. The transmitted power is thus maintained as a constant. The repeaters are assumed to be similar, except that the amplification in each repeater can be controlled to match the losses in each section of fiber.

At the first repeater, the noise from the photodetector and preamplifier stages will be added to the signal and amplified along with the signal. The S/N established in the front end of the receiver will be maintained through the linear amplification, modulation, and transmission through the next section of fiber. In the second repeater, the received signal has associated with it the noise from the first repeater, now attenuated so that it is at the same level as in the input to the first repeater. The total noise power at the output of the receiver in the second repeater is thus twice that produced in a single repeater. In a system having n links, the total noise in the final receiver is n times that due to a single receiver. The S/N for an n-link system is 10 log (n) (dB) less than that for a single link.

It should be pointed out here that the linear amplifier repeater is not an attractive option for PCM systems. It is described to indicate the advantages of the digital regenerator repeater and the advantages of PCM over CW modulation in systems having many repeaters.

In the digital regenerator repeater, the received signal, with the associated detector and preamplifier noise, determines the S/N at the input to the decision circuits. The decision circuits determine, for each signaling interval, whether a binary *0* or *1* was received. The output of the decision circuit is a clean binary signal, with no noise, no dispersion, and a waveshape that can be selected by the system designer. This is the signal that modulates the PCM transmitter. Thus, noise does not accumulate as it did in the system that used linear amplifiers rather than digital regenerators in the repeater.

In the digital repeater there will be some decision errors. These errors will accumulate so that the overall system error rate will be higher than that of any individual link. However, it is possible to sustain high-quality performance over a long total path using digital repeaters. Equivalent performance is not available over corresponding path lengths using linear amplifier repeaters.

For low probabilities of error, the total system error rate is very nearly equal to the sum of the error rates of the individual links in the system. To calculate the total error rate, the correct method is to find the probabilities of errorless transmission over each individual path. The probability that the total path has errorless transmission is the product of the corresponding probabilities for each path. It is easy to show that, for low error rates, the sum of the individual error rates is very close to the total error rate.

EXERCISE 9.4

A binary PCM system uses rectangular pulses of duration $T = 1/R$ and amplitude V or zero to represent the binary *1* or *0* respectively. The overall BER for this system must not exceed 10^{-9}. The system has 19 repeaters (20 links). What peak-signal to rms-noise ratio (V/σ) is required in each regenerator?

Answer
22 dB

9.1.3 *Bandwidth and Data Rate*

In previous chapters, we have established an approximate relationship between dispersion, bit rate, and bandwidth. In Section 2.2.3 it was shown that for PCM a bandwidth of one-half of the pulse repetition rate is adequate. In Section 4.4.3, the relationship $R = \frac{1}{4\sigma}$ between the bit rate R and the rms fiber dispersion σ was established. Thus, for binary PCM, we have

$$R = \tfrac{1}{4\sigma} = 2\Delta f \tag{9.1.7}$$

We will now be concerned with the complete system rather than only with the fiber dispersion. The transmitter and photodiode will each have a characteristic rise time and the receiver amplifier will have a maximum bandwidth; all of these limit the speed of response, or the rise time, of the signal at the output of the receiver. By thinking in terms of the rms width of the pulse as it propagates through the system, it is possible to devise a method for calculating the rms width of the pulse at the output of the receiver. The rms width is used rather than some other measure of pulse width because it leads to a very simple method for combining the effects of different parts of the system on the shape of the output pulse. It is also easily related to bit rate or effective bandwidth through Equation (9.1.7).

To find the rms pulse width, σ, for a pulse $h(t)$, we first normalize $h(t)$ so that

$$\int_{-\infty}^{\infty} h(t)\, dt = 1 \tag{9.1.8}$$

The rms width of $h(t)$ is defined as follows:

$$\sigma^2 = \langle t^2 \rangle - \langle t \rangle^2 \tag{9.1.9}$$

where

$$\langle t \rangle = \int_{-\infty}^{\infty} t\, h(t)\, dt \tag{9.1.10}$$

and

$$\langle t^2 \rangle = \int_{-\infty}^{\infty} t^2 h(t)\, dt \tag{9.1.11}$$

For many pulse shapes the calculation of σ is straightforward.

The rms widths of several common pulse shapes are given in Table 9.1.1. Derivations of the results summarized here, and formulas for other pulse shapes, are given by Gowar [5.1].

When two or more linear systems are connected in tandem, the overall impulse response of the combination can be found by the convolution of the impulse responses of the individual systems. If $h_0(t)$ represents the input pulse shape, $h_1(t)$ impulse response of the system, and $h_2(t)$ the output pulse shape,

Table 9.1.1. Rms pulse width, σ, and spectral width, Δf, of three pulse shapes.

	$h(t)$*	σ	$H(f)$	Δf
Exponential:	$\dfrac{1}{\tau} \exp\left(-\dfrac{t}{\tau}\right)$ $t > 0$	τ	$\dfrac{1}{1 + j\omega\tau}$	$\dfrac{0.159}{\sigma}$
Rectangular:	$\dfrac{1}{\tau}$ $\dfrac{-\tau}{2} < t < \dfrac{\tau}{2}$	$\dfrac{\tau}{2\sqrt{3}}$	$\dfrac{\sin(f\tau)}{f\tau}$	$\dfrac{0.402}{\sigma}$
Gaussian:	$\dfrac{1}{\sqrt{2\pi}\tau} \exp\left[-\dfrac{t^2}{2\tau^2}\right]$	τ	$\exp\left[-2(\pi f\tau)^2\right]$	$\dfrac{0.133}{\sigma}$

* *The h(t) are normalized to unit area.*

then

$$h_2(t) = h_0(t) * h_1(t)$$
$$= \int_{-\infty}^{\infty} h_1(\tau)\, h_0(t - \tau)\, d\tau \tag{9.1.12}$$

Let us now derive expressions for $\langle t \rangle$ and $\langle t^2 \rangle$ for $h_2(t)$ in terms of properties of $h_0(t)$ and $h_1(t)$.

We assume that all of the $h(t)$ functions have been normalized according to Equation (9.1.8) above and that the origin of time is chosen such that $\langle t \rangle$ for the input pulse is zero.

$$\langle t \rangle_0 = \int_{-\infty}^{\infty} t\, h_0(t)\, dt = 0 \tag{9.1.13}$$

To find $\langle t \rangle_2$

$$\langle t \rangle_2 = \int_{-\infty}^{\infty} t\, h_2(t)\, dt \tag{9.1.14}$$

substitute Equation (9.1.12) for $h_2(t)$:

$$\langle t \rangle_2 = \int_{-\infty}^{\infty} t \int_{-\infty}^{\infty} h_1(\tau)\, h_0(t - \tau)\, d\tau\, dt \tag{9.1.15}$$

We can rearrange this double integral to integrate first with respect to t.

$$\langle t \rangle_2 = \int_{-\infty}^{\infty} h_1(\tau) \int_{-\infty}^{\infty} t\, h_0(t - \tau)\, dt\, d\tau \tag{9.1.16}$$

The inside integral can be evaluated by a change of variable. Let $x = t - \tau$;

$$\int_{-\infty}^{\infty} t\, h_0(t - \tau)\, dt = \int_{-\infty}^{\infty} (x + \tau)\, h_0(x)\, dx$$
$$= \int_{-\infty}^{\infty} x\, h_0(x)\, dx + \tau \int_{-\infty}^{\infty} h_0(x)\, dx = \tau \tag{9.1.17}$$

We recognize the first of the integrals in the last line as $\langle t \rangle_0$, which was defined to be zero, and the second integral as unity, due to normalization as defined by Equation (9.1.8). By substituting this result into (9.1.16) we find

$$\int_{-\infty}^{\infty} t\, h_2(t)\, dt = \int_{-\infty}^{\infty} \tau\, h_1(\tau)\, d\tau \tag{9.1.18}$$

which is equivalent to

$$\langle t \rangle_2 = \langle t \rangle_1 \tag{9.1.19}$$

To derive the rms width of $h_2(t)$ we follow a similar procedure.

$$\int_{-\infty}^{\infty} t^2\, h_2(t)\, dt = \int_{-\infty}^{\infty} t^2 \int_{-\infty}^{\infty} h_1(\tau)\, h_0(t-\tau)\, d\tau\, dt$$
$$= \int_{-\infty}^{\infty} h_1(\tau) \int_{-\infty}^{\infty} t^2\, h_0(t-\tau)\, dt\, d\tau \tag{9.1.20}$$

With the change of variable, $x = (t - \tau)$, the inside integral can be written

$$\int_{-\infty}^{\infty} (x^2 + 2x\tau + \tau^2)\, h_0(x)\, dx = \int_{-\infty}^{\infty} x^2 h_0(x)\, dx$$
$$+ 2\tau \int_{-\infty}^{\infty} x h_0(x)\, dx + \tau^2 \int_{-\infty}^{\infty} h_0(x)\, dx \tag{9.1.21}$$

The first of these integrals is recognized σ_0^2, the second zero, and the third unity. Then

$$\int_{-\infty}^{\infty} t^2 h_0(t-\tau)\, dt = \sigma_0^2 + \tau^2 \tag{9.1.22}$$

and

$$\int_{-\infty}^{\infty} t^2 h_2(t)\, dt = \int_{-\infty}^{\infty} h_1(\tau)\left[\sigma_0^2 + \tau^2\right] d\tau$$
$$= \sigma_0^2 \int_{-\infty}^{\infty} h_1(\tau)\, d\tau + \int_{-\infty}^{\infty} \tau^2 h_1(\tau)\, d\tau \tag{9.1.23}$$

The rms width of $h_2(t)$ is

$$\sigma_2^2 = \langle t^2 \rangle_2 - \langle t \rangle_2^2$$
$$= \int_{-\infty}^{\infty} t^2 h_2(t)\, dt - \left[\int_{-\infty}^{\infty} t h_2(t)\, dt\right]^2 \tag{9.1.24}$$

With Equations (9.1.23), (9.1.19), and (9.1.9) this can be reduced to

$$\sigma_2^2 = \sigma_0^2 + \sigma_1^2 \tag{9.1.25}$$

This relationship between the σ's is independent of the shapes of the various $h(t)$ pulses. Each σ is related to the shape of the pulse it characterizes, but the combination of σ's found here is simply the sum of the squares, independent of the pulse shapes.

If there are several subsystems in tandem, each represented by an impulse response $h(t)$ with rms width σ, the overall rms width of the impulse response is

$$\sigma^2 = \sigma_1{}^2 + \sigma_2{}^2 + \cdots + \sigma_n{}^2 \tag{9.1.26}$$

This is an important and useful result. Since we can relate σ to the pulse rate R, the pulse shape, or the bandwidth Δf, this equation provides a convenient way to relate system parameters to performance parameters. When we know the rms pulse width somewhere in the communication system, and we know the impulse response for the system from that point to the output, we can use this equation to find the rms width of the output pulse. For example, in the design of the optical receiver for a PCM communication system, the bit rate R in effect defines the rms width of the output pulse. If we have defined the transmitter pulse shape and know the dispersion of the fiber, we can use these data to calculate the rms width of the optical pulse into the receiver. Equation (9.1.7) can then be used to find the impulse response and bandwidth required in the receiver.

A similar relationship for rise times is used by some system engineers. If the shapes of the impulse responses are all the same, or are such that the ratio of σ to rise time is essentially the same for all $h(t)$'s, then the mean squared rise time for the complete system is equal to the sum of the squares of all of the component rise times. When the constraint concerning similarity of shapes is satisfied,

$$\tau^2 = \tau_1{}^2 + \tau_2{}^2 + \cdots + \tau_n{}^2 \tag{9.1.27}$$

The time domain characteristics of the transmitter, that is, LED or laser, are often given in terms of rise and fall times of the output pulse. If the leading edge of the pulse is assumed to be approximately exponential, then the time constant is half the rise-time. From Table 9.1.1 we see that for an exponential pulse, the rms width is equal to the time constant. Then $\sigma = \tau_r/2$. It can be shown that a pulse with exponential rise and decay, both with time constant τ, and with duration T between the start of the rise and the start of the decay, will have rms pulse width $\sigma^2 = \tau^2 + T^2/12$. For T of the order of τ, $\sigma \approx \tau \approx \tau_r/2$ (see problem 9.8).

The photodiode is also often characterized by the rise and decay times of its impulse response. Then $\sigma_{ph} = \tau_r/2$.

The effect of the bandwidth of the receiver amplifier can be expressed in terms of a σ by using the relationships repeated as Equation (9.1.7) above. If the bandwidth Δf is known, its effect can be represented by a $\sigma = \frac{1}{8\Delta f}$.

All of these effects, each of which represents a dispersion of the leading and trailing edges of the PCM pulse, are combined on a mean-squared basis. The overall system σ is found from the sum of the squares of the rms widths that must be considered.

$$\sigma^2 = \sigma_{tx}^2 + \sigma_f^2 + \sigma_{ph}^2 + \sigma_a^2 \qquad\qquad (9.1.28)$$

The terms on the right-hand side of Equation (9.1.28) represent the rms widths of the impulse responses of the transmitter, the fiber, the photodetector, and the receiver amplifier.

EXAMPLE 9.1

The rise times of the transmitter laser and the receiver photodiode are 5 ns and 3 ns, respectively. The bandwidth of the receiver amplifier is 25 MHz. The total dispersion in the optical fiber is 1.3 ns. What is the maximum PCM bit rate for this system?

Solution

$$\sigma_{tx} = \tfrac{5}{2} = 2.5 \text{ ns} \qquad \sigma_{ph} = 1.5 \text{ ns} \qquad \sigma_f = 1.3 \text{ ns}$$

$$\sigma_a = \frac{1}{(8 \cdot 25 \text{ MHz})} = 5 \text{ ns}$$

The total rms width of the pulse in the output of the receiver is

$$\sigma^2 = (2.5)^2 + (1.5)^2 + (1.3)^2 + (5)^2 = 35.19$$

$$\sigma = 5.93 \text{ ns}$$

The maximum bit rate is $R = \frac{1}{4\sigma} = 42.1 \text{ Mb/s}$.

In a classic paper on the design of receivers for optical communication systems, Personick [4.18] has studied the optimum pulse shape in the receiver and the penalty one must pay for departures from the optimum design. An important property of the optimum pulse shape is that it should have zero magnitude at all integral multiples of T, the pulse repetition period; this property reduces the effects of intersymbol interference. The $(\sin x)/x$ pulse has this property. Thus, an ideal low-pass filter has an impulse response that would produce this shape. A still better pulse shape is that corresponding to the impulse response of a raised-cosine frequency response. The pulse shape and its Fourier transform are:

$$H(f) = \tfrac{1}{2}[1 + \cos(\pi f T)] \qquad -\frac{1}{T} < f < \frac{1}{T} \qquad \text{(9.1.29a)}$$

$$h(t) = \frac{1}{1 - \left(\dfrac{2t}{T}\right)^2} \frac{\sin \dfrac{2\pi t}{T}}{\dfrac{2\pi t}{T}} \qquad \text{(9.1.29b)}$$

Personick gives graphs of the excess input power required for a given signal-to-noise ratio as a function of the rms width of the input pulse. He shows that the lowest input power corresponds to very short rms pulse width. The penalty for longer input pulses is small for $\sigma/T < 0.25$, but increases rapidly and also becomes strongly dependent on pulse shape as the pulse width increases. To keep this power penalty below 1 dB, the normalized pulse width, σ/T, must be less than 0.25. On this basis, the widely used relationship between bit rate and rms width, $R = \frac{1}{4\sigma}$, is established. This relationship was derived in Chapter 2 on the basis of the step response of an ideal low-pass filter and repeated as Equation (9.1.7).

To design the receiver amplifier/equalizer from the Personick point of view, one can follow these steps:

1. Plot the raised-cosine output spectrum corresponding to the bit rate B. The spectrum would normally be plotted on log-log scales, that is, a Bode plot.
2. Determine the width and, if possible, the shape of the photocurrent pulse. Make the width of this pulse as small as possible. Estimate its spectrum. If the pulse width is $\sigma \ll T$, the spectrum will be practically flat over the pass band of the amplifier-equalizer. If the pulse width is not short and if the effects of transmitter rise time, photodiode transit time, and fiber dispersion are significant in determining the width of the photocurrent pulse, the pulse shape may be assumed to be Gaussian.
3. The amplifier-equalizer circuits are designed to provide the transfer function determined by the output and input spectra that were found in steps 1 and 2. This transfer function includes the RC load impedance of the photodetector as well as the amplifier and equalizer circuits.

In a multilink PCM system, the effects of dispersion in limiting the bandwidth or data rate are not cumulative from one link to the next. At each repeater, a clean digital signal is regenerated with neither pulse-shape distortion nor pulse stretching. The only effect of dispersion that is not removed by regeneration is any effect it may have on the probability of error in the binary decisions in each repeater.

In the linear-amplifier repeater system, the repeater does not compensate or correct for the dispersion or bandwidth limitations inherent in the received signal. The overall bandwidth of such a system can be estimated in either of two ways.

The bandwidth of one length of fiber can be found by first determining the impulse response of the fiber (not a simple task) and finding $H(\omega)$, the Fourier transform of the impulse response. The width of $H(\omega)$ is defined to be the bandwidth of the fiber. For n such links in tandem, the overall frequency response is $H_n(\omega) = H_1{}^n(\omega)$. The bandwidth of $H_n(\omega)$ is less than that of $H_1(\omega)$ but how much less is not clear in the general case.

In many cases, the impulse response of the fiber link, including transmitter and receiver, can be approximated with a Gaussian pulse shape. When several networks are connected in tandem, with each represented by an impulse response, the impulse response of the complete cascade tends to become Gaussian. For the Gaussian pulse shape

$$h(t) = \exp\left[-\frac{t^2}{2\tau^2}\right] \tag{9.1.30a}$$

the Fourier transform has the form

$$H(\omega) = \exp\left[-\frac{(\omega\tau)^2}{2}\right] \tag{9.1.30b}$$

The bandwidth of this $H(\omega)$ is found by setting $|H(\omega)| = 1/\sqrt{2}$ and solving for ω. Thus, we find that $\omega_{3\,dB} \approx 0.83/\tau$. For n such lengths of fiber in tandem,

$$H_n(\omega) = \exp\left[-\frac{n(\omega\tau)^2}{2}\right] \tag{9.1.31}$$

By solving $|H_n(\omega_n)| = 1/\sqrt{2}$ for ω_n, we can find that $\omega_n = \omega_1/\sqrt{n}$.

If we assume an exponential impulse response,

$$h(t) = \exp(-\alpha t) \tag{9.1.32a}$$

the Fourier transform is

$$H(\omega) = \frac{1}{\alpha + j\omega} \tag{9.1.32b}$$

The 3-dB bandwidth of this $H(\omega)$ is $\omega_1 = 1/\alpha$. The bandwidth of n such lengths in tandem can be shown to be $\omega_n = [2^{1/n} - 1]^{\frac{1}{2}}\omega_1$. The simpler $\omega_n \approx \omega_1/\sqrt{n}$ gives an estimate that is approximately 20 percent high for large n.

Bandwidth can also be estimated from the dispersion parameters of the fiber. Each link has rms dispersion of σ seconds. Subsequent links produce similar rms dispersion of the input signal, which itself includes the accumulated dispersion of the previous links. The total rms dispersion is estimated to be

$\sigma_n{}^2 = n\sigma_1{}^2$. Since the bandwidth is inversely proportional to σ, we again have $\omega_n = \omega_1/\sqrt{n}$.

EXAMPLE 9.2

A PCM system is to transmit 10 Mb/s a distance of 200 km. The optical fiber to be used has average attenuation of 0.7 dB/km and bandwidth of 200 MHz-km. The system will consist of five links of 40 km each. The transmitter peak power is 0.1 mW. The sensitivity of the receiver is -35 dBm (peak) for an average probability of error of 10^{-10}. Compare the linear amplifier and digital regenerator type repeaters for this system. Assume that both type repeaters use the same photodetector and preamplifier.

Solution

The attenuation over a 40-km length of fiber is 28 dB. The bandwidth is $200/40 = 5$ MHz. Since the PCM bit rate can be twice the bandwidth, 5 MHz is adequate for the 10 Mb/s bit rate.

The S/N can be related to the probability of error by using the erfc graph. For $P(e) = 10^{-10}$, $\text{erfc}(V_t/\sqrt{2}\sigma) = 2 \cdot 10^{-10} = 10^{-9.7}$. From the *erfc* graph, Figure 9.1.2, $V_t/\sigma = 4.5\sqrt{2} = 6.3\,(16\text{ dB})$. The peak signal is $V_1 = 2V_t$; the S/N, using peak signal power, is 6 dB greater than V_t/σ, or 22 dB.

Since the receiver peak input power for this condition is -35 dBm, the equivalent noise power, referred to the optical input of the receiver, is -57 dBm or 2 nW. We assume for this example that this noise is independent of the input signal power; this is equivalent to assuming that the photodiode shot noise is small in comparison to other receiver noise sources.

Consider first the digital-regenerator repeaters. The transmitted power is 0.1 mW, that is, -10 dBm. With 28 dB attenuation, the received optical power will be -38 dBm. Since the noise power is -57 dBm, the S/N is $-38 + 57 = 19$ dB. With 19 dB, $V_1/\sigma = 8.9$, $V_t/\sigma = 4.45$, and $V_t/\sqrt{2}\sigma = 3.15$. The probability of error is $1/2\ \text{erfc}(3.15) = 4 \cdot 10^{-6}$. With five such links in tandem, the overall probability of error is $P(e) = 2 \cdot 10^{-5}$.

The bandwidth for each link is adequate, though barely so, for the 10 Mb/s PCM data rate.

Consider now the use of the linear-amplifier type repeater. At the end of the first link, the signal level is -38 dBm and the noise, referred to the same point, is -57 dBm. After amplification of $+28$ dB, to compensate for the fiber attenuation, and attenuation of -28 dB in the next fiber, both signal and noise are at the -38 and -57 dBm levels. At this point, in the second amplifier an additional 2 nW of noise is added to that received. At the end of five transmissions, the signal level is still -38 dBm and the noise is $5 \cdot 2$ nW, or -50 dBm. The S/N is now $-38 + 50 = 12$ dB. With $S/N = 12$ dB with peak signal power, we can find the probability of error to be $P(e) = 0.023$.

The bandwidth of each 40 km link is 5 MHz, barely adequate for the 10 Mb/s PCM signal. Any further bandwidth reduction would reduce the system bandwidth below that required. Since cascading similar $H(\omega)$ functions will narrow the overall bandwidth, the bandwidth of the five 40-km links will be less than that required for the 10 Mb/s PCM signal. Estimation techniques described in Section 9.1.3 give $f_{3\,dB} \approx 5/\sqrt{5} = 2.2$ MHz for the 200-km system. The PCM data rate would be limited to 4.4 Mb/s and would have a very high error rate.

9.1.4   *PCM System Design*

System design is a complex process. It can be described in terms of the stages of the design process. The stages can be broadly defined as (1) preliminary design, (2) detailed design, (3) prototype development and evaluation, and (4) final design. The process is iterative in that at any point in the evolution of the ultimate system it may be desirable to return to an earlier stage to reconsider a preliminary decision. The four stages defined here may overlap in time; they may merge to the extent that a particular activity may not fit clearly into one of these categories; they may each be broken into stages, tasks, and subtasks.

The level of system design that we are prepared to consider at this stage in our study of optical communication systems is preliminary design. Preliminary design consists of evaluating the specifications for the proposed system, identifying alternative means for meeting these specifications, making tentative selections among the alternatives, and preparing more detailed specifications that will be required for the more detailed design to follow.

In this section, we will use two example design problems to illustrate some principles of optical fiber communication system design.

EXAMPLE 9.3

An optical communication system is needed to meet the following specifications: data rate = 100 Mb/s; BER = 10^{-9}; distance between terminals = 500 km. Make tentative selections for the transmitter, optical fiber, and receiver specifications and indicate the number of repeaters that will be required.

Solution

Preliminary considerations:

The pulse repetition period is $T = 1/R = 10$ ns. The maximum total system dispersion, including effects of receiver band-limiting, is $\sigma = \frac{1}{4R} = 2.5$ ns max. The minimum overall system bandwidth is $f = R/2 = 50$ MHz min.

Because the data rate is high and the total distance is long, a laser transmitter and a single mode fiber will probably be required. A PIN photodetector may be adequate for the receiver.

Select for the wavelength $\lambda = 1.3$ μm. High performance components are more readily available for this wavelength than for the 1.55-μm option.

Subsystem specifications:

The following specifications for the transmitter, fiber, and receiver are tentatively selected. The individual subsystems will then be examined in more detail. Preliminary selections of subsystem parameters can be changed if they do not lead to a satisfactory system.

After study of manufacturers' data sheets describing devices for 1.3-μm systems, the following specifications are selected:

Transmitter:

$\tau_r = \tau_f = 2$ ns
$P_t = -5$ dBm, peak output power into a fiber pigtail
laser spectral width = 3 nm.

Optical fiber:

Single mode: $\Delta = 0.0037$
$\alpha = 0.6$ dB/km
$D = 3.2$ ps/nm-km

PIN Photodiode:

$C = 1$ pF
$\tau_r = \tau_f = 1$ ns
$\mathcal{R}_0 = 0.8$ A/W @ 1.3 μm
100-μm-dia active area
$I_d = 15$ nA

Preliminary system design:

Receiver:

Assume that receiver sensitivity of 2000 photons per bit can be achieved.

$$P_r = 2000 \cdot R \cdot \frac{hc}{\lambda} = 3.06 \cdot 10^{-5} \text{ mW } (-45.1 \text{ dBm})$$

Attenuation:

Total losses = -5 dBm + 45.1 dBm = 40.1 dB
For fiber sections of 3 km between splices and 0.1-dB average loss per splice, the average loss due to splices is 0.1 dB/3km = 0.033 dB/km. Other losses are a 2-dB coupling loss at the receiver and a 3-dB system margin. These total 5 dB.

The maximum fiber attenuation losses, including splice losses, can total $40.1 - 5 = 35.1$ dB; the attenuation per km, including splice loss, is 0.633 dB/km.
The maximum length of fiber is 35.1/0.633 = 55.5 km.

Dispersion:

The fiber dispersion per unit length is

$\sigma/L = 3.2$ ps/nm-km $\times$ 3 nm = 9.6 ps/km

The maximum dispersion is 2.5 ns. The maximum length of fiber is $2.5/0.0096 =$ 260 km. Since attenuation limits the fiber length to 55 km, the dispersion in the fiber will be $9.6 \times 55 = 528$ ps or 0.53 ns. The equivalent bandwidth of the fiber is

$$\Delta f = \frac{1}{8\sigma} = 237 \text{ MHz}$$

Bit Error Rate:

The BER for the 500-km system cannot exceed 10^{-9}. Since the maximum distance between repeaters is 55 km, there must be ten lengths averaging 50 km each. The maximum BER for each link is 10^{-10}.

The S/N necessary to realize this BER can be found by using the *erfc* graph, Figure 9.1.2, and Equations (9.1.5). An equivalent graph, showing $P(e)$ versus S/N, is given in Figure 9.1.4. From Figure 9.1.4 we see that the required S/N is

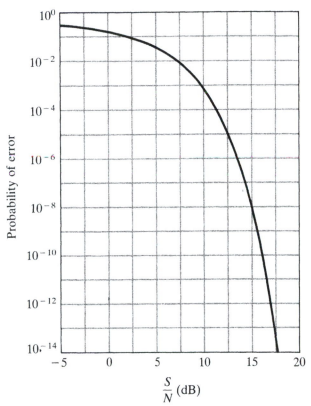

Figure 9.1.4 *Probability of error versus signal-to-noise ratio. The probability of error can be found from the erfc(x) function by Equation (9.1.5). The effective signal is the amount by which the signal level differs from the threshold level. For the on-off binary PCM signal, S/N is 20 log (V_t/σ) or 20 log $[(V_1 - V_t)/\sigma]$, and $P(e) = \frac{1}{2}$ erfc $\left[\dfrac{V_t}{\sigma\sqrt{2}}\right]$ or $\frac{1}{2}$ erfc $\left[\dfrac{(V_1 - V_t)}{\sigma\sqrt{2}}\right]$.*

16 dB. This is based on the signal amplitude being interpreted as the amount by which the peak pulse amplitude exceeds the threshold. If the threshold is placed at one-half the peak amplitude, the peak signal is twice the magnitude of the amplitude above threshold. Thus, the peak-signal to rms-noise ratio is 6 dB higher than the S/N read from the probability-of-error graph, or 22 dB.

Receiver Bandwidth:

The receiver bandwidth can be found by using the rms widths of the transmitter pulse and of the impulse responses of the fiber and photodetector.

The speed of response of the transmitter laser and the photodiode are specified in terms of rise and fall times. Following Equation (9.1.27), the relationship between rms width and exponential rise time or fall time was shown to be approximately $\sigma = \tau_r/2$. We then have estimates for the system σ and the σ's of all parts of the system except the amplifier. Then

$$\sigma_s^2 = \sigma_{tx}^2 + \sigma_f^2 + \sigma_{ph}^2 + \sigma_a^2$$

$$(2.5)^2 = (1)^2 + (0.53)^2 + (0.5)^2 + \sigma_a^2$$

from which

$$\sigma_a = 2.17 \text{ ns}$$

and

$$\Delta f = \tfrac{1}{8\sigma} = 57.5 \text{ MHz}$$

The bandwidth of the receiver is 57.5 MHz. For the preliminary design, we will interpret this as the cutoff frequency of an ideal low-pass filter characteristic.

Receiver Sensitivity:

We have already selected a PIN photodiode and have assumed a receiver sensitivity of 2000 photons per bit. We must now examine the receiver in more detail to see that this sensitivity is a reasonable design objective.

The photodetector will be followed by a transimpedance amplifier having the following parameters:

$$C_a = 2 \text{ pF}$$

$$R_f = ?$$

$$\{i_a^2\} = 6 \cdot 10^{-25} \text{ A}^2/\text{Hz}$$

$$\{v_a^2\} = 6 \cdot 10^{-20} \text{ V}^2/\text{Hz}$$

If we assume the parasitic capacitances in the photodetector circuit to be 1 pF, then the total effective C in the photodetector load impedance is 4 pF. The load resistance should be made as high as is feasible. The equivalent resistance is made up of this R and the R_f in parallel. The magnitude of this equivalent resistance is limited by practical considerations; as it is made very large, the range of frequencies over which equalization is required, and hence the difficulty of designing the amplifier, increases. We will limit this range to one decade in frequency. The RC corner frequency is then approximately $\Delta f/10$, or 5.75 MHz.

$$R_f \approx \frac{1}{2\pi\Delta fC} = \frac{1}{2\pi \cdot 5.75 \cdot 10^6 \cdot 4 \cdot 10^{-12}} = 6920 \text{ ohms}$$

Let $R_f = 6500$ ohms.

The receiver sensitivity can now be calculated. Equation (8.1.20), with $M = 1$ and $F = 1$, is applicable. For the PCM system, the signal current I is either zero or that due to the peak transmitted power, that is, the signal current I and the average photocurrent I_{ph} are the same current.

$$\frac{S}{N} = \frac{I^2}{\left\{ 2q(I_{ph} + I_d) + \dfrac{4kT}{R_f} + \{i_a^2\} + \{v_a^2\}\left[\dfrac{1}{R_f^2} + \dfrac{(2\pi\Delta fC)^2}{3} \right] \right\} \Delta f}$$

The values of the individual noise terms in the denominator are:

$$\frac{4kT}{R_f} = 2.55 \cdot 10^{-24} \text{ A}^2/\text{Hz}$$

$$2qI_d = 0.005 \cdot 10^{-24}$$

$$\{i_a^2\} = 0.60 \cdot 10^{-24}$$

and

$$\{v_a^2\}[\ldots] = 0.04 \cdot 10^{-24}$$

These total $3.19 \cdot 10^{-24}$ A²/Hz. The other term in the denominator is $2qI = 3.2I \cdot 10^{-19}$. It is evident that the thermal noise is the dominant one and that the shot noise due to dark current is negligible.

We found above that for the BER required for each link, the S/N must be 22 dB or larger. With this figure, the noise terms, and $I_{ph} = I$, the sensitivity equation can be reduced to a quadratic equation in I.

$$I^2 - 2.92 \cdot 10^{-9}I - 2.91 \cdot 10^{-14} = 0$$

This equation has one positive and one negative solution. The negative solution is meaningless. The positive one gives the sensitivity of this receiver.

$$I = 0.169 \ \mu\text{A}$$

$$P_{in} = \frac{I}{\mathcal{R}_0} = 0.21 \ \mu\text{W} \ (-36.8 \text{ dBm})$$

We assumed initially a receiver sensitivity of −45.1 dBm. We have calculated the sensitivity of this receiver to be −36.8 dBm, or 8.3 dB poorer than that required for the BER specified. We must reexamine the design to find a way either to improve the sensitivity or to provide larger input optical power to the receiver.

Since the receiver noise is dominated by thermal noise, it could be reduced by increasing R_f. However, we cannot get 8 dB improvement from this alone.

Other options that can be considered include

1. An APD photodetector.
2. A higher power laser for the transmitter.
3. A fiber with lower attenuation.
4. A shorter distance between repeaters.

We will consider first an APD photodetector having the same specifications as the PIN and with $k = 0.1$.

Equation (8.1.20) can be used with M and F as unknowns. They are to be determined to give the maximum sensitivity. From Figure 7.3.5 we see that for $k = 0.1$ and $M < 100$, $F \approx M^{0.5}$. The equation to be solved for the optimum M is

$$I^2 - 2.92 \cdot 10^{-9}(I + I_d)M^{0.5} - \frac{2.91 \cdot 10^{-14}}{M^2} = 0$$

The optimum M is the M that gives minimum I. In principle, it can be found by differentiating this equation with respect to M and setting $dI/dM = 0$. The resulting equation gives the value of M for which $dI/dM = 0$, which can be shown to be a minimum. Another option, often easier, is to plot I versus M. A graph of the equation above is shown in Figure 9.1.5. The minimum I and the corresponding M can then be read from the graph.

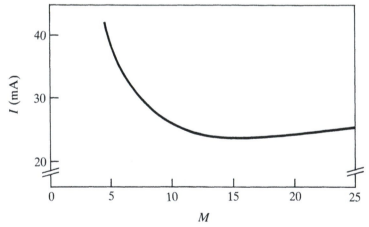

Figure 9.1.5 *Graph of sensitivity versus M. The sensitivity is highest for $M \approx 16$.*

By using the graphical technique, the solution is found to be $M = 16$ and $I = 23.8$ nA. The corresponding input optical power, which is the receiver sensitivity, is

$$P_{in} = \frac{I}{\mathcal{R}_0} = \frac{23.8}{0.8} = 29.75 \text{ nW}$$

The receiver sensitivity is 29.75 nW (-45.3 dBm), or 1945 photons per bit.

This sensitivity is slightly better than the assumed value. The APD receiver, with the laser and optical fiber characteristics selected initially, gives a system that satisfies the performance specifications.

To summarize, the 500-km system will have ten links of average length 50 km, a maximum BER of 10^{-10} for each link, and receiver sensitivity of -45.3 dBm and transmitter power of -5 dBm at each repeater.

In the example above, the performance objectives, and little else, were specified. Many system and subsystem parameters were left for the designer to choose in evolving a design to meet the performance objectives. After one tentative design proved to be unsatisfactory, a second effort, based on improving the "weak link" of the first design, was successful.

In the following example, most system parameters are specified. The problem is to evaluate trade-offs between data rate and distance within the constraints imposed. As we define the limits on system performance and identify the system parameters that lead to these limits, the most likely routes for a productive attack on the performance limitations are apparent.

EXAMPLE 9.4

For the systems specified below, plot graphs of the maximum distance between repeaters versus data rate for $\lambda = 0.85$ and $\lambda = 1.3$ μm. Both the attenuation limits and the dispersion limits must be included.

Specifications

Optical fiber:

Graded-index fiber with the profile parameter α optimized to minimize intermodal dispersion. The index $\Delta = 0.015$ and the group index $N = 1.5$. The attenuation of the fiber is (a) $\alpha_{0.85} = 4$ dB/km and (b) $\alpha_{1.3} = 0.6$ dB/km at $\lambda = 0.85$ μm and $\lambda = 1.3$ μm, respectively. These attenuation constants include splicing and coupling losses not otherwise specified.

Transmitters:

(a) At 0.85 μm, an LED delivers 100 μW peak power into the fiber. Its spectral width is $\Delta\lambda = 50$ nm. (b) At 1.3 μm, a single-mode laser delivers 1 mW peak power into the fiber. Its spectral width is 5 nm.

Receivers:

The photodetector is a PIN diode, with responsivity of (a) 0.4 at 0.85 μm and (b) 0.7 at 1.3 μm.

The preamplifier is an FET transimpedance amplifier with $R_f = 1000$ ohms and $g_m = 5000$ μS. I_G and the noise-current source at the input of the amplifier are zero. The total shunt capacitance in the amplifier input circuit is 5 pF.

The S/N at the output of the preamplifier should be 20 dB.

Solution

Attenuation:

To find the total attenuation in the fiber, it is necessary first to calculate the sensitivity of the receiver. From equation (8.19), with $R \gg R_F$, M and $F = 1$, and $I_{ph} = \mathscr{R}_0 P$,

$$\frac{S}{N} = \frac{\mathscr{R}_0^2 P^2}{\left\{ 2q(\mathscr{R}_0 P + I_d) + \dfrac{4kT}{R_F} + \{i_a^2\} + \{v_a^2\}\left[\dfrac{1}{R_F^2} + \dfrac{(2\pi CB)^2}{3} \right] \right\} B}$$

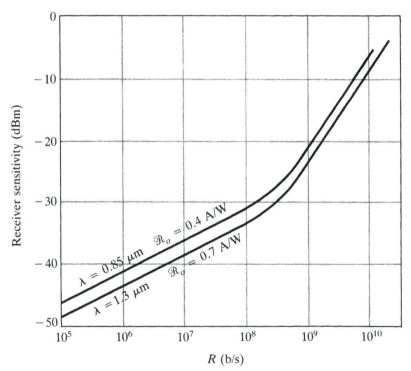

Figure 9.1.6 *Receiver sensitivity versus data rate for 0.85- and 1.3-μm receivers. The decrease in sensitivity for higher data rates is due to the higher bandwidths, with a concomitant increase in noise for higher data rates. The noise bandwidth is proportional to R at low data rates and to R^3 at high data rates.*

From [4.9, p.136]

$$\{v_a{}^2\} = \tfrac{2}{3}\left[\frac{4kT}{g_m}\right]$$

With the numerical values given in the specification, and $R = 2B$, the receiver sensitivity equation for $\lambda = 1.3 \ \mu m$ can be reduced to

$$P^2 = (2.26 \cdot 10^{-17}P + 1.88 \cdot 10^{-21})R + 1.816 \cdot 10^{-38}R^3$$

This equation is plotted in Figure 9.1.6. It can be plotted with straight-line asymptotes, using the second term for small R and the third term for large R. The first term is very small for all values of P of interest.

For $\lambda = 0.85 \ \mu m$, the sensitivity equation is the same except that the responsivity is reduced from 0.7 to 0.4. The resulting equation is

$$P^2 = (4.00 \cdot 10^{-17}P + 5.87 \cdot 10^{-21})R + 5.68 \cdot 10^{-38}R^3$$

This equation is also plotted in Figure 9.1.6.

The total attenuation is equal to the difference between the transmitted power and the receiver sensitivity. The maximum distance between repeaters, so far as the power levels are concerned, is the total attenuation in dB divided by the fiber attenuation in dB per km. Curves showing the maximum distance between repeaters versus data rate for the attenuation limit are shown in Figure 9.1.7

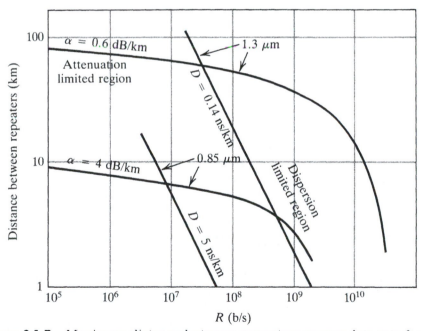

Figure 9.1.7   *Maximum distance between repeaters versus data rate for 0.85- and 1.3-μm receivers. The maximum distance between repeaters is limited by signal attenuation at low data rates and by dispersion at high data rates.*

Dispersion:

The graded-index fiber has $\Delta = 0.015$ and $N = 1.5$. The modal dispersion is given by Equation (4.3.22b):

$$\frac{\tau}{L} = \frac{N\Delta^2}{8c} = \frac{1.5(0.015)^2}{8c} = 1.4 \cdot 10^{-13} \text{ s/m}$$
$$= 0.14 \text{ ns/km}$$

The material dispersion will be different for the two wavelengths. For $\lambda = 0.85$ μm, we estimate from Figure 3.4.3 that $D = 100$ ps/(nm-km). The material dispersion is then

$$\frac{\Delta\tau}{L} = 100 \text{ ps/(nm-km)} \times 50 \text{ nm}$$
$$= 5 \text{ ns/km}$$

The combination of modal and material (intramodal) dispersion is

$$\left(\frac{\Delta\tau}{L}\right)_{0.85} = [0.14^2 + 5^2]^{\frac{1}{2}} \approx 5 \text{ ns/km}$$

To find material dispersion for 1.3 μm, we will assume $D = 3$ ps/(nm-km). The material dispersion is

$$\frac{\Delta\tau}{L} = 3 \text{ ps/(nm-km)} \times 5 \text{ nm} = 15 \text{ ps/km}$$

The total dispersion is

$$\left(\frac{\Delta\tau}{L}\right)_{1.3} = [0.14^2 + 0.015^2]^{\frac{1}{2}} \approx 0.14 \text{ ns/km}$$

From Equation (4.4.6), we have $R = \frac{1}{4\sigma}$, where σ is equivalent to $\Delta\tau$, the dispersion in seconds. The relationship between R and L is then $RL = 1/(4 \cdot 5 \text{ ns})$ for $\lambda = 0.85$ μm and $RL = 1/(4 \cdot 0.14 \text{ ns})$ for $\lambda = 1.3$ μm. These dispersion limits are plotted on Figure 9.1.7.

If we assume for the moment that the parameters used in developing the data shown in Figure 9.1.7 are realistic, then we can deduce from this figure some interesting and useful generalizations concerning such systems. For example, we see that the 0.85 μm/LED system is limited to maximum repeater spacing of approximately 10 km and to data rates of 10 Mb/s. Similarly, the 1.3 μm/laser system is limited to 60 km and 50 Mb/s. Trade-offs of distance for data rates are available. The 1.3-μm system could function at 100 Mb/s if the distance is limited to 20 km or over a distance of 65 km if the data rate is limited to 10 Mb/s. The system is dispersion-limited at 100 Mb/s and is attenuation-limited at 65 km.

Graphs such as those of Figure 9.1.7 are useful for selecting the most likely wavelengths, source devices, and other system parameters to meet data rate and distance specifications. They are also useful for extending the limits shown by indicating the specific parameters that impose limits.

Consider the means for designing a system to transmit data at rates of 1 Gb/s with distances up to 100 km between repeaters. It is clear from Figure 9.1.7 that neither the attenuation nor the dispersion characteristics of the 1.3-μm system are adequate for the proposed system. We then must consider the available alternatives for upgrading both.

To raise the attenuation limits, there are several possible approaches; we could consider increasing the transmitter power, decreasing the transmission losses, increasing the receiver sensitivity, and extending to higher frequencies the corner frequency at which the "f" noise becomes significant. The transmitted power can be increased if we can find a higher-power laser or if we are willing to degrade lifetime and reliability by operating an available laser at higher power levels. The losses can be reduced by a factor of two by changing the wavelength to 1.55 μm, at which the fiber attenuation constant can be less than 0.3 dB/km. The receiver sensitivity can be increased by using an APD detector. The corner frequency can be increased by reducing the total input-circuit capacitance. Each of these options has potential for achieving some of the improvements that we require. All would probably be considered and evaluated in some detail.

We must decrease the dispersion in order to move the dispersion limit to higher data rates. The dispersion per km is the product of the fiber dispersion properties, expressed as ps/(nm-km), and the spectral width of the signal being propagated through the fiber; here we should include effects of the photodetector response time and the receiver bandwidth in the overall system dispersion, in addition to the dispersion in the fiber. The fiber dispersion can be reduced by selecting a single-mode fiber rather than the graded-index fiber chosen for this example. That will eliminate the modal dispersion and reduce the total fiber dispersion to 15 ps/(nm-km). This material dispersion can be reduced by (1) using a fiber with its wavelength of zero dispersion shifted to the operating wavelength, or by (2) using a laser with a smaller line width.

Evidence that the 1-GHz–100-km objectives can be achieved is found in descriptions of such systems in the periodical technical literature. For example, reference [5.2] describes a 4-Gb/s–117-km system developed in a research laboratory. It uses a 1.55-μm, compound-cavity laser with 9.5-mW output power and an external traveling-wave modulator. The receiver sensitivity was −30.6 dBm, which corresponds to 1424 photons per bit. The fiber dispersion was 18 ps/(nm-km). The sensitivity and dispersion figures suggest that it should be possible to extend the performance to even higher distances and data rates.

The state-of-the-art, circa 1987, extends to 4 Gb/s-100$^+$ km system in development laboratories and 500 Mb/s-50 km in commercial application. Both of these pairs of numbers are approximate, order-of-magnitude indications of current achievements. Higher capacity components and/or systems are announced regularly at technical conferences and in almost every issue of the *Journal of Lightwave Technology* and other technical journals.

9.2 Analog Systems

Unlike digital signals, analog signals are continuous functions of time. The range of amplitudes is continuous rather than quantized. Communication systems for analog signals are expected to be linear so that the analog waveforms at the output of the receiver are faithful reproductions of the modulating signals.

Analog signals can transmit video, audio, or data waveforms. An analog signal can be a sinusoidal waveform that is itself modulated with an analog or digital waveform.

In this section, we will briefly examine analog communication systems to identify ways in which they differ from digital systems. These differences, and their impact on system design, will be discussed. Many considerations are essentially the same for both analog and digital systems; these, having been covered in Section 9.1, will not be repeated.

9.2.1 *Signal-To-Noise Ratio and Bandwidth Requirements*

The signal-to-noise ratio required for an analog signal is often higher than that for a digital signal. In the case of the digital signal, the S/N is determined by the probability of error in digital decisions in the receiver; in binary PCM, 20 dB or less is adequate for many systems. For analog signals, there is some performance measure that will determine the S/N.

The S/N required at the output of the receiver depends on the application. In coding the data from an instrumentation system, for example, the S/N would be related to the accuracy of the data being transmitted or perhaps to the accuracy required in the use of the data. For example, if the instrument generating the data has a probable error of $\pm$ 5 percent, then $S/N = 30$ dB should be adequate. As another example, studio-quality video signals should have S/N of 60 dB; a high-quality broadcast system design can require 50 dB S/N at the receiver.

There is little to be said about transmission bandwidth requirements that was not summarized in Section 2.1. The analog signal has a frequency spectrum that is specified; the communication system designer is given the signal bandwidth or some signal characteristics from which bandwidth can be found. The transmission-channel bandwidth is determined by the signal bandwidth and the modulation method selected. The bandwidth requirements for AM, IM, FM, and PCM were discussed in Chapter 2.

9.2.2 *Microwave Subcarrier Systems*

In some cases, it is advantageous to transmit a modulated microwave carrier over an optical fiber system rather than through a coaxial cable or other microwave channel. In such cases, the microwave carrier is an analog signal input to the optical fiber system. This is an example of a compound modulation system.

Intensity modulation of optical carriers with modulating frequencies in the GHz range is feasible [5.3][5.4]. The use of the optical carrier for the microwave signal can have advantages over conventional waveguides and coaxial cables. The optical fiber may be simply an alternative transmission channel for the microwave signal, with advantages in attenuation, bandwidth, size, weight, and/ or cost. It can offer options not available with more conventional transmission methods.

It is feasible to intensity-modulate an optical carrier with the entire VHF broadcast television spectrum. The gaps within this spectrum, reserved in the radio frequency allocations for other broadcast services, can have additional video or nonvideo channels. This offers an example of how the combination of frequency-division multiplexing and compound modulation enables us to begin to use the very broad spectrum capabilities of optical fiber systems.

9.2.3 *Video Signal Transmission*

Transmission of video signals over optical fiber systems has received some attention. Transmission over short, unrepeatered links can be done by an analog modulation method. Long-distance video transmission is most likely done with PCM.

Transmission of the baseband video signal by intensity modulation of an optical carrier must include consideration of linearity as well as the usual attenuation and bandwidth constraints. The laser, or LED, is not strictly linear, as our introductory analysis in Chapters 5 and 6 suggested. The laser characteristic has some curvature near the threshold current, rather than a sharp cutoff, and it has one or more kinks in the current versus power (I-P) characteristic for high current levels. The designer must determine where the I-P characteristic is reasonably linear and bias the laser so that the video signal falls within the linear region. It may be reasonable to allow the sync pulses, or black-level components of the composite signal, to extend into regions on the I-P characteristic that are not sufficiently linear for the picture component of the composite signal.

Optical fiber transmission of video signals by compound modulation methods, described in Section 9.2.2, can have advantages. If a sinusoidal subcarrier is frequency-modulated with the video signal, the linearity requirements for the optical system are eased and the noise advantage of frequency modulation can be realized.

9.2.4 *Analog Signal Transmission by PCM*

PCM is probably the most common system for the transmission of analog signals over distances that require the use of repeaters. The analog signal must be sampled, quantized, and coded for digital transmission. In such cases, digital transmission is preferred because of the problems of managing the cumulative effects of noise and dispersion over long distances.

In a PCM system, a limited number of discrete amplitudes can be transmitted. The PCM code thus conveys an approximation of the true amplitude. The larger the number of quantizing levels, the better the approximation. The difference between the true magnitude and the PCM representation is a kind of distortion. Its effect is much like noise, and it is called quantizing noise.

Quantizing noise is determined by the number of discrete levels into which the analog signal is quantized. The PCM code, with a discrete number of bits, represents an approximate value of the magnitude of the analog signal at each sampling time. If the separation between the possible discrete levels is ΔA, then the error in the magnitude transmitted with the PCM code is in the range $-\Delta A/2 < \text{error} < \Delta A/2$. This rms error is defined as quantizing noise. It can be shown to be

$$N_Q = \frac{A^2}{12M^2} \tag{9.2.1}$$

where M is the number of quantizing levels. If the signal can have any value in the range $-A/2 < s(t) < A/2$ with equal probability, the rms signal is $A^2/12$ and the signal-to-quantizing-noise ratio is [1.11, Section 7.7]

$$\frac{S}{N_Q} = M^2 \tag{9.2.2}$$

When the range of signal magnitudes is defined in a different way, for example, as $0 < s(t) < A$, then the signal power and the S/N can differ from Equation (9.2.2) by a constant factor.

We will assume that the additional noise, caused by an occasional binary decision error, is small compared to the quantizing noise. The number of quantizing levels is then M and the bits per sample $D = \log_2 M$. Sampling must be at greater than the Nyquist rate, giving a PCM data rate

$$R > 2B \log_2 M \tag{9.2.3}$$

and a PCM system bandwidth (at baseband) of at least $R/2$.

In transmitting analog signals by PCM, we trade greater bandwidth for lower receiver S/N. As we have seen in Sections 9.1.2 and 9.1.3, PCM also has the great advantage that its performance over a long-distance multilink transmission system is much better than conventional intensity modulation for the analog signal.

9.3 Coherent Optical Communications

Coherent optical communication systems have the potential for significantly higher levels of performance than that available from intensity modulation with direct detection. This potential has been demonstrated in research laboratories but has not yet been used in practical systems.

Coherent communication techniques rely on the coherence of the optical carrier in order to use modulation and demodulation methods more complex, and correspondingly more powerful, than those used with elementary intensity modulation. Frequency and phase modulation and heterodyne detection are examples of coherent modulation and detection methods. In each case, the frequency or phase of the carrier and the receiver's local oscillator must be known and controlled.

In this section, we will examine several coherent communication methods to compare their advantages and identify some of the problems in implementing such methods.

9.3.1 Quantum Limits and the Ideal Detector

This section does not deal directly with coherent detection methods. It is provided because the quantum limits or ideal detectors provide a useful and commonly used basis for comparing coherent and other modulation/detection techniques and for determining whether a specific principle or circuit approaches an ideal level of performance.

The quantum limit, or another form of ideal detector, is used as a basis for evaluating the quality of a communication system and for comparing the performance of various systems. The concept of the quantum limit was introduced in Section 9.1.1. It was shown that if the average number of photons received in a signaling interval is greater than 20, the probability that no photons will be received in any specific signaling interval is less than $2 \cdot 10^{-9}$. For PCM systems using on-off intensity modulation and a decision threshold of one photon, the average number of photons per bit is $20/2 = 10$ and the average probability of error is 10^{-9}. The quantum limit for this probability of error is 10 photons per bit.

A detector can be made ideal by assigning to its critical parameters their optimum values, even though these values may be unrealistic. For example, in the basic PIN detector, the ideal detector has a quantum efficiency of unity and the only noise present is the shot noise associated with the photocurrent. The output signal-to-noise ratio of this ideal detector can be calculated and used as a measure against which practical designs can be evaluated.

In this section, we will use binary PCM for illustrative discussion and numerical examples. For direct detection, a pulse of power P_1 is transmitted to represent a binary *1* and nothing is transmitted for a binary *0*. This is on-off intensity modulation. This type of PCM system is known as on-off keying (OOK) or amplitude-shift keying (ASK). The ASK designation has become the one in most common use.

For the P_1 signal, the photocurrent is $I_{ph} = \mathcal{R}_0 P_1$. If this is taken to be the signal current and the shot noise associated with this current is the only noise, then the signal-to-noise ratio is

$$\frac{S}{N} = \frac{I_{ph}{}^2}{2qI_{ph}B} = \frac{I_{ph}}{2qB} \tag{9.3.1}$$

To express the photocurrent in terms of the responsivity, we substitute $\mathscr{R}_0 = q\eta/hf$, and let $\eta = 1.0$; then

$$\frac{S}{N} = \frac{P_1}{2hfB} \tag{9.3.2}$$

If the signaling interval $T = 1/R$, $R = 2B$, and $P_1/hf = \phi$ photons per second, then

$$\frac{S}{N} = \phi T \quad \text{photons per bit} \tag{9.3.3}$$

This is the S/N for an ideal PIN detector.

For a given received power, wavelength, and bit rate, we can calculate the average received energy in photons per bit; it is a useful way to normalize the power/energy relationships for many problems.

For an on-off intensity-modulated PCM system, a pulse of average power P_1 is received for a binary 1 and no signal power is received for a 0. There will be some noise associated with each. The peak signal to rms noise ratio for the binary 1 is that given by Equation (9.3.2) and (9.3.3). For the binary 0, there is no signal; the rms noise, σ_0, can be found when the noise sources are identified. For small signal and noise powers, that is, a few photons per bit, Gaussian statistics and the *erfc* are not applicable. The probabilities of error, $P(e|1)$ and $P(e|0)$, must be calculated independently using the Poisson probability distribution (see Example 9.5.).

In some cases, the Poisson distribution can be approximated by the Gaussian distribution function. However, this approximation is not valid in all cases of interest. If N is the average number of photons per signaling interval, the Gaussian approximation is valid when n, the number of photons satisfies two conditions: $|n - N| \ll N$ and $n \gg 1$. The approximation is valid only near the peak of the distribution and only if the number of photons per bit is large. It must be used with caution unless these conditions are satisfied. In comparing a real detector with the quantum limit of 10 photons per bit, the Gaussian approximation is not applicable.

EXAMPLE 9.5

In an on-off PCM system, the average received photons per bit is 35 for a binary 1 and 0.1 for a 0. The small received signal for the binary 0 may be due to background light or to some small residual light from the transmitter during the 0 intervals. Find the optimum decision threshold and the average probability of decision error.

Solution

There will be two types of error, interpreting a *1* as a *0* and interpreting a *0* as a *1*. They must be treated separately and averaged.

The Poisson distribution when a *1* was transmitted is

$$P(n|1) = \frac{35^n \exp(-35)}{n!}$$

If a *1* is decided when $n > n_{th}$, then the probability of an error is the sum of the $P(n|1)$ for $n = 0$ through n_{th}.

The Poisson distribution when a *0* was transmitted is

$$P(n|0) = \frac{0.1^n \exp(-0.1)}{n!}$$

The probability of error is the sum of the $P(n|0)$ from n_{th} to infinity. The calculation for small n_{th} is easier by summing from $n = 0$ through n_{th}, then subtracting this sum from unity. We can calculate and tabulate these probabilities.

n_{th}	$P(e|1)$	$P(e|0)$	$P(e)$
0	6.31E − 16	9.52E − 2	4.76E − 2
1	2.27E − 14	4.68E − 3	2.34E − 3
2	4.09E − 13	1.55E − 4	7.75E − 5
3	4.91E − 12	3.58E − 6	1.93E − 6
4	4.43E − 11	7.67E − 8	3.84E − 8
5	3.20E − 10	1.27E − 9	7.95E − 10
6	1.93E − 9	1.6E − 11	9.73E − 10
7	9.98E − 9	*	4.99E − 9

The lowest probability of error would be achieved with $n_{th} = 5$. The BER is less than 10^{-9} for thresholds of either 5 or 6.

This is not an ideal detector, because some noise is assumed when no signal power is received. However, it is otherwise ideal, and its sensitivity, 17.5 photons per bit, is close to the quantum limit.

The Gaussian approximation for the Poisson distribution and probability-of-error calculations based on the error function are not valid with only 35 photons per pulse.

─────────────────────────────

9.3.2 *Homodyne Detection*

Homodyne detection resembles heterodyne detection in most respects. The fundamental difference is that the signal frequency and the local oscillator frequency are the same in homodyne detection.

We will study the on-off intensity-modulated binary PCM system with homodyne detection. The notation will follow that used for the heterodyne except that the $mf(t)$ term, representing general amplitude modulation, will be dropped. The electric field for the binary *1* is $e_1(t) = E_1 \cos \omega_0 t$; for the *0* it is zero.

Because the optical signal and the local oscillator have the same frequency, the relative phase angle becomes significant. The local oscillator field is $e_0(t) = E_0 \cos (\omega_0 t + \theta)$. The total field at the photodetector is the sum of the signal and local oscillator fields:

$$e_d(t) = e_s(t) + e_0(t) \tag{9.3.4}$$

where $e_s(t)$ is $e_1(t)$ or zero for the binary *1* or *0*, respectively.

For the binary *1*, the field at the photodetector is

$$\begin{aligned} e_d(t) &= E_1 \cos \omega_0 t + E_0 \cos (\omega_0 t + \theta) \\ &= (E_1 + E_0 \cos \theta) \cos \omega_0 t - E_0 \sin \theta \sin \omega_0 t \end{aligned} \tag{9.3.5}$$

The magnitude of the total field is given by

$$\begin{aligned} |E_d|^2 &= (E_1 + E_0 \cos \theta)^2 + (E_0 \sin \theta)^2 \\ &= E_1{}^2 + E_0{}^2 + 2E_1 E_0 \cos \theta \end{aligned} \tag{9.3.6}$$

Since the power in the light wave is proportional to the square of the electric field strength,

$$P_{d1} = P_1 + P_0 + 2[P_1 P_0]^{\frac{1}{2}} \cos \theta \tag{9.3.7}$$

For the binary *0*, $P_s = 0$ and $P_{d0} = P_0$.

To calculate the probabilities of error, $P(e|1)$ and $P(e|0)$, we can justify and use the Gaussian approximation to the Poisson distribution. For the binary *1* and *0*, we have total input power P_{d1} and P_{d0} and the photocurrents are I_1 and I_0.

$$I_1 = \mathscr{R}_0[P_0 + 2[P_1 P_0]^{\frac{1}{2}} \cos \theta] \tag{9.3.8}$$

$$I_0 = \mathscr{R}_0 P_0 \tag{9.3.9}$$

Each of these currents will generate shot noise. Because we assume P_0 to be large, the Gaussian approximation will be valid for $n \approx N$. Furthermore, because $P_0 \gg P_1$, the shot noise for both signal currents is $\langle i_{sh}{}^2 \rangle = 2qI_0B$. This is the variance, σ^2, for the Gaussian density functions. If the thermal and other noises are not negligible, then it must be added to the shot noise to find the total noise and the correct σ.

The distributions for I_1 and I_0 are illustrated in Figure 9.3.1. The probability of error, $P(e|0)$, can be found by integrating the probability density function.

$$P(e|0) = \frac{1}{[2\pi\sigma^2]^{\frac{1}{2}}} \int_{I_{th}}^{\infty} \exp\left[-\frac{(i - I_{th})^2}{2\sigma^2} \right] di$$

$$= \tfrac{1}{2} erfc\left[\frac{(I_0 - I_{th})}{\sqrt{2}\sigma} \right] = \tfrac{1}{2} erfc\left[\frac{y}{\sqrt{2}} \right] \tag{9.3.10}$$

where y is defined as the ratio of the signal current, measured with respect to the threshold, to the rms noise current.

The threshold is normally set halfway between the two signal levels. If this is the case, $I_1 - I_{th} = I_{th} - I_0 = (I_1 - I_0)/2$. From equations (9.3.8) and (9.3.9),

$$I_1 - I_0 = 2\mathscr{R}_0[P_1 P_0]^{\frac{1}{2}} \cos\theta \tag{9.3.11}$$

The shot noise current is

$$\sigma = [2q\mathscr{R}_0 P_0 B]^{\frac{1}{2}} \tag{9.3.12}$$

and the ratio of signal current to noise current

$$y = \frac{I_1 - I_0}{2\sigma} \tag{9.3.13}$$

$$= \left[\frac{\mathscr{R}_0 P_1}{2qB} \right]^{\frac{1}{2}} \cos\theta = [\phi_1 T]^{\frac{1}{2}} \cos\theta \tag{9.3.14}$$

By symmetry, $P(e|1) = P(e|0) = P(e)$. The probability of error is

$$P(e) = \tfrac{1}{2} erfc \frac{y}{\sqrt{2}} \tag{9.3.15}$$

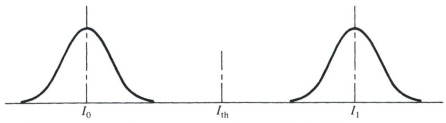

Figure 9.3.1 *Gaussian distribution functions for PCM-ASK signal with homodyne detection. Even though there is no optical signal present for the binary 0, there is a voltage at the detector output caused by the local oscillator power input to the detector. The rms noise voltages for the two Gaussian distributions are approximately equal.*

as in Equation (9.3.10). If the distributions and the probabilities of *1* and *0* were not completely symmetrical, it would be necessary to find $P(e|0)$ and $P(e|1)$ independently and calculate the weighted average to find $P(e)$.

The sensitivity of a PCM detector, or receiver, is the input optical power required to achieve a specified probability of error. A common basis for comparison of detectors, receivers, or modulation methods is to compare the sensitivities for a BER of 10^{-9}. An example calculation is given below.

EXAMPLE 9.6

Calculate the sensitivity of the homodyne detector for a binary PCM ASK system with BER $= 10^{-9}$. The data rate is 250 Mb/s and the wavelength is 1.3 μm. It uses a PIN photodiode that has responsivity of 0.8 and dark current of 10 nA. The PIN load resistance is 1000 ohms and the receiver noise bandwidth is 130 MHz. The receiver preamplifier noise is twice the thermal noise. The probabilities of *1* and *0* are equal. The relative phase between the received light wave and the local oscillator is not larger than 20°.

Solution

The thermal noise in the load resistance is

$$\langle i_{th}^2 \rangle = \frac{4kTB}{R_L} = 2.153 \cdot 10^{-15} \ (A^2)$$

and the sum of thermal and amplifier noise is

$$\sigma_0^2 = 6.46 \cdot 10^{-15} \ (A^2)$$

To realize the advantages of the homodyne detector, we should make the shot noise dominant. Let

$$\langle i_{sh}^2 \rangle = 6.46 \cdot 10^{-14} \ (A^2)$$
$$2q\mathscr{R}_0 P_0 B = 6.46 \cdot 10^{-14} \ (A^2)$$
$$P_0 = 1.94 \ \text{mW}$$

Let $P_0 = 2$ mW. Then $I_0 = \mathscr{R}_0 P_0 = 1.6$ mA. The dark current is negligible. The shot noise is then

$$\langle i_{sh}^2 \rangle = 2qI_0 B = 6.66 \cdot 10^{-14}$$

and the total noise

$$\sigma^2 = 6.66 \cdot 10^{-14} + 6.46 \cdot 10^{-15}$$
$$= 7.3 \cdot 10^{-14} \ (A^2)$$

and

$$\sigma = 2.7 \cdot 10^{-7} \text{ (A)}$$

We can justify using the Gaussian distribution for the shot noise by showing that the number of electrons is large and that the relative width of the distribution of interest to us is small. The number of electrons per bit is

$$N = \frac{I_0 T}{q} = \frac{1.6 \cdot 10^{-3} \cdot 4 \cdot 10^{-9}}{1.6 \cdot 10^{-19}}$$

$$= 4.0 \cdot 10^7 \text{ electrons per bit}$$

The spread of the distribution is several σ, that is, relative to the peak,

$$\frac{\sigma}{I_{ph}} = \frac{2.7 \cdot 10^{-7}}{1.6 \cdot 10^{-3}} = 1.68 \cdot 10^{-4}$$

It was shown in Exercise 9.1 that for an average BER of 10^{-9}, the *erfc* graph reads $x = 6/\sqrt{2}$. Because the noise spectra for *1* and *0* are the same, the decision threshold will be placed in the center of the range. Then, by Equations (9.3.13) and (9.3.11),

$$6 = \frac{I_1 - I_0}{2\sigma}$$

from which

$$I_1 - I_0 = 6 \cdot 2 \cdot 2.7 \cdot 10^{-7}$$

$$= 3.24 \ \mu A$$

$$= 2\mathscr{R}_0 [P_0 P_1]^{\frac{1}{2}} \cos \theta$$

and

$$P_1 = 2.32 \text{ nW } (-56.3 \text{ dBm})$$

9.3.3 *Phase Stability and Phase Noise*

Homodyne detection and other coherent detectors require that the phase difference between the received optical signal and the local oscillator be stable. It is clear from Equation (9.3.11) that if θ is far from zero (or π), the magnitude of the detector signal current will be reduced, perhaps to zero.

The phase difference will be stable if both the transmitted and local oscillator lasers are stable. The phase difference can be kept small by using automatic frequency control (AFC) of the local oscillator. In practice we will usually need both stable lasers and automatic frequency control. Automatic frequency control is necessary because absolute frequency stability of the lasers is not possible. Stable lasers are necessary because the range and speed of automatic frequency control systems are limited. Coherent systems are not widely used in current practical optical fiber communication systems because the stability of semiconductor lasers is not adequate. Coherent systems have been operated sucessfully in research laboratories where extraordinary means can be used to achieve the stability required, fine-tuning and close monitoring during operation are acceptable, and cost is a secondary consideration.

Phase instability in lasers is the result of several factors. Spontaneous emission of photons into the lasing mode(s) will add a small step function to the electromagnetic field propagating in the laser. The phase of this added component, relative to the lasing field, is random and uniformly distributed. The total field will have a phase that is different from that of the original laser field. The natural regeneration and nonlinear saturation in the laser will establish the new phase as a stable condition until another random disturbance changes it once again. Additional spontaneous photons will cause further steps, random in both magnitude and direction, in the phase of the laser field. The laser phase angle exhibits a random-walk behavior.

If the phase of the laser at some arbitary time, $t = 0$, is defined as the reference phase, the mean-squared deviation from this reference, for $t > 0$, is

$$\langle \theta^2(t) \rangle = (2\pi)^2 N_0 t \tag{9.3.16}$$

where N_0 can be related to the laser linewidth, B_L. B_L is the full width between 3-dB points in Hz. In most cases, both transmitter and local oscillator lasers will have line widths that must be included in the calculation; in such cases, B_L is the sum of the linewidths of the individual lasers. In terms of total laser line width,

$$N_0 = \frac{B_L}{2\pi} \tag{9.3.17}$$

and

$$\langle \theta^2(t) \rangle = 2\pi B_L t \tag{9.3.18}$$

The mean-squared phase deviation increases with time. In a coherent detector, if the time period required for the measurement of the phase of a signal element is short, the phase will not change significantly during the measurement. If this period is the signaling interval, $T = 1/R$, then the mean-squared phase deviation at the end of the interval can be expressed

$$\langle \theta^2(T) \rangle = \frac{2\pi B_L}{R} \tag{9.3.19}$$

We can conclude that the phase deviation is smaller for higher data rates. The ratio B_L/R must be small if the effects of phase noise are to be minimized.

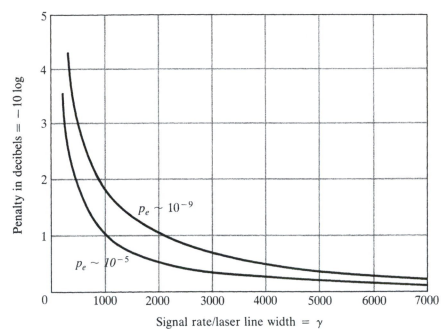

Figure 9.3.2 *Power penalty versus signaling rate parameter for homodyne PSK detectors with two different BERs. The power penalty is the amount by which the signal power must be increased to maintain the BER. The abscissa, γ, is $2R/B_L$, where R is the data rate and B_L is the sum of the line widths of the transmitter and local oscillator lasers. From Salz, AT&T Technical Journal, vol. 64. pp. 2153–2209; copyright © 1985, Bell Telephone Laboratories, Incorporated, reprinted by permission [5.5].*

Salz [5.5] summarizes the performance of several coherent communication systems with phase noise and AFC. He expresses system degradation in terms of a dB penalty, the amount by which the receiver optical power must be increased to compensate for the degradation of detector performance because of phase noise. He calculates the dB penalty as a function of B_L/R. The graphs of Figures 9.3.2 and 9.3.3 show examples of these results.

EXERCISE 9.5

Calculate the power penalty in dB in a homodyne ASK detector for a phase difference of 20°.

Answer
0.54 dB.

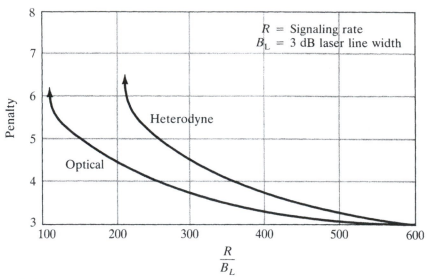

Figure 9.3.3 *Power penalty versus R/B_L due to phase noise in DPSK detectors. The BER is 10^{-9} and the signal power level is 20 photons per bit. The optical detector is one in which the phase comparison between successive pulses is done at the optical frequency. In the heterodyne detector, the phase comparison is done at the intermediate frequency. From Salz, AT&T Technical Journal, vol. 64, pp. 2153–2209; copyright © 1985, Bell Telephone Laboratories, Incorporated, reprinted by permission [5.5].*

In addition to spontaneous emission, several factors can degrade the phase and frequency stability of the laser. Among these factors are temperature and charge density in the laser resonator. Some effects are slow changes in the natural frequency of the resonator; these effects can be interpreted as a slow drift of the center frequency of the laser line. Some have the effect of broadening the laser line width. Although this identification of two classes of effects is somewhat arbitrary, it corresponds to two ways in which the system designer responds to the instabilities. Slow changes can be tracked with automatic frequency and phase control with minimum degradation of detector performance. Faster changes cannot be compensated with automatic control systems and will result in degradation of performance.

Commercially available single-mode lasers have line widths of the order of 1 nm (10^{11} Hz) and smaller. Line widths are typically much larger than the minimum line widths found from theoretical considerations [3.22]. Line widths as small as 10 kHz have been reported [5.7].

9.3.4 *Phase Shift Keying*

The phase-shift keyed (PSK) PCM system transmits a carrier of one phase for the binary *1* and a different phase for the binary *0*. The amplitude and frequency should be the same for both. The electric field for the binary PCM PSK light wave has the form

$$e_s(t) = E_1 \cos \omega_0 t \qquad \text{or} \qquad E_1 \cos (\omega_0 t + \pi)$$
$$= \pm E_1 \cos \omega_0 t \tag{9.3.20}$$

If the local oscillator light wave is, as before, represented by

$$e_0(t) = E_0 \cos (\omega_0 t + \theta) \tag{9.3.21}$$

then the electric field of the light wave incident on the photodetector is

$$e_d(t) = E_0 \cos (\omega_0 t + \theta) \pm E_1 \cos \omega_0 t \tag{9.3.22}$$

and its magnitude is given by

$$|E_d|^2 = E_0{}^2 + E_1{}^2 \pm 2E_0 E_1 \cos \theta \tag{9.3.23}$$

Because the power is proportional to the square of the electric field intensity,

$$P_d = P_0 + P_1 \pm 2[P_0 P_1]^{\frac{1}{2}} \cos \theta \tag{9.3.24}$$

The two currents, representing the binary *1* and *0*, are

$$I_1 = \mathcal{R}_0[P_0 + P_1 + 2[P_0 P_1]^{\frac{1}{2}} \cos \theta] \tag{9.3.25a}$$

and

$$I_0 = \mathcal{R}_0[P_0 + P_1 - 2[P_0 P_1]^{\frac{1}{2}} \cos \theta] \tag{9.3.25b}$$

The difference between the two detector currents is

$$I_1 - I_0 = \mathcal{R}_0 4[P_0 P_1]^{\frac{1}{2}} \cos \theta \tag{9.3.26}$$

If the decision threshold is placed halfway between I_1 and I_0, then the probability of decision error is

$$P(e) = \tfrac{1}{2} erfc \frac{y}{\sqrt{2}} \tag{9.3.27}$$

where

$$y = \frac{I_1 - I_0}{2\sigma} \tag{9.3.28}$$

The standard deviation, σ, is given by Equation (9.3.12). Thus

$$y_1 = \frac{4\mathscr{R}_0[P_0 P_1]^{\frac{1}{2}}\cos\theta}{2[2q\mathscr{R}_0 P_0 B]^{\frac{1}{2}}}$$

$$= \left[\frac{2\mathscr{R}_0 P_1}{qB}\right]^{\frac{1}{2}}\cos\theta \tag{9.3.29}$$

For the PSK signal, the same power is transmitted for both binary symbols. Then $\phi_1 = \phi$ and

$$y = 2[\phi T]^{\frac{1}{2}}\cos\theta \tag{9.3.30}$$

where ϕT is the received optical energy in photons per bit.

For the ideal PSK system, $\theta = 0$. A probability of error of 10^{-9} requires that $y = 6$. The sensitivity of the ideal PSK detector is then

$$\phi T = \left[\frac{6}{2}\right]^2 = 9 \text{ photons per bit} \tag{9.3.31}$$

For $\eta < 1$ and $\theta \neq 0$, the sensitivity is increased to

$$\phi T = \frac{9}{(\eta \cos^2\theta)} \text{ photons per bit} \tag{9.3.32}$$

The rms phase error, given by Equation (9.3.19), is a measure of how heavy a penalty we must pay because of the inherent line width, that is, phase instability, of the laser. Salz [5.5] has analyzed the PSK with automatic frequency control to evaluate this penalty in terms of the increase in received power required to maintain a specified $P(e)$ in the presence of phase instability due to the laser line width. Figure 9.3.2 is the power penalty as evaluated by Salz.

A variation of PSK uses each pulse as the phase reference for the following pulse. A pulse with the same phase as the previous pulse represents a binary *1*; a pulse with a change of phase represents a *0*. This variation of PSK is differential phase-shift keying (DPSK).

The sensitivity of an ideal DPSK detector is the same as that for PSK; both have a perfect phase reference. The advantage of DPSK is its better performance in the presence of phase instabilities. The phase error in the phase reference can increase with time with PSK; the growth in phase error increases indefinitely or until it is reduced by automatic frequency control. With DPSK, the phase error can get no larger than that corresponding to one signaling interval; the rms phase error is no larger than $(2\pi B_L/R)^{\frac{1}{2}}$. Graphs illustrating the power penalty for phase noise in DPSK are shown in Figure 9.3.3.

9.3.5 *Frequency Shift Keying*

To complete our review of coherent modulation and detection methods, we will now examine the characteristics of frequency shift keying (FSK). In binary FSK, the transmitted signal consists of one of two frquencies, one representing the binary *1* and the other the binary *0*. The intensities of the two possible signals are the same.

The block diagram of a heterodyne FSK receiver is shown in Figure 9.3.4. The intermediate-frequency signal is applied to two band-pass filters, one tuned to each of the possible signals. The envelope detectors produce baseband signals, one of which is the sum of a signal and noise and the other noise alone. The comparator and decision circuits must determine which of the two channels most likely includes the signal. The output is a binary signal generated from the *1-0* decisions. Because of noise in both channels, there will be some decision errors in the output binary signal.

There is an inherent 3-dB improvement over ASK in S/N at the decision circuit that results from transmitting a signal for each of the binary symbols. In ASK, the decision threshold must be placed between the peak signal level and zero. In the FSK decision circuit, the two detector outputs are subtracted and the difference voltage applied to a zero-threshold decision circuit; the effective signal, that is, the difference between the signal level and the threshold level, is twice that for a comparable ASK receiver. The mean-squared noise in the FSK decision circuit is the sum of the noises from the two detectors; it is twice that of an ASK receiver. The rms noise is $\sqrt{2}$ times that of a single detector. The S/N for the FSK receiver is therefore increased by $\sqrt{2}$, or 3 dB.

The reader should note that the FSK receiver consists of an optical heterodyne detector followed by an i-f amplifier and a conventional electronic FSK detector. In principle, optical FSK detection is conceivable. It could use a beam splitter and two homodyne detectors; the local lasers for the two homodyne detectors would have frequencies equal to the two optical frequencies that constitute the optical FSK signal. However, stability of laser frequencies is not yet adequate to support this mode of operation.

Salz [5.5] indicates that the FSK performance is 6 dB poorer than the quantum limit, 3 dB of which is attributed to the heterodyne function and the other 3 dB to the properties of FSK detection. FSK has poorer receiver sensitivity than PSK but has the advantage of being less sensitive to phase errors in the transmitter and local oscillator lasers. It is often attractive for lower data rates and where receiver sensitivity is not critical.

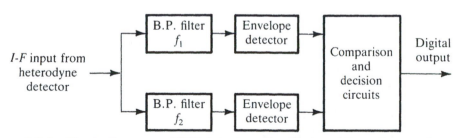

Figure 9.3.4 *Block diagram of PSK detector. The IF signal from the receiver heterodyne detector is applied to two band-pass filters. The comparison and decision circuits must determine which channel is most likely to contain the signal.*

9.3.6 *Comparison of Coherent Modulation and Detection Methods*

We have examined the characteristics of several coherent communication systems. Our objective has been to build an understanding of what a coherent system is and why it can be attractive in communication system design. In this section, we will compare the receiver sensitivities that could be achieved with these coherent modulation and detection techniques.

Comparisons can be made by calculating the signal-to-noise ratio for each detection method. Some care must be taken in interpreting such calculations, however, because there are several different ways in which the comparisons can be structured. For example, the peak and average signal powers are the same for PSK but differ by 3 dB for ASK. Our comparisons will be made on the basis of the same peak power for each method; comparisons based on the same average power could differ by 3 dB from our results. There are also other ways in which the bases for comparison may differ.

We will compare the S/N at the detector output for three modulation-detection methods: (1) ASK with shot-noise-limited direct detection, (2) ASK with homodyne detection, and (3) PSK with homodyne detection. In each case, the signal at the output of the detector is defined as $(I_1 - I_0)/2$. Method (1) is not a coherent technique; it is included as a reference for comparison.

For ASK with shot-noise-limited direct detection, the optical power is P_1 for the binary *1* and 0 for the binary *0*. Then $I_1 = \mathscr{R}_0 P_1$ and $I_0 = 0$. The signal current, as defined above, is

$$I_s = \frac{\mathscr{R}_0 P_1}{2} \tag{9.3.33}$$

The noise current is

$$\langle i_n^2 \rangle = 2q\mathscr{R}_0 P_1 B$$

when the input power is P_1 and zero otherwise. The signal-to-noise ratio for binary *1* decisions is

$$\frac{S}{N} = \frac{I_s^2}{\langle i_n^2 \rangle} = \frac{(\mathscr{R}_0 P_1/2)^2}{2q\mathscr{R}_0 P_1 B} = \frac{\mathscr{R}_0 P_1}{8qB} \tag{9.3.34}$$

For binary *0* decisions there is no shot noise; this decision cannot be shot-noise limited. For the purposes of this comparison we will use Equation (9.3.34) for shot-noise-limited direct detection.

The ideal detector has quantum efficiency $\eta = 1$. Equation (9.3.34) can then be reduced to

$$\frac{S}{N} = \frac{I_s^2}{\langle i_n^2 \rangle} = \frac{P_1}{8hfB} = \frac{\phi T}{4} \tag{9.3.35}$$

This is 6 dB less than the S/N given in Equation (9.3.2). In that equation the signal was defined to be the entire photocurrent, $\mathscr{R}_0 P_1$. In the case considered here the signal current, defined as the difference between the photocurrent and threshold current level, is half of the photocurrent used in Equation (9.3.2).

The use of an average shot noise and a single S/N figure for direct detection is highly artificial and unrealistic; it is done here solely for the purpose of expressing S/N in the same terms as those to be used for homodyne detection. Since the shot noise for the *0* state is zero, the S/N for this state cannot be shot-noise limited. We can conclude that, since thermal and amplifier noises will usually be substantially higher than shot noise, the S/N for direct detection will be substantially poorer than that given by Equation (9.3.35).

For ASK with homodyne detection we have, from Equation (9.3.11)

$$I_s = \mathscr{R}_0 [P_0 P_1]^{\frac{1}{2}} \cos \theta \tag{9.3.36}$$

and, for the noise

$$\langle i_n{}^2 \rangle = 2q\mathscr{R}_0 P_0 B \tag{9.3.37}$$

The signal-to-noise ratio is

$$\frac{S}{N} = \frac{\mathscr{R}_0 P_1 \cos^2\theta}{2qB} = \phi T \cos^2\theta \tag{9.3.38}$$

For PSK with homodyne detection, from Equation (9.3.26),

$$I_s = 2\mathscr{R}_0 [P_0 P_1]^{\frac{1}{2}} \cos \theta \tag{9.3.39}$$

The noise is given by Equation (9.3.37). The S/N is

$$\frac{S}{N} = \frac{2\mathscr{R}_0 P_1 \cos^2\theta}{qB} = 4\phi T \cos^2\theta \tag{9.3.40}$$

This is 6 dB higher than ASK-homodyne.

Coherent techniques are more complex and more difficult to implement than direct detection. They would not be chosen for a system design unless they offered some advantage to justify the costs of their complexity. One important advantage is greater receiver sensitivity. There are also circumstances where coherent systems offer options for signal processing at the intermediate frequency that are not feasible with direct detection. Creative system designers will surely find other new ways to use coherent techniques to advantage.

9.4 Examples of Optical Fiber Communication Systems

In our final section, we will review two specific types of optical fiber communication system. The first, long-distance undersea systems, is described in moderate detail. The second, local area networks, is defined and the special considerations important in such systems are identified. Both are practical and economically attractive applications of optical fiber technology

9.4.1 *Undersea Communication Systems*

As an example of the state of the art in long-distance optical fiber communication systems, we will review one current undersea system and summarize some of the other undersea systems of current interest. The system we will review is the SL Undersea Lightwave System, developed by AT&T Bell Laboratories. The SL system, and several other undersea systems developed in other laboratories, are described in papers in a special joint issue of the IEEE *Journal on Selected Areas in Communications* (November 1984) and the IEEE *Journal of Lightwave Technology* (December 1984) [5.8].

The principles of optical fiber communication systems are applicable to undersea and terrestrial systems alike. The power, attenuation, and sensitivity considerations and the dispersion, bandwidth, and data rate limitations are applicable in essentially the same way to both classes of long-distance systems. However, the technology required for undersea systems is different from that for terrestrial systems in several respects.

A basic difference is the stringent reliability requirements that must be imposed on the undersea system. The reliability target for the first transatlantic light wave system is fewer than three cable-ship repairs during the 25-year service life of the system. The reliability requirement is a major consideration in the planning and detailed design of all parts of the system.

Other special technology problems include the unique light-wave cable requirements and the environmental conditions that the cable and repeaters must accommodate. The cable must have the mechanical strength to survive the deep-sea cable-laying operations without damage; it must carry power to the repeaters; and it must protect the light guide and power supply conductors from the environmental hazards of the sea. The mechanical structure of the repeaters must also have similar strength and protective characteristics.

Although the technologies of design for high reliability, mechanical characteristics, and effects of a hazardous environment in most respects are beyond the scope of this book, the subjects are vital in most system design. Although the communication system specialist on the design team may not have the responsibility for assuring adequate overall system reliability, the optical and electronic design tasks will be influenced by reliability and other nonoptical and nonelectronic considerations. All members of the design team should have some understanding and sympathy for the responsibilities of other specialists.

The first deep-water, repeatered undersea optical fiber communication system will be the TAT-8, a transatlantic system to be placed in service in 1988. The extent of the TAT-8 system is shown in Figure 9.4.1. The SL system will extend 5600 km from the U.S. terminal to a branching repeater in the eastern Atlantic. From the branching repeater, the British Standard Telephones and Cables (STC) NL2 system will extend 500 km to the British terminal and the French Submarcom S280 system 300 km to the French terminal. The SL system

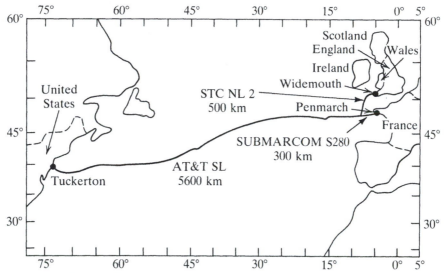

Figure 9.4.1 *The TAT-8 undersea cable system. From Runge and Traschitta, IEEE Journal of Lightwave Technology, vol. LT-2, pp. 744–753; © 1984 IEEE.*

is also proposed for a link between Hawaii and the western U.S. and for part of the transpacific system, TPC-3, linking Hawaii with Japan and Guam.

The SL system will use three pairs of 1.3-μm, single-mode fiber. Transmission and supervisory signaling will be via two of the fiber pairs. The additional fiber pair will serve as standby redundancy, to be switched into service if needed. The cable will also have a copper conductor carrying a constant 1.6 A for power supply to all repeaters.

The total capacity of the TAT-8 system will be 577 Mb/s. Each active fiber pair will carry 295.6 Mb/s. Some of these bits will be parity-check bits for error detection and some will be used for system management data. The overall BER, averaged over any 24 hour period, is $4.8 \cdot 10^{-8}$. Each link is required to have an average BER of 10^{-9}.

Each repeater will have six receiver-regenerators, including two standby regenerators for redundancy. The regenerator, illustrated in Figure 9.4.2, includes the photodetector, amplifier, timing circuits, and decision circuits. Supervisory circuits in the repeater monitor the performance of each regenerator and control redundancy switching.

Specifications for the optical fiber are attenuation of less than 0.48 dB/km and dispersion less than 2.8 ps/(km-nm). There are additional specifications on excess losses due to cabling and various environmental effects; these excess losses are not to exceed a total of 0.09 dB/km over the 25-year life of the system.

The receiver uses an InGaAs PIN photodetector and a transimpedance amplifier. Its sensitivity is -34 dBm at 10^{-9} BER and it has an 18-dB dynamic range. An equalizer-AGC amplifier equalizes the signal pulse shape to a raised

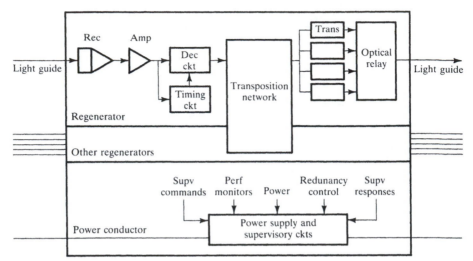

Figure 9.4.2 *The SL repeater block diagram. Each of the six input optical fibers in the cable is connected to a regenerator and each of the six output fibers to a transmitter. Each transmitter has four laser positions, one of which is connected to the output fiber through an optical switch. Any regenerator can be connected to any transmitter through a transposition network. Monitoring of regenerator and transmitter performance and switching in spare units to replace failed units are done through supervisory circuits. From Runge and Traschitta, IEEE Journal of Lightwave Technology, vol. LT-2, pp. 744–753; © 1984 IEEE.*

cosine, and it maintains the output level constant over a 40-dB input voltage-range. The decision circuits provide a retimed and reshaped signal to modulate the transmitter. In case of timing circuit failure, the decision circuit can provide a reshaped but not retimed output; in that case the data would be retimed at the next repeater.

The transmitter is an InGaAsP buried-heterostructure laser with a 5-mW (7-dBm) output. It has a flat wide-band modulation response up to 2 GHz. Because the laser is the least reliable of the active components, provision is made for each regenerator to have as many as four transmitter lasers. The laser performance is monitored through both its bias current and the back-face optical power. When a laser fails, it can be replaced by activating a spare and switching the output fiber to the active transmitter.

The dB budget includes 23 dB for fiber attenuation and a 5-dB margin for aging. The maximum distance between repeaters is approximately $23/0.48 = 48$ km. The 5600-km span of the SL system will require more than 100 repeaters.

9.4.2 Local Area Networks

Local area networks (LANs) have became a major application for optical fiber communication systems. They represent a very different set of problems than those of long-distance communication systems. LAN applications are characterized by short distances, typically from 0.1 to 2 km, with a complex system for the interconnection of many terminals. Two special issues of IEEE technical periodicals provide a view of the state of the art of fiber optics in local-area networks and an indication of the growing importance of optical fiber systems in these networks [5.9][5.10].

A local area network can be defined as a communication network that provides paths for communication among a number of terminals within a reasonably compact service area. The service area may be a building, a manufacturing plant, a hospital, or a university campus. The LAN concept arose in connection with extensive office automation and has spread to many other multiterminal computer-based automation systems. Although it is typically a computer-oriented system, it does not necessarily involve computers.

The design of a local area network must begin with the selection of a network configuration, a network protocol, and a communication technology. To a degree, these selections are interdependent.

Two basic configurations are the star and the loop. Many networks, especially the larger and more complex ones, may use combinations or variations of the star and loop structures. The star, illustrated in Figure 9.4.3, is characterized by one central hub through which all interconnections are made. The hub may consist of nothing more than an $n \times n$ switch through which any node (terminal) can be connected to any other node. It may be a computer that can receive, store, and forward messages either to a single node or to all nodes simultaneously.

In the loop, or ring, configuration, illustrated in Figure 9.4.4, all nodes are connected to a single, shared data line. Each terminal monitors the data on the

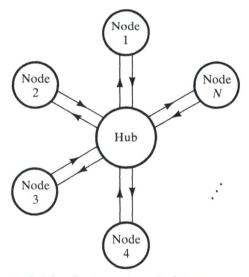

Figure 9.4.3 *A star network. Interconnections between nodes can be made in the hub so that any node can be connected to any other node.*

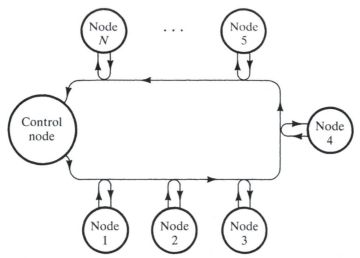

Figure 9.4.4 *A loop network. All nodes can receive any messages propagating in the loop. Any node can put a message on the loop.*

line to identify any messages addressed to that terminal, including authorization to transmit. Opportunities to transmit may be in sequence around the loop, managed by a "token-passing" system, or they may be based on some request-and-authorization system.

It is evident that combinations of loops and stars are possible and, in systems with many terminals, desirable. A node of a star network can be the master terminal of an outlying loop or each node on a loop can be the hub of an outlying star. Hybrid systems in which all terminals are connected in both a loop and a star configuration have advantages of flexibility and transmission-path redundancy, at the cost of added complexity.

Network protocols are the rules under which all terminals on a network must operate. Without a well-defined network discipline, a multiple-access network would be chaotic. The price a user must pay for access to the network includes acceptance of the necessary discipline in its use.

There must be some control function that monitors the network activity, authorizes transmissions, and detects problems. Delays are an inherent problem. The time required for the control entity to receive a request for access and in turn authorize a transmission can constitute a bottleneck. The maximum delay that can be tolerated can determine the maximum number of terminals the system can handle and the maximum distances within the LAN. Propagation delays, rather than attenuation, often limit the maximum distance between terminals in a local area network.

The LAN terminal can consist of a passive directional coupler that couples a small fraction of the power in the line to an off-line receiver, or it can be an active receiver/transmitter (repeater) terminal. The passive coupler is a natural and less expensive choice for a loop system, but has the disadvantage that each node must divert enough energy from the data line to meet the sensitivity level of its receiver, thus reducing the intensity of the signal in the loop and limiting the number of nodes that can be connected in the loop.

When the distances are short and the data rates modest, as are true in many LANs, the demands on the optical fiber communication links are modest. An LED/graded-index fiber/PIN system is common for LANs having data rates up to 5 Mb/s. Systems using lasers and APD receivers are found when data rates exceed 5 to 10 Mb/s. The error rate specification depends on the application, but a BER of 10^{-9} is not unusual.

One design problem that can be more difficult to handle in LANs than in long-distance systems is that of the wide range of input power levels at the receiver. The signal at the receiver may have originated at any of several other terminals on the network. The distance, and the intervening attenuation, may cover a wide range. The receiver must be designed to handle as large a range of input power levels as is feasible, and the system must be designed so that the power level at each receiver will fall within the range that the receiver can detect and process without excessive noise or distortion.

SUMMARY

Long-distance digital systems can maintain good signal quality by the use of regenerative repeaters at suitably spaced intervals. The quality of digital transmission is expressed in terms of the bit error rate, or probability of error in binary decisions. For weak signals, the probability of error can be calculated by using the Poisson distribution function

$$P(n) = \frac{N^n \exp(-N)}{n!}$$

This gives the probability of receiving exactly n photons during a pulse interval T when the average number of photons received during this interval is N.

When the number of photons becomes large, the quantum fluctuations are represented as shot noise and can be represented by the Gaussian probability density function

$$f(x) = \frac{1}{[2\pi\sigma^2]^{\frac{1}{2}}} \exp\left[\frac{-(x-m)^2}{2\sigma^2}\right]$$

The probability of error, when the Gaussian distribution is applicable, can be found from the complementary error function.

$$P(e) = \tfrac{1}{2} erfc\left[\frac{V}{\sigma\sqrt{2}}\right]$$

where V/σ is the voltage signal-to-noise ratio. For this calculation, the signal is defined as the difference between the received signal level and the decision threshold level.

The maximum distance between repeaters is limited by the receiver sensitivity, the transmitter power, and the attenuation between the transmitter and receiver. The maximum permissible attenuation in dB divided by the fiber attenuation in dB/km establishes a maximum path length. The highest possible receiver sensitivity is given by the quantum limit. For BER $= 10^{-9}$, the quantum limit is 10 photons per bit; practical receivers have sensitivities of 1000 or more photons per bit.

Dispersion and bandwidth limitations can also determine a maximum distance between repeaters. The total system, or any part of it, can be characterized by the rms width of its impulse response. The rms width for the complete system includes the effects of fiber dispersion, transmitter and photodetector rise times, and receiver bandwidth. The rms width for the system can be found from the rms widths of its parts by the sum of the squares:

$$\sigma^2 = \sigma_{tx}^2 + \sigma_f^2 + \sigma_{ph}^2 + \sigma_a^2$$

The rms width and the data rate are related by $R = \frac{1}{4\sigma}$. The fiber dispersion, which is equal to the rms width, is proportional to distance. When R is specified, there is a maximum distance for which σ is small enough to permit data transmission at the specified rate.

The maximum distance between repeaters is determined by attenuation limits for low data rates and by dispersion limits for high data rates.

In most respects, design considerations for analog systems are similar to those for digital systems. When the signal-to-noise ratio required for the specific application under consideration has been established, the signal and noise levels can be determined. Because the analog signal is a continuous variable that must be faithfully reproduced in the receiver, the analog system must be linear throughout. Repeaters for an analog system must be linear. Because the linear repeater will amplify the noise as well as the signal, the total noise will continue to increase as the signals and noise are amplified in successive repeaters. Repeaters will not extend the maximum total distance for the analog system in the way they can for digital systems.

In addition to the conventional analog signals, such as audio and video, sinusoidal carriers that are themselves modulated are an important class of analog signals. A modulated microwave carrier can be transmitted through an optical fiber as well as through a coaxial cable or radio system. Analog signals are often transmitted by PCM and reconstituted in analog form after reception.

Coherent optical communication systems, although not yet practical for commercial use, offer realistic expectations for increasing receiver sensitivities and extending maximum distances between repeaters. Coherent modulation and detection methods include heterodyne and homodyne detection, phase modulation, and frequency modulation. In principle, PSK with homodyne detection can approach the quantum limit. Most coherent techniques can offer substantial improvements over intensity modulation with direct detection. Coherent techniques are not yet practical because the frequency and phase stability of practical lasers are not adequate.

PROBLEMS _____

9.1 In a binary decision circuit, the rms noise is 50 nV for both *0* and *1*. The signal voltages are V_1 and zero. The threshold voltage is $V_1/2$. Find V_1 to make the average probability of error 10^{-6}. Note that in this case, because the Gaussian curves are symmetrical about the threshold voltage, the individual probabilities of error are equal.

9.2 For the distributions given in Figure 9.1.1, and with $V_t = V_1/2$, find the probabilities of error, $P(e|0)$ and $P(e|1)$. If $P(0) = 0.65$ and $P(1) = 0.35$, find $P(e)$.

9.3 An LED delivers power of 50 μW, with $\lambda = 0.85$ μm, into an optical fiber. The total attenuation between transmitter and receiver is 40 dB. What is the probability that fewer than three photons will be received in an interval of 1 ns?

9.4 In a PCM receiver with signal voltages of V_1 and 0 for the binary *1* and *0*, and with $P(1) = P(0)$, the rms noise voltages are $\sigma_1 = 0.20V_1$ and $\sigma_0 = 0.05V_1$. What threshold voltage, V_t, will give the lowest probability of error?

9.5 Plot a graph of sensitivity versus data rate similar to those of Figure 9.1.3, for
(a) 1000 photons per bit, with $\lambda = 0.85$, 1.3, and 1.55 μm, and
(b) 100, 1000, and 10,000 photons per bit, with $\lambda = 1.55$ μm.

9.6 A receiver with sensitivity of -45 dBm receives its input signal through an optical fiber having an attenuation of 0.4 dB/km. The coupling losses are 3 dB from the transmitter laser to the fiber and 2 dB from the fiber to the photodetector.
(a) Plot the minimum laser output power, P_t (mW), versus distance.
(b) Plot P_t (dBm) versus distance.

9.7 The transmitted pulse has rectangular shape with duration 2 ns. It propagates through 70 km of single-mode fiber having dispersion $D = 10$ ps/km. What is the rms width of the pulse at the input to the receiver?

9.8 A pulse has leading and trailing edges with the same time constant:

$$h_1(t) = 1 - \exp\left(-\frac{t}{\tau}\right) \quad 0 < t < \tau$$

$$h_2(t) = (e - 1) \exp\left(-\frac{t}{\tau}\right) \quad \tau < t$$

Find the rms pulse width.

9.9 A multilink optical fiber PCM system transmits data at the rate $R = 30$ Mb/s. The transmitter output power is 0 dBm. The transmitted pulse is approximately triangular in shape, with rms width $\sigma_{tx} = 3.5$ ns. The optical fiber has average attenuation of 0.5 dB/km, including splice losses; the total connector losses and design margin are 5 dB and 3 dB, respectively. The fiber is a multimode graded-index fiber having dispersion of $\sigma_f/L = 20$ ps/km. The receiver has bandwidth $\Delta f = 25$ MHz and sensitivity -35 dBm. The rms width of the impulse response of the photodetector is $\sigma_{ph} = 5$ ns. What is the maximum distance between repeaters?

9.10 The pulse at the input to the receiver has a Gaussian shape, with rms width $\sigma = 1.0$ ns and pulse repetition rate $R = 100$ Mb/s.
   (a) Calculate the duration of the Gaussian pulse between half-amplitude points.
   (b) The pulse at the output of the amplifier has a raised-cosine spectrum, with nulls in the pulse amplitude at intervals of $1/R$ from the peak amplitude (see Equation 9.1.29). Plot the pulse shape, $h(t)$, versus time.
   (c) On a Bode plot, sketch the frequency response of the amplifier.

9.11 An analog system of total length 150 km has linear-amplifier repeaters spaced at intervals of 30 km. The overall impulse response of one 30-km link, including the transmitter, fiber and receiver, is

$$h(t) = \exp\left[-\frac{t^2}{2\tau^2}\right]$$

where $\tau = 15$ ns. What is the bandwidth (electrical bandwidth) in MHz of the total 150-km system?

9.12 A PCM-ASK system, with $R = 150$ Mb/s and $\lambda = 1.3$ μm, must have an S/N in the detected signal of 15 dB. The quantum efficiency of the PIN photodiode is 0.6. Neglect amplifier noise. Calculate the sensitivity for
   (a) an ideal (shot-noise-limited) direct detector,
   (b) direct detection, including thermal noise from a load resistance of 500 ohms, and
   (c) a homodyne detector, with local oscillator power of -3 dBm and phase error of 10 degrees.

APPENDIX A

List of Symbols and Acronyms

Symbols

a	radius of core of optical fiber	m
$\mathbf{a}_x$	unit vector, in x direction	
A_{21}	coefficient of spontaneous emission	1/s
b	normalized propagation constant	
B	bandwidth	Hz
B	coefficient of stimulated emission	$m^3/J \cdot s^2$
$\mathbf{B}$	magnetic flux density	Wb/m^2
c	velocity of electromagnetic waves in vacuum	m/s
D	dispersion parameter	s/m^2
$\mathbf{D}$	electric flux density	C/m^2
dB	decibel	
e	2.71828..., the base for natural logarithms	
$\mathbf{E}$	electric field intensity	V/m
$\mathscr{E}$	electron energy level	J, eV
F	excess noise factor (APD)	
$F(\omega)$	Fourier transform	
g	rate of generation of EHP	$1/m^3 s$
h	Planck's constant, $6.626 \cdot 10^{-34}$	Js
$h(t)$	impulse response	
$\mathbf{H}$	magnetic field intensity	A/m
$H(j\omega)$	network transfer function	
I_{ph}	photocurrent	A
j	$\sqrt{-1}$	
$\mathbf{J}$	current density	A/m^2
$J_n(\beta)$	Bessel function of the first kind of order n	
k	Boltzmann's constant, $1.38 \cdot 10^{-23}$	J/K
$\mathbf{k}$	vector propagation constant	1/m
$K_n(x)$	modified Bessel function of the second kind of order n	
m	amplitude modulation index	
m	mean value, of voltage or current	
M	current multiplication factor (APD)	
n	electron concentration	$1/m^3$
n	index of refraction	
$\mathbf{n}$	unit vector, normal to a surface or wavefront	

Δn	excess electron density	$1/m^3$
n_i	intrinsic carrier concentration	$1/m^3$
N	group index	
N	noise power	
NA	numerical aperture	
N_A, N_D	doping levels in the p,n regions	$1/m^3$
p	hole concentration	$1/m^3$
$P(e)$	probability of error	
q	charge of an electron	C
r	reflection coefficient	
r	recombination rate	$1/m^3 s$
R	data rate	b/s
R	reflectivity	
$\mathscr{R}$	responsivity of a photodetector	A/W
S	signal power	W
$\mathbf{S}$	Poynting vector	W/m^2
S/N	signal-to-noise ratio	
t	transmission (refraction) coefficient	
T	transmissivity	
v	velocity	m/s
v_g	group velocity	m/s
v_p	phase velocity	m/s
V_d	diffusion potential	V
V	normalized frequency in dielectric waveguide	
$\langle v \rangle$	average value of v	V
$\{v^2\}$	spectral density	V^2/Hz
α	attenuation propagation constant	$1/m$
α	profile parameter in graded-index fiber	
β	frequency modulation index	
β	phase propagation constant	$1/m$
γ	complex propagation constant; $\gamma = \alpha + j\beta$	$1/m$
Δ	normalized index difference	
$\Delta\tau$	dispersion	s
ϵ	permittivity, dielectric constant	F/m
η	intrinsic impedance	Ω
η	quantum efficiency	
λ	wavelength	m, μm
μ	permeability	H/m
ρ	charge density	C/m^3
σ	conductivity	S/m
σ	rms dispersion	s
σ	rms pulse width	s
σ	standard deviation	
ϕ	photon density	$1/m^3$
ω	angular frequency (radians/second)	$1/s$
ω_B	radian bandwidth, $2\pi B$	$1/s$

Acronyms

APD	avalanche photodiode
ASK	amplitude shift keying
BER	bit error rate, probability of error
DPSK	differential phase-shift keying
EHP	electron-hole pair
erf	error function
erfc	complementary error function
FSK	frequency-shift keying
IEEE	Institute of Electrical and Electronics Engineers
I/O	input-output
LAN	local-area network
LED	light-emitting diode
OOK	on-off keying
PCM	pulse code modulation
PIN	*p*-intrinsic-*n*, a three-layer semiconductor diode structure
PSK	phase-shift keying
S/N	signal-to-noise ratio

APPENDIX B

Physical Constants

Symbol	Value	Remarks
c	$2.998 \cdot 10^8$ m/s	
μ_o	$4\pi \cdot 10^{-7}$ H/m	
ϵ_0	$8.854 \cdot 10^{-12}$ F/m	$\approx 1/36\pi \cdot 10^{-9}$
q	$1.602 \cdot 10^{-19}$ C	electronic charge
h	$6.626 \cdot 10^{-34}$ J $\cdot$ s	Planck constant
k	$1.381 \cdot 10^{-23}$ J/$^\circ$K	Boltzmann constant
m_o	$9.110 \cdot 10^{-31}$ kg	electron rest mass

APPENDIX C

Properties of Semiconductor Materials

Material	Band gap energy $\mathscr{E}_g$, eV	Lattice spacing nm	Intrinsic carrier concentration n_i, $1/m^3$	Relative permittivity ϵ_r
Si	1.11	0.5431	$1.5 \cdot 10^{16}$	11.8
Ge	0.67	0.5646	$2.5 \cdot 10^{19}$	16.0
GaAs	1.43	0.5653	$1.8 \cdot 10^{12}$	13.2
GaP	2.26	0.5451	10^6	11.1
GaSb	0.70	0.6096	$2 \cdot 10^{19}$	15.7
AlAs	2.16	0.5661	$6 \cdot 10^6$	10.9
AlP	2.45	0.5451	10^5	9
AlSb	1.60	0.6136	$3 \cdot 10^{11}$	14.4
InAs	0.36	0.6058	10^{21}	14.6
InP	1.28	0.5869	$2 \cdot 10^{13}$	12.4
InSb	0.18	0.6479	$2 \cdot 10^{22}$	17.7

The values in the table are representative of these materials at room temperature; however, these data can vary with temperature, frequency, and other parameters. More extensive data on III-V compound semiconductor materials are given in Neuberger, M., *III-V Semiconducting Compounds*, Plenum Publishing Company, New York, 1971 [3.16].

APPENDIX D

SI Units

The International System of Units (SI) provides a coherent system of units that has become the standard for science and engineering publications. It is based on seven primary units and quantities; these base units are listed in Table D1. Other secondary units can be expressed in terms of the base units; secondary units for quantities used in electrical and optical systems are listed in Table D2. Standard prefixes that are used to indicate decimal multiples or submultiples of SI units are listed in Table D3.

Table D1 Base Units

Quantity	Unit	Symbol
length	meter	m
mass	kilogram	kg
time	second	s
current	ampere	A
temperature	kelvin	°K
amount of substance	mole	mol
luminous intensity	candela	cd

Table D2 Secondary Units

Quantity	Unit	Symbol	Dimension
frequency	hertz	Hz	$1/s$
force	newton	N	$kg\text{-}m/s^2$
energy	joule	J	N-m
power	watt	W	J/s
electric charge	coulomb	C	A-s
potential	volt	V	J/C
conductance	siemens	S	A/V
resistance	ohm	Ω	V/A
capacitance	farad	F	C/V
magnetic flux	weber	Wb	V-s
inductance	henry	H	Wb/A

Table D3 Standard Prefixes

Multiple	SI Prefix	Symbol
10^{18}	exa	E
10^{15}	peta	P
10^{12}	tera	T
10^{9}	giga	G
10^{6}	mega	M
10^{3}	kilo	k
10^{-3}	milli	m
10^{-6}	micro	μ
10^{-9}	nano	n
10^{-12}	pico	p
10^{-15}	femto	f
10^{-18}	atto	a

APPENDIX E

Greek Alphabet

Letter	Lowercase	Uppercase
Alpha	α	A
Beta	β	B
Gamma	γ	Γ
Delta	δ	Δ
Epsilon	ϵ	E
Zeta	ζ	Z
Eta	η	H
Theta	θ	Θ
Iota	ι	I
Kappa	κ	K
Lambda	λ	Λ
Mu	μ	M
Nu	ν	N
Xi	ξ	Ξ
Omicron	o	O
Pi	π	Π
Rho	ρ	P
Sigma	σ	Σ
Tau	τ	T
Upsilon	ϑ	Υ
Phi	ϕ	Φ
Chi	χ	X
Psi	ψ	Ψ
Omega	ω	Ω

Answers to Selected Problems

Chapter 2

2.1 (a) f_c: 100 MHz,   1 kV;
      $f_c \pm f_1$:   98, 102 MHz,   225 V,   1012.5 W;
      $f_c \pm 2f_1$:   96, 104 MHz,   150 V,   450.0 W;
      $f_c \pm 3f_1$:   94, 106 MHz,   75 V,   112.5 W;
      (b) $V_{peak} = 1.9$ kV; $P_{peak} = 36.1$ kW.

2.3 The intensity modulation is changed from $1.002 + 0.752 \cos \omega_1 t$ to $1.012 + 0.797 \cos \omega_1 t$.

2.5 (a) 12 MHz   (b) 7 MHz   (c) 3 MHz   (d) 1 MHz.

2.7 $1.05 \cdot 10^5$ bits/s.

2.12 (a) 36.7 μV   (b) 33.4 μV.

2.14 50.2 dB.

Chapter 3

3.1 $2.23 - j0.0016$.

3.3 (a) $0.0544 + j2.56 \cdot 10^7$ (b) $0.0544 + j1.67 \cdot 10^7$
    (c) $0.0544 + j1.40 \cdot 10^7$.

3.6 $-2.26a_x + 3.87a_y$.

3.11 $N = 1.4819$; $dN/d\lambda = -6.9 \cdot 10^{-4} \mu m^{-1}$.

3.13 $\theta_c = 83.4°$, $r_E = \exp(j144.64°)$, $t_E = 0.608 \exp(j72.32°)$.

3.15 98 mW;   0.601 fJ.

Chapter 4

4.2 3.

4.4 (a) 1.26 μm   (b) 2.30 μm.

4.6 even: 0.28, 0.836, 1.373×10^6 m^{-1}.
    odd: 0.498, 1.109, 1.610×10^6 m^{-1}.

4.8 $7.28 \cdot 10^6$ m^{-1}.

4.11 $E_1 = A_1 J_2(5.0r/a) \exp(j2\theta) \exp[j(\omega t - 7.078z)]$ in the core, and
     $E_1 = C_1 K_2(13.7r/a) \exp(j2\theta) \exp[j(\omega t - 7.078z)]$ in the cladding

where all linear dimensions are in μm, and

$$C_1 = A_1 \frac{J_2(5.0)}{K_2(13.7)}$$

E_2, E_3, and E_4, have similar forms but different numerical parameters.

4.13   2.02 μs.

4.15   2.5 μm.

4.17   (a)   33.5 ns/km   (b)   28.9 ps/km.

Chapter 5

5.3   1.46 mW.

5.5   (a)   0.66   (b)   0.91.

5.9   2.3 ns (recombination); 3.7 ps (propagation).

5.12   (a) $5 + 0.998 \cos(\omega t - 3.6°)$ (b) $5 + 0.847 \cos(\omega t - 32.14°)$
       (c) $5 + 0.157 \cos(\omega t - 80.96°)$.

Chapter 6

6.1 $g > 1873$ m^{-1}.

6.3   1.01 nm.

6.4 $P(x)/P(0) = \exp(2772x)$.

6.7   4.8 ps.

6.8   (c) $7 + 6 \cos \omega t$ mW.

6.10   41.7 μW.

6.13   651 m^{-1}, 11 modes.

6.18 $2.78 \cdot 10^5 \cos[\omega t - (5\pi/3) \cos \omega_m t]$ V/m.

Chapter 7

7.2   0.72.

7.4   5.7 nm.

7.6   (a) $3.9 \cdot 10^{19}$ m^{-3} (b)   30.8 μm   (c) -20.2 V.

7.10   (a) $8.27 \cdot 10^5$ V/m   (b)   14.9 μm   (c) -8.7 V.

7.14   (a) $2.56 \cdot 10^{-25}$ A^2/Hz   (b)   55.6 dB.

7.15 $M_e = 13.8$; $M_h = 2.3$.

7.17   5.8 GHz.

Chapter 8

8.1   73.4 μV.

8.3   APD: $S/N = 46.7$ dB;   PIN: $S/N = 31.4$ dB.

8.5 (a) 14.5 MHz (b) 22.7 MHz (c) 0.11 mV.
8.12 21.1 dB, 19.7 dB, 23.3 dB.
8.13 7.0 dB.

Chapter 9

9.1 0.47 μV.
9.2 0.0063; 0.0195; 0.0109.
9.3 $1.3 \cdot 10^{-7}$.
9.7 0.91 ns.
9.9 54 km.
9.12 -56.2 dBm; -33.5 dBm; -59.1 dBm.

Bibliography

I. General

1.1   Li, T., "Advances in optical fiber communications: An historical perspective," *IEEE Journal on Selected Areas in Communications*, vol. SAC-1, n. 3, pp. 356–372, Apr. 1983.

1.2   O'Neill, E. F., ed., *A History of Science and Engineering in the Bell System*, AT&T Bell Laboratories, 1985.

1.3   Hecht, E., *Optics*, 2nd ed. Addison-Wesley, Reading, Mass. 1987.

1.4   Ditchburn, R. W., *Light*, 3d ed., Academic Press, London, 1976.

1.5   Jenkins, F. H., *Fundamentals of Optics*, 4th ed., McGraw-Hill, New York, 1976.

1.6   Born, M., and E. Wolf, *Principles of Optics*, 6th ed., Pergamon Press, 1980.

1.7   Ziemer, R. E., W. H. Tranter, and D. R. Fannin, *Signals and Systems: Continuous and Discrete*, Macmillan, New York, 1983.

1.8   Haykin, S., *Communication Systems*, 2d ed., Wiley, New York, 1983.

1.9   Dwight, H. B., *Tables of Integrals and Other Mathematical Data*, 4th ed., Macmillan, New York, 1961.

1.10 Jahnke, E., and F. Emde, *Tables of Functions*, 4th ed., Dover, New York, 1945.

1.11 Couch, L. W., II, *Digital and Analog Communication Systems*, Macmillan, New York, 1983.

1.12 ANSI/IEEE Std 280-1985, *IEEE Standard Letter Symbols for Quantities Used in Electrical Science and Electrical Engineering*, The Institute of Electrical and Electronics Engineers, New York, 1984.

1.13 Stremler, F. G., *Introduction to Communication Systems*, 2d ed., Addison-Wesley, Reading, MA, 1982.

1.14 Schwartz, M., *Information Transmission, Modulation, and Noise*, 3d ed., McGraw-Hill, New York, 1980.

1.15 Henry, P. S., "Lightwave primer," *IEEE Journal of Quantum Electronics*, vol. QE-21, pp. 1862–1879, Dec. 1985.

1.16 ANSI/IEEE Std 812-1984, *IEEE Standard Definitions of Terms Relating to Fiber Optics*, IEEE, New York, 1984.

II. Dielectric Waveguides and Optical Fibers

2.1   Goldin, E., *Waves and Photons: An Introduction to Quantum Optics*, Wiley, New York, 1982.

2.2   Haus, H. A., *Waves and Fields in Optoelectronics*, Prentice-Hall, Englewood Cliffs, N.J., 1984.

2.3   Feynman, R. P., R. B. Leighton, and M. Sands, *The Feynman Lectures on Physics*, Addison-Wesley, Reading, Mass., 1977.

2.4   Yariv, A., *Optical Electronics*, 3d ed., Holt, Rinehart and Winston, New York, 1985.

2.5   Fleming, J. W., "Material dispersion in lightguide glasses," *Electronics Letters*, vol. 14, no. 11, pp. 326–328, 1978.

2.6   Marcuse, D., *Light Transmission Optics*, 2d ed., Van Nostrand Reinhold, New York, 1982.

2.7   Cheo, P. K., *Fiber Optics, Devices and Systems*, Prentice-Hall, Englewood Cliffs, N.J., 1985.

2.8   Gloge, D., "Weakly Guiding Fibers", *Applied Optics*, vol. 10, no. 10, pp. 2252–2258, October, 1971.

2.9   Gloge, D., "Dispersion in weakly guiding fibers," *Applied Optics*, vol. 10, no. 11, pp. 2442–2245, Nov. 1971.

2.10   Synder, A. W., "Asymptotic expressions for eigenfunctions and eigenvalues of a dielectric or optical waveguide," *IEEE Transactions on Microwave Theory and Techniques*, vol. MTT-17, no. 12, pp. 1130–1138, Dec. 1969.

2.11   Cohen, L. G., W. A. Mammel and S. J. Lang, "Low-loss quadruple-clad single-mode lightguides with dispersion below 2 ps/km · nm over the 1.28 μm–1.65 μm wavelength region," *Electronics Letters*, vol. 18, no. 24, pp. 1023–1024, 25 No. 1982.

2.12   Gloge, D., and E. A. J. Marcatili, "Multimode theory of graded-core fibers," *Bell System Technical Journal*, vol. 52, pp. 1563–1578, Nov. 1973.

2.13   Olshansky, R., and D. B. Keck, "Pulse broadening in graded-index optical fibers," *Applied Optics*, vol. 15, pp. 483–491, Feb. 1976.

2.14   Li, Tingye, ed., *Optical Fiber Communications, Volume 1, Fiber Fabrication*, Academic Press, Orlando, 1985.

2.15   Marcatili, E. A. J., "Bends in optical dielectric guides," *Bell System Technical Journal*, vol. 48, pp. 2103–2132, Sept. 1969.

2.16   Arnaud, J. A., *Beam and Fiber Optics*, Academic Press, Orlando, 1976.

2.17   Gloge, D., ed., *Optical Fiber Technology*, IEEE Press, New York, 1976.

2.18   Kao, C. K., ed., *Optical Fiber Technology II*, IEEE Press, New York, 1981.

2.19   Kapany, N. S., and J. J. Burke, *Optical Waveguides*, Academic Press, Orlando, 1972.

2.20   Midwinter, J. E., *Optical Fibers for Transmission*, Wiley, New York, 1979.

2.21   Okoshi, T., *Optical Fibers*, Academic Press, Orlando, 1982.

2.22   Unger, H.-G., *Planar Optical Waveguides and Fibers*, Clarendon Press, Oxford, 1977.

2.23   Giallorenzi, T. G., "Optical communications research and technology: Fiber optics," *Proc. IEEE*, vol. 66, pp. 744–780, July, 1978.

III.   Optical Sources and Transmitters

3.1   Sze, S. M., *Physics of Semiconductor Devices*, 2d ed., Wiley, 1981.

3.2   Wada, O., et al., "Performance and reliability of high-radiance InGaAsP/InP DH LED's operating in the 1.15–1.5 μm wavelength region," *IEEE Journal of Quantum Electronics*, vol. QE-18, pp. 368–373, March 1982.

3.3 Anderson, R. L., "Experiments on Ge-GaAs heterojunctions," *Solid State Electronics*, vol. 5, pp. 341–351, 1962.

3.4 Kressel, H., ed., *Semiconductor Devices for Optical Communication*, vol. 39, Topics in Applied Physics, 2d ed., Springer-Verlag, New York, 1982.

3.5 Burrus, C. A., and B. I. Miller, "Small area double-heterostructure aluminum-gallium-arsenide electroluminescent sources for optical-fiber transmission lines," *Optics Communications*, vol. 4, pp. 307–309, Elsevier, 1971.

3.6 Ettenberg, M., G. H. Olsen, and F. Z. Hawrylo, "On the reliability of 1.3 μm InGaAsP/InP edge-emitting LED's for optical fiber communication," *IEEE Journal of Lightwave Technology*, vol. LT-2, no. 6, pp. 1016–1023, Dec. 1984.

3.7 Marathay, A. S., *Elements of Optical Coherence Theory*, Wiley, New York, 1982.

3.8 Thompson, G. H. B., *Physics of Semiconductor Laser Devices*, Wiley, New York, 1980.

3.9 Lau, K. Y., and A. Yariv, "High-frequency current modulation of semiconductor injection lasers," Chapter 2, *Semiconductors and Semimetals*, Vol. 22, Part B, Academic Press, Orlando, 1985.

3.10 Lau, K. Y., and A. Yariv, "Ultra-high speed semiconductor lasers," *IEEE Journal of Quantum Electronics*, vol. QE-21, pp. 121–138, Feb. 1985.

3.11 Adler, R., "A study of locking phenomena in oscillators," *Proceedings of the IRE*, vol. 34, pp. 351–357, June 1946. (Reprinted in *Proceedings of the IEEE*, vol. 61, pp. 1380–1386, Oct. 1973.)

3.12 Mogensen, F., H. Olsen, and G. Jacobsen, "Locking conditions and stability properties for a semiconductor laser with external light injection," *IEEE Journal of Quantum Electronics*, vol. QE-21, pp. 784–793, July 1985.

3.13 Taylor, H. F., "Optical switching and modulation in parallel dielectric waveguides," *Journal of Applied Physics*, vol. 44, pp. 3257–3262, July 1973.

3.14 Alferness, R. C., and R. V. Schmidt, "Tunable optical waveguide directional coupler filter," *Applied Physics Letters*, vol. 33, pp. 161–163, 15 July 1978.

3.15 Alferness, R. C., "Waveguide electrooptic modulators," *IEEE Transactions on Microwave Theory and Techniques*, vol. MTT-30, pp. 1121–1137, Aug. 1982.

3.16 Neuberger, M., *III-V Semiconducting Compounds*, Plenum, New York, 1971.

3.17 Becker, R. A., "Broad-band guided-wave electrooptic modulators," *IEEE Journal of Quantum Electronics*, vol. QE-20, pp. 723–727, 1984.

3.18 Gee, C. M., and G. D. Thurmond, "Wideband traveling-wave electrooptic modulator," *Proceedings of the Society of Photo-Optical Instrumentation Engineers*, vol. 477, pp. 17–22, 1984.

3.19 Goodfellow, R. C., B. T. Debney, G. J. Reese, and J. Buus, "Optoelectronic components for multigigabit systems," *IEEE Journal of Lightwave Technology*, vol. LT-3, pp. 1170–1179, Dec. 1985.

3.20 Hawes, M. J., and D. V. Morgan, *Optical Fibre Communications: Devices, Circuits, and Systems*, Wiley, New York, 1980.

3.21   Wilson, J., and J. F. B. Hawkes, *Optoelectronics, An Introduction*, Prentice-Hall International, London, 1983.

3.22   Kressel, H., and J. K. Butler, *Semiconductor Lasers and Heterojunction LEDs*, Academic Press, Orlando, 1977.

3.23   Butler, J. K., ed., *Semiconductor Injection Lasers*, IEEE Press, New York, 1980.

3.24   Panish, M. B., and I. Hayashi, "Heterostructure junction lasers," *Applied Solid State Science*, vol. 4, Academic Press, Orlando, 1974.

3.25   Kressel, H., M. Ettenberg, and I. Ladany, "Accelerated step-temperature aging of Al_xGa_{1-x} as heterojunction laser diodes," *Applied Physics Letters*, vol. 32, pp. 305–308, 1978.

IV.   Photodetectors and Optical Receivers

4.1   Bar-Lev, A., *Semiconductors and Electronic Devices*, 2d ed., Prentice-Hall, 1984.

4.2   Streetman, B. G., *Solid State Electronic Devices*, 2d ed., Prentice-Hall, Englewood Cliffs, N.J., 1980.

4.3   Sze, S. M., *Semiconductor Devices: Physics and Technology*, Wiley, New York, 1985.

4.4   Stillman, G. E., V. M. Robbins, and N. Tabatabaie, "III-V compound semiconductor devices: Optical detectors," *IEEE Transactions on Electron Devices*, vol. ED-31, p. 1643, 1984.

4.5   Stillman, G. E., and C. M. Wolfe, "Avalanche photodiodes," Chapter 5 in *Semiconductors and Semimetals*, Volume 12, edited by R. K. Willardson and A. C. Beer, Academic Press, New York, 1977.

4.6   Miller, S. L., "Ionization rates for holes and electrons in silicon," *Physical Review*, vol. 105, p. 1246, Feb. 15, 1957.

4.7   Emmons, R. B., "Avalanche photodiode frequency response," *Journal of Applied Physics*, vol. 38, p. 3705, 1967.

4.8   McIntyre, R. J., "Multiplication noise in uniform avalanche diodes," *IEEE Transactions on Electron Devices*, vol. ED-13, p. 164, 1966.

4.9   Buckingham, M. J., *Noise in Electron Devices and Systems*, Wiley, New York, 1983.

4.10   van der Ziel, A., *Solid State Physical Electronics*, Prentice-Hall, Englewood Cliffs, N.J., 1976.

4.11   Personick, S. D., "Receiver design for optical fiber systems," *Proceedings of the IEEE*, vol. 65, pp. 1670–1678, 1977.

4.12   Ambrozy, A., *Electronic Noise*, McGraw-Hill, Budapest, Hungary, 1982.

4.13   Muoi, T. V., "Receiver design for high-speed optical-fiber systems," *IEEE Journal of Lightwave Technology*, vol. LT-2, no. 3, pp. 243–267, June, 1984.

4.14   Smith, R. G., and S. D. Personick, "Receiver design for optical fiber communication systems," Chapter 4 in *Semiconductor Devices for Optical*

Communications, 2d ed., Volume 39, Topics in Applied Physics, Springer-Verlag, New York, 1982.

4.15 Siegman, A. E., S. E. Harris, and B. J. McMurtry, "Optical heterodyning and optical demodulation at microwave frequencies," in *Optical Masers*, edited by J. Fox, Polytechnic Press, Polytechnic Institute of Brooklyn, Brooklyn, N.Y., 1963.

4.16 Webb, P. P., R. J. McIntyre, and J. Conradi, "Properties of avalanche photodiodes," *RCA Review*, vol. 35, pp. 234–278, 1974.

4.17 Brain, M., and T. P. Lee, "Optical receivers for lightwave communication systems," *IEEE Journal of Lightwave Technology*, vol. LT-3, pp. 1281–1300, Dec. 1985.

4.18 Personick, S. D., "Receiver design for digital fiber optic communication systems," I and II, *Bell System Technical Journal*, vol. 52, pp. 843–874 and 875–886, July-Aug. 1973.

V. Optical Fiber Communication Systems

5.1 Gowar, J., *Optical Communication Systems*, Prentice-Hall International, London, 1984.

5.2 Korotky, S. K., et al., "4-Gbit/s transmission experiment over 117 km of optical fiber using a $Ti-LiNbO_3$ external modulator," *IEEE Journal of Lightwave Technology*, vol. LT-3, pp. 1027–1031, Oct. 1985.

5.3 Bechtle, D. W., and S. A. Siegel, "An optical communications link in the 2.0–6.0 GHz band," *RCA Review*, vol. 43, pp. 277–309, 1982.

5.4 Stephens, W. E., and T. R. Joseph, "A 1.3 μm microwave fiber-optic link using a direct-modulated laser transmitter," *IEEE Transactions of Lightwave Technology*, vol. LT-3, pp. 308–315, Apr. 1985.

5.5 Salz, J., "Coherent lightwave communications," *AT&T Technical Journal*, vol. 64, pp. 2153–2209, Dec. 1985.

5.6 Salz, J., "Modulation and detection for coherent lightwave communications," *IEEE Communications Magazine*, vol. 24, pp. 38–49, June 1986.

5.7 Wyatt, R., and W. J. Devlin, "10 kHz linewidth 1.5 μm InGaAsP external cavity laser with 55 nm tuning range," *Electronics Letters*, vol. 19, pp. 110–112, Feb. 3, 1983.

5.8 Runge, P. K., and P. R. Trischitta, "The SL undersea lightwave system," *IEEE Journal on Selected Areas in Communications*, vol. SAC-2, pp. 784–793, Nov. 1984, or *IEEE Journal of Lightwave Technology*, vol. LT-2, pp. 744–753, Dec. 1984.

5.9 Special issue on fiber optics for local communications, *IEEE Journal on Selected Areas in Communications*, vol. SAC-3, June 1985.

5.10 Special issue on fiber-optic local area networks, *IEEE Journal of Lightwave Technology*, vol. LT-3, June 1985.

5.11 Eng, S. T., R. Tell, T. Anderson, and B. Eng, "250 Mb/s ring local computer network using 1.3 μm single-mode optical fibers," *IEEE Journal of Lightwave Technology*, vol. LT-3, pp. 820–823, Aug. 1985.

5.12 Special issue on undersea lightwave communications, *IEEE Journal of Lightwave Technology*, vol. LT-2, Dec. 1984.

5.13 Barnoski, M. K., *Fundamentals of Optical Fiber Communications*, 2d ed., Academic Press, Orlando, 1981.

5.14 Bickers, L., L. C. Blank, and S. D. Walker, "Long-span optical transmission experiment over 222.8 km of commercial monomode fibre at 140 Mbits/s and 1.525 μm," *Electronics Letters*, vol. 21, p. 7, Mar. 28, 1985.

5.15 Blank, L. C., L. Bickers, and S. D. Walker, "Long-span optical transmission experiments at 34 and 140 Mb/s," *IEEE Journal of Lightwave Technology*, vol. LT-3, pp. 1017–1026, Oct. 1985.

5.16 Gnauk, A. H., J. E. Bowers, and J. C. Campbell, "8 Gb/s transmission over 30 km of optical fiber," *Electronics Letters*, vol. 22, pp. 600–602, 22 May 1986.

5.17 Maylon, D. J., et al., "PSK homodyne receiver sensitivity measurements at 1.5 μm," *Electronics Letters*, vol. 19, pp. 144–146, 1983.

5.18 Gagliardi, R. M., and S. Karp, *Optical Communications*, Wiley, New York, 1976.

5.19 Kao, C. K., *Optical Fiber Systems: Technology, Design, Applications*, McGraw-Hill, New York, 1982.

5.20 Keiser, G., *Optical Fiber Communications*, McGraw-Hill, New York, 1983.

5.21 Miller, S. E., and A. G. Chynoweth, *Optical Fiber Telecommunications*, Academic Press, Orlando, 1979.

5.22 Palais, J. C., *Fiber Optic Communications*, Prentice-Hall, Inc., Englewood Cliffs, N.J., 1984.

5.23 Personick, S. D., *Optical Fiber Transmission Systems*, Plenum Press, New York, 1981.

5.24 Sandbank, C. P., ed., *Optical Fibre Communication Systems*, Wiley, New York, 1980.

5.25 Senior, J., *Optical Fiber Communications, Principles and Practice*, Prentice-Hall International, London, 1985.

5.26 Sharma, A. B. R., S. J. Halme, and M. M. Batusov, *Optical Fiber Systems and Their Components*, vol. 24, Springer-Verlag Series in Optical Sciences, Springer-Verlag, New York, 1981.

5.27 Suematsu, Y., and K. I. Iga, *Introduction to Optical Fiber Communications*, Wiley, New York, 1982.

5.28 Taylor, H. F., ed., *Fiber Optics Communications*, Artech House, Dedham, Mass., 1983.

Index

Lightning Source UK Ltd.
Milton Keynes UK
UKOW05n1010121215

3620UKLV00017B/255/P

9 780195 107265